ATOMIC FLUORESCENCE SPECTROSCOPY

VAN NOSTRAND REINHOLD
SERIES IN ANALYTICAL CHEMISTRY

Edited by

DR. R. A. CHALMERS

Department of Chemistry
University of Aberdeen

This series is designed as a coverage of reliable analytical information of value to chemists in research, industry and teaching. Each volume is carefully selected and planned as a modern treatment of a topic of importance to analytical chemists today. New volumes will be added from time to time.

Chemistry of Complex Equilibria	M. T. BECK
Solvent Extraction of Metals	A. K. DE, S. M. KHOPKAR AND R. A. CHALMERS
Modern Methods in Organic Microanalysis	JEAN P. DIXON
Errors, Measurement and Results in Chemical Analysis	K. ECKSCHLAGER
Laboratory Handbook of Chromatographic Methods	O. MIKEŠ
Flame Photometry Theory	E. PUNGOR
Thermometric and Enthalpimetric Titrimetry	G. A. VAUGHAN
Gas-Liquid Chromatography	D. W. GRANT
Solution Equilibria in Analytical Chemistry	L. ŠŮCHA AND S. KOTRLÝ
Thermal Analysis	A. BLAŽEK
Inorganic Chromatographic Analysis	J. MICHAL

Additional titles will be listed and announced as published

ATOMIC FLUORESCENCE SPECTROSCOPY

by

Dr. VÁCLAV SYCHRA, CSc.,
PRAGUE INSTITUTE OF CHEMICAL TECHNOLOGY

Ing. VRATISLAV SVOBODA, CSc.,
INSTITUTE FOR RESEARCH, PRODUCTION AND USE OF RADIOISOTOPES, PRAGUE

Dr. IVAN RUBEŠKA, CSc.,
GEOLOGICAL SURVEY OF CZECHOSLOVAKIA, PRAGUE

Translation Editor

Dr. MALCOLM CRESSER,
UNIVERSITY OF ABERDEEN

VNR

VAN NOSTRAND REINHOLD COMPANY
LONDON

NEW YORK CINCINNATI TORONTO MELBOURNE

VAN NOSTRAND REINHOLD COMPANY LTD
25–28 Buckingham Gate, London SW1E 6LQ

INTERNATIONAL OFFICES
New York Cincinnati Toronto Melbourne

Translated by the authors
First published in 1975
by Van Nostrand Reinhold Company Ltd,
25—28 Buckingham Gate, London SW1E 6LQ
in co-edition with SNTL—Publishers of Technical Literature, Prague

Library of Congress Catalog No. 74-3596
ISBN 74-22264

Library of Congress Cataloging in Publication Data

Sychra, Václav
Atomic fluorescence spectroscopy.

(Van Nostrand Reinhold series in analytical chemistry)

Bibliography: p.
Includes index.
1. Atomic spectra, 2. Fluorescence spectroscopy.
I. Svoboda, Vratislav, joint author. II. Rubeška, Ivan, joint author.
III. Title.
Qd96.A8S95 1975 543.085
ISBN 74-22264

Printed in Czechoslovakia by SNTL, Prague

Contents

PREFACE

Although atomic fluorescence spectrometry as a general analytical method has now been in use for more than ten years no monograph has so far been published. Those wishing to get acquainted with and start working in this field must rely mostly on review articles published in scientific journals or on chapters devoted to atomic fluorescence spectrometry in spectroscopic compendia dealing with atomic spectroscopic methods in general, which, due to their limited scope, cannot and do not bring together all the information on specific problems and elements.

Our book is intended to provide spectroscopists and analytical chemists with a general up-to-date knowledge of the theory and instrumentation for atomic fluorescence spectrometry as well as with a knowledge of the analytical potential of the method and possible practical applications.

In the writing of monographs on a particular subject complete coverage and reasonable length are competing interests. In atomic fluorescence spectrometry the existing material has just reached the point when practically all of it may be assembled in one single volume. In a few years a selection would already be necessary. We therefore chose the first alternative despite the fact that the outcome is somewhat but not excessively bulky.

When treating subjects common to all three "flame spectroscopic methods and associated procedures", e.g. instrumentation, atomizers, we tried to lay emphasis on properties which we think important for atomic fluorescence spectrometry, dealing with the rest as briefly as possible consistent with easy understanding. We expect the reader to look for more complete information on these topics in one of several existing excellent monographs on atomic emission and/or absorption spectrometry.

We have quoted freely from the work of many authors whose results and experiences have contributed to the present state of the art of atomic fluorescence spectrometry. We are particularly indebted to all who have permitted us to use their data and have promptly provided us with their publications. The second of the authors is specially indebted to J. D. Winefordner for constant support and vivid interest during his stay at University of Florida.

It is a pleasure for us to acknowledge the help and support we have received from our institutes and close co-workers, during the time of writing this book. We particularly wish to thank Miss N. Dudová, who assembled the bibliography and took care of many editorial tasks in addition to typing the manuscript.

Prague, January 1975 The Authors

Chapter 1

HISTORY AND INTRODUCTION

Atomic fluorescence flame spectrometry – the most recent flame analytical method – is based upon the absorption of radiation of a certain wavelength by an atomic vapour and subsequent radiational deactivation of the excited atoms, a phenomenon known as atomic fluorescence; both the absorption and the measured atomic emission processes occur at wavelengths which are characteristic of the atomic species present.

The basic layout of such measurements remained unchanged for more than seventy years and may be seen in Fig. 1.1. Radiation of a primary light

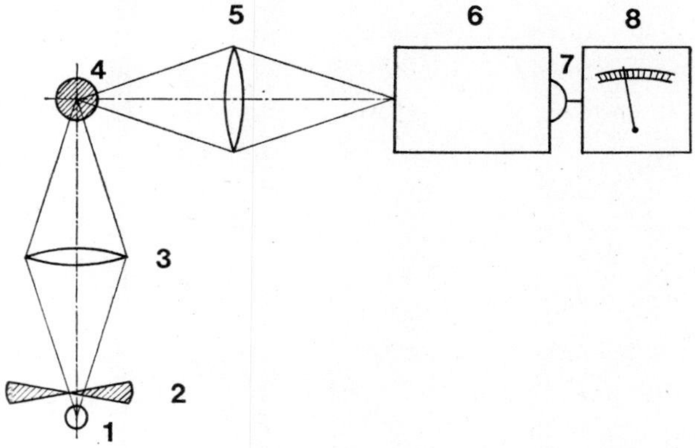

FIG. 1.1. Basic layout of fluorescence measurement: (1) primary light source; (2) chopper; (3) illuminating optics; (4) flame; (5) entrance optics; (6) monochromator; (7) detector; (8) amplifier and read-out system.

source strikes a cloud of free atoms. Part of it is absorbed and subsequently re-emitted in all directions as fluorescence radiation. This is usually observed at a right angle to the exciting beam.

The fluorescence of atomic vapours had already been studied by a number of physicists in the late 19th and early 20th centuries. These investigations have been summarized by Mitchell and Zemansky [289] and Pringsheim [331].

Wood [468, 469] observed for the first time the emission of the sodium D-lines after radiative excitation of sodium vapour in a sealed glass tube by the sodium D-doublet emission from a flame. He called this resonance radiation. The atomic fluorescence of sodium vapour was then investigated by other spectroscopists [155, 156, 471, 473, 474]. Dunoyer [155, 471] showed that the time-interval between absorption and subsequent emission of fluorescence radiation is very short and that only the centre of a D-line is efficient for producing fluorescence.

The fluorescence of lithium atoms excited by a lithium flame was described by Bogros [53]. In 1912, Wood [470, 472] illuminated mercury vapour with radiation from a quartz mercury arc lamp and observed intense fluorescence emission at 2537 Å. Terenin [396, 397] investigated the atomic fluorescence of cadmium, thallium, lead, bismuth, arsenic, and antimony: in the case of thallium vapour excited with the 3776 Å line from a quartz thallium arc, he obtained both the 3776 and 5350 Å lines as fluorescence. Similarly, Strutt [377, 378] and later Christensen and Rollefson [86] excited sodium vapour with the 3303 Å doublet and measured fluorescence radiation at 3303, 5890, and 5896 Å. Boeckner [51] used the helium 3888 Å line to excite caesium fluorescence at both 3888.6 and 8521 Å.

Fuchtbauer [178] studied the absorption of radiation by excited atoms and their subsequent fluorescence. He excited mercury vapour with the 2537 Å line and subsequently with the 3125 and 3132 Å lines and observed fluorescence lines at 5770 and 3654 Å in addition to those at 2537, 3125, and 3132 Å. A similar effect was described by Bender [44] for cadmium and zinc.

The fluorescence of calcium, strontium, barium, lithium, and sodium atoms was observed for the first time in flames by Nichols and Howes [303, 304]. Later Badger [25] reported the atomic fluorescence of thallium, mercury, magnesium, copper, silver, cadmium, and sodium in a Bunsen flame.

The next major advance came in 1956, when Boers, Alkemade, and Smit [52] used atomic fluorescence to study basic physical and chemical processes in flames. At the International Spectroscopy Colloquium in 1962, Alkemade [10] summarized the methods by which atoms can be excited in flames and described apparatus for measuring quantum efficiencies of resonance fluorescence of the sodium D-doublet in various flames: he also suggested the possible use of atomic fluorescence in analytical chemistry.

Meanwhile, Robinson [342] attempted to excite atomic fluorescence of sodium, magnesium, and nickel with hollow-cathode lamps; he was successful only in the case of magnesium, for which he observed weak fluorescence at the main resonance line, 2852 Å.

In spite of the fact that the principles of atomic fluorescence have been known for many years, its use in chemical analysis dates back only to 1964, when Winefordner and Vickers [444] proposed atomic fluorescence flame spectroscopy as a new analytical method. Since then, atomic fluorescence spectroscopy has gone through a period of very rapid growth, similar to that of atomic absorption between 1957 and 1965. Much of the credit for this is attributable to two research groups in particular, Professor Winefordner's group at Florida State University, and Professor West's group at Imperial College, London.

Chapter 2

THEORY

2.1. TYPES OF ATOMIC FLUORESCENCE

Atomic fluorescence is a radiational deactivation process which occurs after the excitation of free atoms by the absorption of radiation of characteristic wavelength from an appropriate external excitation source. Basically, a fluorescent line can have a wavelength that is the same as that of the exciting line, or one that is longer or (very rarely) shorter. There are two basic types of atomic fluorescence, resonance fluorescence and non-resonance fluorescence.

2.1.1. RESONANCE FLUORESCENCE

Resonance fluorescence occurs when excited atoms absorb and re-emit radiation of the same wavelength. Lines corresponding to transitions between excited electronic states and the ground state of an atom (resonance transitions) constitute most analytically useful examples of resonance fluorescence, since the transition probabilities for resonance transitions are usually much greater than those for other transitions (Fig. 2.1A)*. Absorption and re-emission of the 2138.6, 2320.0, and 2833.1 Å resonance lines by zinc, nickel, and lead atoms, respectively, are examples of resonance fluorescence.

Provided an atom has low-lying metastable energy levels which can be thermally populated (as in the case of indium, gallium, lead, tin, etc.), resonance fluorescence may originate from the metastable levels (Fig. 2.1B).

---

* Note that the expression "resonance fluorescence" refers to "resonance" in wavelength between an exciting and a fluorescent line, and not to the emission of a fluorescent line corresponding to a resonance transition only.

Atoms populating these levels after thermal excitation are further excited by the absorption of radiation corresponding to an appropriate non-resonance line from the excitation source; they can re-emit the same line

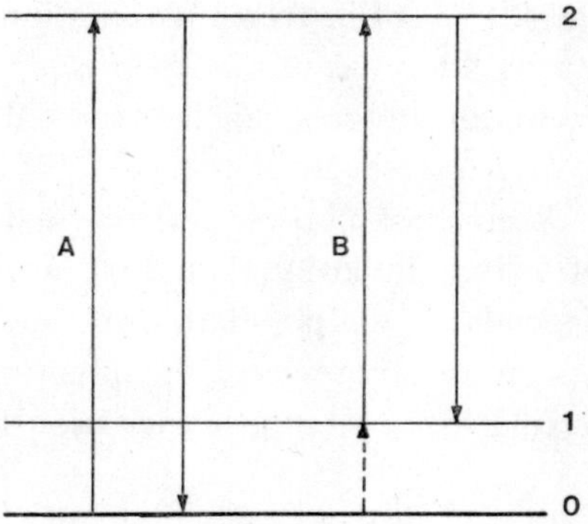

FIG. 2.1. Types of resonance fluorescence: (A) resonance fluorescence originating from the ground state of an atom (resonance transition); (B) resonance fluorescence originating from a metastable state; (0 — ground state of an atom; 1, 2, 3 — excited electronic states). The solid lines represent radiational processes, the dashed line represents a non-radiational process.

as resonance fluorescence. The flame temperature and the energy of the metastable level above the ground state are parameters which govern the contribution of this type of fluorescence, the existence of which was established by Omenetto and Rossi [313] in explaining the excitation mechanism of indium and gallium fluorescence: they called the process "thermally-assisted" resonance fluorescence.*

In Table 2.1, the ratios of the number of atoms populating the $^2P_{3/2}$ level and the ground state are given for thallium, indium, and gallium. This ratio is negligible for thallium, but it is significant for the other two. Thus,

TABLE 2.1. Population of $^2P_{3/2}$ level for thallium, indium and gallium in an air-hydrogen flame at a temperature of 2100 °C [313]

Reprinted from Ref. [313] by permission of Pergamon Press Ltd.

Element	$N_{^2P_{3/2}}/N_{^2P_{1/2}}$	E (eV) of $^2P_{3/2}$ level
Tl	0.02	0.966
In	0.52	0.274
Ga	1.2	0.103

* Recently [526], the term "excited state resonance fluorescence" was recommended for this type of fluorescence transition.

absorption and re-emission of the 4511.3 and 4172.1 Å lines by indium and gallium atoms, respectively, are examples of resonance fluorescence corresponding to non-resonance transitions (cf. Sections 5.12 and 5.15). For indium, this excitation mechanism was experimentally confirmed by isolating the 4511.3 Å line from an excitation source with a suitable filter. A relatively strong fluorescence radiation could still be observed at 4511.3 Å. The same excitation process can be expected for indium and gallium at 3250.6 and 2944.2 Å, respectively.

Resonance fluorescence originating from the ground state is generally of higher intensity than that originating from metastable states, which depends on the population of these states. It will be shown later that this resonance fluorescence is often mixed with non-resonance fluorescence having the same upper excitation level.

2.1.2. NON-RESONANCE FLUORESCENCE

Non-resonance fluorescence occurs when the exciting line and the observed fluorescent line are of different wavelengths. The two main categories here, direct-line fluorescence and stepwise-line or stepwise fluorescence, are represented schematically in Figs. 2.2 and 2.3.

2.1.2.1. Direct-Line Fluorescence

Direct-line fluorescence is non-resonance fluorescence which results when the exciting and the fluorescent line originate from optical transitions with a common upper level. An atom is radiatively excited (usually from the ground state) to a higher excited electronic state, and then undergoes a direct radiational transition to a metastable level above the ground state (see Fig. 2.2A).

Absorption of the 2833.1 Å line by lead atoms in the ground state and subsequent emission of the 4057.8 and 7229.0 Å lines are typical cases of simple direct-line fluorescence [390] (cf. Section 5.19); similar examples are the excitation of thallium, indium, and gallium ground-state atoms by absorption at 3775.5, 4101.8, and 4033.0 Å followed by emission at 5350.5, 4511.3, and 4172.1 Å, respectively (cf. Sections 5.12, 5.15, and 5.35). This excitation mechanism has been confirmed experimentally by eliminating primary resonance lines from the excitation source; only very weak fluorescence emission (due to resonance fluorescence of atoms in the metastable state) is then observed. The smaller the energy difference between the ground state of an atom and the lower energy level of the direct-line fluorescence transition, the larger the contribution of resonance fluorescence to the apparent direct-line fluorescence intensity.

Direct-line fluorescence excited by absorption at a non-resonance line (see Fig. 2.2B) is probably responsible for the fluorescence of tin at 3330.6 Å (cf. Section 5.36).

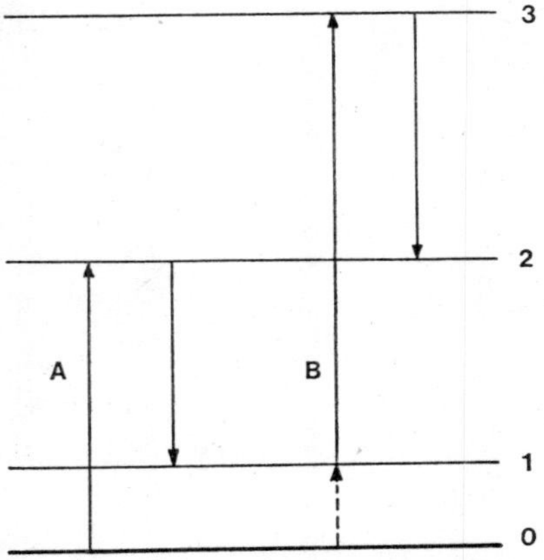

FIG. 2.2. Direct-line fluorescence transitions originating from the ground state of an atom (A) and from a metastable state (B). (See footnotes to Fig. 2.1.)

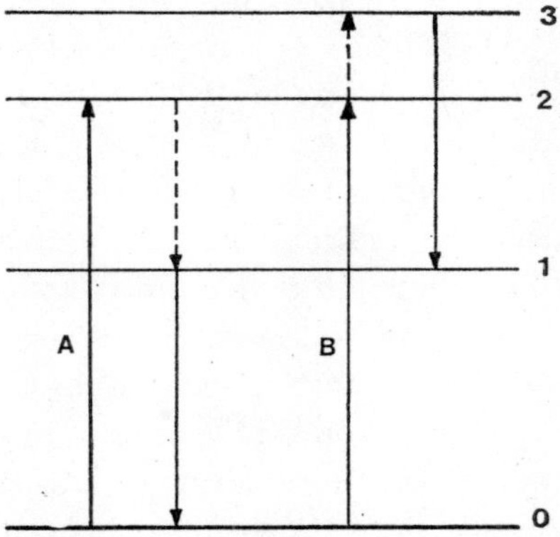

FIG. 2.3. Types of stepwise-line fluorescence: (A) normal stepwise-line fluorescence; (B) thermally-assisted stepwise-line fluorescence. (See footnotes to Fig. 2.1.)

2.1.2.2. Stepwise-Line Fluorescence

When the upper levels of the exciting and of the emitted line are different, stepwise-line fluorescence is said to occur. Radiatively excited atoms lose part of their energy, usually by collisional deactivation, before emitting fluorescence radiation of longer wavelength. Two cases can be distinguished here (Fig. 2.3).

Normal stepwise fluorescence (Fig. 2.3A) involves excitation to an electronic state above the first excited state and subsequent two-stage deactivation; the first stage is a radiationless transition to a lower excited state, induced by collisions with flame molecules; the second state is a fluorescence transition, usually down to the ground state. For example, normal stepwise-line fluorescence occurs in the emission of the sodium 5890.0 Å line after the excitation of sodium atoms by the sodium 3303 Å doublet [86] (cf. Section 5.32).

An example of normal stepwise fluorescence when the lower level of the fluorescence transition is not the ground state is the emission of lead fluorescence at 3683.5 Å [390]. The mechanism, which has been confirmed experimentally [388], involves excitation at 2833.1 Å to the $7s\,^3P_1$ state, followed by a radiationless transition to the $7s\,^3P_0$ state, and then a fluorescence transition to the metastable $6p^2\,^3P_1$ state (cf. Section 5.19).

A radiatively excited atom can be further excited thermally to higher electronic states; it can then become deactivated by emitting fluorescence radiation (see Fig. 2.3B). This type of atomic fluorescence was termed "thermally-assisted direct-line fluorescence" when it was first postulated by Dagnall and co-workers to explain some fluorescence transitions of antimony [107] and bismuth [108]. According to the "Nomenclature for analytical flame spectroscopy and associated procedures" suggested by IUPAC, thermally-assisted stepwise-line fluorescence is a more appropriate term, since this indicates that the difference in upper levels of the exciting and fluorescent line is due to a thermal energy contribution from the flame (i.e. due to collisional activation of the excited atoms, and not due to collisional deactivation, as in the case of normal stepwise fluorescence). Any possibility of ambiguity or confusion is best avoided by simply outlining the energy transitions involved, and stating whether these are radiational or thermal, or by the use of partial Grotrian diagrams. Thermally-assisted stepwise fluorescence only occurs when the energy separating two or more excited electronic states is small enough for thermal energy to cause transfer from a lower to a higher state.

2.1.2.3. Anti-Stokes Fluorescence

Anti-Stokes fluorescence occurs when the wavelength of the fluorescent line is shorter than that of the exciting line. The deficit in photon energy is usually supplied thermally and the term "thermally-assisted fluorescence" can be used here.

Thermally-assisted anti-Stokes fluorescence occurs when an atom populating an energy level lying near the ground state is excited to a higher energy level and then undergoes a radiational transition to the ground state (Fig. 2.4A), or when an atom in the ground state is excited to a certain electronic state, then raised to a slightly higher electronic state by absorbing thermal energy from the flame, and is finally radiationally deactivated to the ground state (Fig. 2.4B).

It is evident that anti-stokes fluorescence shown in Fig. 2.4A is a special case of direct-line fluorescence while that shown in Fig. 2.4B is a special case of stepwise-line fluorescence.

An example of the mechanism shown on Fig. 2.4A which has been confirmed experimentally [316] occurs for indium, which has a low-lying metastable energy level. Atoms thermally populating this level can be

FIG. 2.4. Types of thermally-assisted anti-Stokes fluorescence. (See footnotes to Fig. 2.1.)

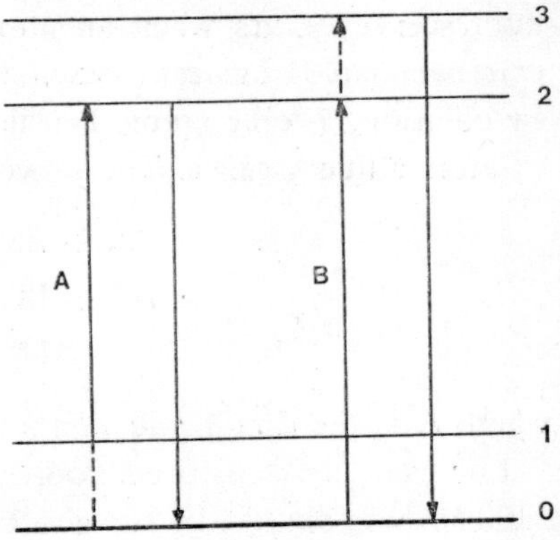

excited by absorption at 4511.3 Å and then deactivated to the ground state by emission of fluorescence at 4101.8 Å.

An example of case 2.4B is the emission of the most intense component of the chromium triplet at 3578.7 Å after excitation at 3593.5 Å (cf. Section 5.9). It should be noted that anti-Stokes fluorescence is always accompanied by resonance fluorescence at a particular resonance wavelength [45].

Recent investigations of non-resonance fluorescence [45, 245, 316, 390, etc.] have demonstrated that atomic fluorescence involves more complicated processes than originally presumed (especially if high-temperature flames are used for excitation) and other possible excitation mechanisms therefore cannot be excluded.

All types of non-resonance fluorescence, particularly direct-line fluorescence, may be analytically useful. Fluorescence intensities corresponding to direct-line fluorescence are in some favourable cases (lead, gallium, indium) higher than those of resonance fluorescence. In practical analytical applications non-resonance fluorescence offers considerable advantages over resonance fluorescence: scatter can be eliminated by removing the exciting line from the fluorescence spectrum. When excitation occurs at a wavelength where the flame exhibits strong thermal background emission, it is usually advantageous to measure the fluorescence at a line where the flame background is lower, if such a line is available. Finally, self-absorption problems may be overcome by measuring fluorescence at a line that does not correspond to deactivation to the ground state.

2.1.3. SENSITIZED FLUORESCENCE

In addition to the types of atomic fluorescence which have already been discussed, another type exists, called "sensitized fluorescence". Sensitized fluorescence occurs when an atom or a molecule (donor) excited by an external source transfers its excitation energy to the sample atom (acceptor) by collision. The acceptor then undergoes radiative deactivation, resulting in atomic fluorescence. The process can be represented as follows:

$$\begin{aligned} A + h\nu &\rightarrow A^* \\ A^* + M &\rightarrow A + M^* + \Delta \\ M^* &\rightarrow M + h\nu' \end{aligned}$$

where A is the donor and M the acceptor.

For example, sensitized fluorescence of thallium atoms at 3775.7 and 5350.5 Å was observed when a mixture of thallium and a high concentration of mercury vapour was excited with the 2536.5 Å mercury line [81]. This type of fluorescence requires a higher concentration of donor species than can be obtained in flame cells, where the concentration of atoms is low and where atoms are primarily deactivated by collisional means. This phenomenon is, therefore, of academic interest only. However, this type of atomic fluorescence could be observed in some non-flame cells.

Recently an interesting paper classifying in more detail the different types of atomic radiational transitions has appeared [526]. In addition to those mentioned above, multiphoton fluorescence was postulated. This type assumes excitation by two (or more) identical photons.

2.2. THEORY OF ANALYTICAL CURVES

It is essential that the analyst is aware of the form that the growth curve (calibration curve) is expected to take for any analytical method he chooses to use; otherwise, measurements may be unintentionally made under conditions which may be far from ideal.

In atomic fluorescence spectrometry, the lack of detailed and precise information concerning the parameters of the experimental system restricts knowledge of the growth curve shape to a semi-quantitative level. However, even this information should make it less likely that measurements will be accidentally made with poorer precision than is necessary. In extreme cases, ambiguity, due to the use of an unsuitable portion of the growth curve, may be avoided.

It is also important that the significance of atomizer cell geometry and spectrometer optics in atomic fluorescence measurements is appreciated, particularly in the design stage of atomic fluorescence spectrometers. As little equipment is commercially available at the present time (and none is available for single-channel analysis), the majority of atomic fluorescence spectrometers are assembled in the laboratory. The choice of cell geometry and the type of cell illumination best suited to the particular application depends upon many factors, including the range of concentration to be covered and whether a continuous source or a narrow atomic-line source is to be used.

Unfortunately, a thorough treatment of all the variables which must be considered to obtain even an approximate expression for the analytical signal which results from a real measurement situation, is far more complex in atomic fluorescence than in either atomic absorption or atomic emission spectrometry.

Ideally, at *low atomic concentrations (low optical densities)* the integrated radiance of the analytical signal, whether it is from atomic absorption, emission, or fluorescence, will be linearly related to the concentration of ground-state atoms which give rise to the signal. It is only at *higher atomic concentrations (hence, higher optical densities)* that the relationship between integrated radiance and atomic concentration becomes complex and the most complex situation occurs in atomic fluorescence spectrometry. This arises primarily because the optical path followed by the incident radiation, during the absorption process, differs from that of the fluorescence radiation which it excites, and which is measured. This is in marked contrast to the situation in atomic absorption or emission spectrometry: in both of these techniques, the region of the atom cell which can modify the analytical signal is confined to the region which itself produces the signal along the optical axis of measurement. To take a specific example, in flame emission spectrometry the only re-absorption of the emitted radiation which can alter the analytical signal is that which occurs along the optical path of measurement, between the primary point of emission and the detector.

In atomic fluorescence spectrometry, on the other hand, the concentration of atoms *outside* the optical path of measurement will still influence the analytical signal. Therefore, experimental growth curves in atomic fluorescence spectrometry are a direct and complex function of the atom cell geometry (whether it be a flame or a furnace) in a way that experimental atomic absorption or atomic emission growth curves are not.

The parameters of an atomizer cell for atomic fluorescence spectrometry can only be optimised by careful consideration of the shapes of growth curves for several possible cell geometries and methods of cell illumination.

It is important here to distinguish between those factors which actually alter the shape of the analytical curve and those which simply displace the curve, while leaving its form unchanged. The former category includes the cell geometry, the method of cell illumination and the atomic damping constant (which describes the atomic line profile in terms of its collisional and Doppler half-widths): the relationship between the concentration of atoms in the cell capable of absorbing radiation and the concentration of atoms introduced to the atomizer, will in most instances fall into the latter category.

There are certain important exceptions to this, however, such as the point at which ionization becomes significant (see Section 2.2.2).

In the following sections of this monograph, factors which relate the theoretical growth curves to the experimental curves will be considered in detail. It is hoped that the approach adopted will give to those working in the field of atomic fluorescence spectrometry sufficient insight to appreciate how optimization of experimental parameters can lead to improved analytical performance.

2.2.1. THEORETICAL GROWTH CURVES

2.2.1.1. General Comments and Assumptions

Winefordner, Parsons, Mansfield and McCarthy [452] first attempted to predict the shape of growth curves (plots of the logarithm of fluorescence intensity *vs* logarithm of the analyte atomic concentration) in atomic fluorescence flame spectrometry. Hooymayers [200] more recently made a much more thorough study of shape of growth curves in atomic fluorescence flame spectrometry for a rectangular flame excited by narrow line or continuum sources, for resonance fluorescence measured at right angles to the excitation-flame axis; he derived expressions for growth curves based on the assumption that the exciting radiation traversed the flame as a parallel beam of rectangular cross section, and that all radiation emitted along the exciting beam and at right angles to it was measured (see Fig. 2.5a).

Typical experimental systems described in the analytical literature [463] may differ substantially from the models employed by Winefordner *et al.* or Hooymayers. In this monograph, a general expression is evolved for the fluorescence radiant flux emitted from atoms in any shape of atomizer cell, and is valid for different geometric arrangements of the excitation and emission beams. This is then used to construct theoretical growth curves as a function of these parameters and of atomic characteristics.

To make the results compatible with those of Hooymayers [200] the following assumptions from his paper are retained: the solid angle of measured fluorescence resonance radiation is small; the resonance absorption line of the analyte atoms is single and isolated (multiplets and hyperfine structure are not taken into account); the distribution of ground-state analyte atoms is uniform throughout the region of the atomizer cell (flame) under consideration, and the atomizer cell temperature is uniform throughout the same region; re-emission of self-absorbed fluorescence radiation is ignored; the profile of the emission coefficient e_ν is identical to the profile of the absorption coefficient k_ν and is described by the Voigt profile. In this section, fluorescence radiant flux expressions will be given only for resonance absorption-fluorescence; however, the general approach used is valid for non-resonance fluorescence processes: the general expressions are identical to those given here. If more than one absorption line excites fluorescence and/or more than one fluorescence line is measured, summation techniques must be employed.

2.2.1.2. Basic Expressions

In addition to the basic assumptions given above, it will also be assumed that the atomizer cell has well-defined but not necessarily regularly-shaped boundaries (e.g., refer to Fig. 2.5) and that the $x-y$ cross-section of the atomizer cell (in a horizontal plane) and of the excitation beam are constant within the entire height (z-axis) of the excitation beam. Therefore, the observed fluorescence radiant flux $\Delta\Phi_{\mathrm{Fi}}$, in watts (W), due to the analyte atoms within an elementary cell (which constitutes a small part of the total cell) (see Fig. 2.5) of dimensions Δx, Δy, and Δz is approximately

$$\Delta\Phi_{\mathrm{Fi}} = \left(\frac{\Delta\Omega}{4\pi}\right)\Delta\Phi_{\mathrm{A}} f_{\mathrm{S}} Y \tag{2.1}$$

where $\Delta\Omega$ is the solid angle, in steradians (sr), of the fluorescence radiant flux measured by the measurement system, $\Delta\Phi_{\mathrm{A}}$ is the radiant flux absorbed by the analyte atoms (within the elementary cell) from the exciting beam, in W, Y is the fluorescence quantum yield, i.e. the ratio of the number of quanta emitted as fluorescence to the number of quanta absorbed capable of producing the fluorescence, f_{S} accounts for the fluorescence radiance lost within the flame gases due to reabsorption, and 4π is the number of steradians subtended from the centre to the surface of a sphere (fluorescence is isotropic). If a non-resonance process, e.g. direct-line or stepwise-line fluorescence is of interest, Eq. (2.1) is modified by replacing the quantum yield by the radiant flux yield (of course $\Delta\Phi_{\mathrm{A}}$ and f_{S} will then correspond to processes having different wavelengths). (For non-resonance fluorescence, the fluorescence quantum yield must be replaced by the fluorescence power or flux yield, i.e. the ratio of the number of W emitted as fluorescence to the number of W absorbed and producing the fluorescence).

If the dimensions of the elementary cell are small enough, then for every single elementary cell, the function $\Delta\Phi_A$ depends only upon the distance b of the cell from the boundary of the flame cell in the direction of the exciting beam (see Fig. 2.5), the atomic constants of the particular analyte atomic absorption line, the analyte atomic concentration within the flame cell (analyte atoms in the ground state), n_0, and the spectral distribution of exciting radiation; the function f_S, on the other hand, is a function of the distance b' from the cell to the border of the atomizer cell in the direction of observation, of the atomic constants of the particular fluorescence line, and of the analyte atomic concentration (n_0 for resonance fluorescence). If the lower state involved in the absorption process is not the ground state e.g. electronic state i, then

$$n_i = n \left[\frac{g_i e^{-E_i/kT}}{\sum_{j=0}^{l} g_j e^{-E_j/kT}} \right]$$

where the g's are statistical weights of electronic states i and j, E_i and E_j are the energies of levels i and j, n is the total concentration of analyte atoms in all states, k is the Boltzmann constant, T is the atomizer temperature (in K), and the summation is over all electronic states of the analyte atoms. If the ground state ($i = 0$) is involved in the absorption process and if there are low-lying electronic states, then

$$n_0 = n \left[\frac{g_0}{\Sigma g_j e^{-E_j/kT}} \right]$$

The denominator of this expression is called the electronic partition function. In our discussion, the lower-level in the absorption process will be assumed to be the ground state.

The variation of the absorption coefficient of a single atomic line of an atomic vapour is described by the Voigt profile:

$$k_v = k_0 \delta_v \tag{2.2}$$

where k_v is the atomic absorption coefficient (dimensionless) at any relative frequency, v, k_0 is the peak atomic absorption coefficient for pure Doppler broadening (at frequency $\nu = \nu_0$), in cm^{-1}, and δ_v is the Voigt profile function (no units) and is given by

$$\delta_v = \frac{a}{\pi} \int_{-\infty}^{+\infty} \frac{\exp(-y^2)\,dy}{a^2 + (v - y)^2} \tag{2.3}$$

where a is the classical damping constant (dimensionless), v is the relative frequency (dimensionless) at any point on the spectral line from the peak frequency ν_0 in Doppler half-width frequency intervals, and y is the relative frequency (dimensionless) of any point on the spectral line from the point v in Doppler half-width frequency intervals. The classical damping constant a is defined by

$$a = \sqrt{\ln 2}\,\frac{\Delta\nu_L}{\Delta\nu_D} \tag{2.4}$$

where $\Delta\nu_L$ and $\Delta\nu_D$ are the Lorentzian half-width and Doppler half-width, respectively, in Hz. The Lorentzian half-width is composed of the collisional half-width (collisional broadening due to adiabatic collisions of analyte atoms with foreign species and similar analyte species), and natural half-width (broadening due to finite lifetime of an atom in the excited state). Refer to reference [319] for estimates of atomic line half-widths.

The relative frequency v is defined as

$$v = \frac{2(\nu - \nu_0)\sqrt{\ln 2}}{\Delta\nu_D} \tag{2.5}$$

The integration variable, y, has the same dimensions as v. The peak absorption coefficient for pure Doppler broadening, k_0 (in cm^{-1}), is given by

$$k_0 = \frac{2\sqrt{\ln 2}\,X n_0 f}{\sqrt{\pi}\,\Delta\nu_D} \tag{2.6}$$

where X is a constant [i.e. $X = \pi e^2/mc$, where m and e are the mass (in g) and charge (in esu) of the electron and c is the speed of light, in $cm\ s^{-1}$; units of X are $cm^2\ s^{-1}$], and f is the absorption oscillator strength of the atomic transition (dimensionless).

In terms of wavelength units, k_0, in cm^{-1}, is given by

$$k_0 = \frac{2\sqrt{\ln 2}\,X\lambda_0^2 n_0 f}{c\sqrt{\pi}\,\Delta\lambda_D} \tag{2.7}$$

where $\Delta\lambda_D$ is the Doppler half-width in cm, and λ_0 is the peak absorption wavelength, in cm. The absorption oscillator strength is related to the Einstein coefficient of spontaneous emission, A (the reverse process to that involved in the absorption transition) by the equation:

$$f_{lu} = \left(\frac{mc}{8\pi^2 e^2}\right)\left(\frac{g_u}{g_l}\right)\lambda_0^2 A_{ul} = 1.51\left(\frac{g_u}{g_l}\right)\lambda_0^2 A_{ul}$$

where the g's are the statistical weights of the upper (u) and lower (l) electronic states involved in the absorption process, and the subscripts of f and A represent the transitions involved. For simplicity, in this monograph, the subscripts designating the transitions will be omitted.

It is convenient here to define two more functions, related to the optical absorption transitions in atomic spectroscopy, which are used in further computations. For the absorption of continuum radiation, the function α_C is defined as

$$\alpha_C = \int_{-\infty}^{+\infty} [1 - \exp(-tk_0\delta_v)]\, dv \tag{2.8}$$

where t is the total distance travelled by the beam of radiation through the homogeneous absorbing medium (atomizer cell in the present case). The function α_C corresponds to the fraction of radiation absorbed from a continuum source, i.e. a source which emits a constant spectral radiance over the region of $\sim \nu_0 \pm 15\Delta\nu_0$ (ν_0 is the peak frequency of the analyte atomic absorption line and $\Delta\nu_0$ is the half-width of the analyte atomic absorption line).

For the absorption of radiation from a line source, a similar function α_L is defined as

$$\alpha_L = \int_{-\infty}^{+\infty} \delta_v [1 - \exp(tk_0\delta_v)]\, dv \tag{2.9}$$

With a *line* source ($\Delta\nu_s \sim \Delta\nu_0$, where $\Delta\nu_s$ is the half-width of the line source) it is necessary to convolute the fraction of radiation absorbed by analyte atoms and the source line profile (it is assumed for simplicity that the source line and the absorption line have similar profiles; if the two lines have different profiles, the expression for α_L becomes more difficult to solve; fortunately, the assumptions made for Eq. (2.9) are approximately valid for most analytical situations, and α_L is not an extremely sensitive function of the source line profile).

The function α_C increases monotonically with increasing t, but α_L approaches asymptotically a limiting value of

$$\alpha_L(t \to \infty) = \frac{\sqrt{\pi}}{2} \tag{2.10}$$

For the case of a continuum source, the radiant flux absorbed by the atoms within the elementary cell is

$$\Delta\Phi_{A,C} = E_0 \left(\frac{\Delta\Omega}{4\pi}\right) \Delta x \Delta y \Delta z \left(\frac{\Delta\alpha_C}{\Delta x}\right) \tag{2.11}$$

where E_0 is the incident radiance in W cm^{-2} sr^{-1} per unit of v, $\Delta\Omega$ is the solid angle of the beam of exciting radiation, in sr^{-1}, Δx, Δy and Δz are the dimensions of the elementary cell in Cartesian coordinates, in cm, the differential $\Delta\alpha_C/\Delta x$, in cm^{-1}, corresponds to the fraction of radiation absorbed along the light path per unit of length (in passing through the elementary cell), and 4π is the number of solid angles in a sphere (it is assumed the source radiation is isotropic); it is also assumed here that the light comes in the direction of the x coordinate, perpendicularly to the $z - y$ plane).

Similarly, for the case of a *line source*, the radiant flux absorbed is

$$\Delta\Phi_{A,L} = \frac{2}{\sqrt{\pi}} E'_0 \left(\frac{\Delta\Omega}{4\pi}\right) \Delta x \Delta y \Delta z \left(\frac{\Delta\alpha_L}{\Delta x}\right) \tag{2.12}$$

Here, E'_0 is the radiance of the single spectral line, i.e. the integrated spectral radiance of the line.

The re-absorption factor, f_s, in Eq. (2.1) is the same for lines excited by either line or continuum sources and is given by

$$f_S = 1 - \frac{2}{\sqrt{\pi}} \int_{-\infty}^{+\infty} \delta v'[1 - \exp(-b'k'_0\delta_v)]\, dv' \tag{2.13}$$

where δ_v is the Voigt profile factor defined by Eq. (2.3) except that v' is the relative frequency interval for the fluorescence rather than the exciting radiation, b' is the path length through which the fluorescence must pass before reaching the atomizer cell boundary, and k_0 is the peak absorption coefficient for absorption of the fluorescence line, assuming pure Doppler broadening, and is defined by Eq. (2.2). For resonance fluorescence, $v = v'$, (for non-resonance fluorescence $v \neq v'$, Y will differ from the power yield for the resonance process, and the number of atoms in the lower state involved in the absorption process may not be n_0, i.e. $n_i \neq n_0$). If the primary absorption and observed fluorescence lines are identical, i.e. resonance fluorescence, then $k_0 = k'_0$, which is the most common case in analytical atomic fluorescence spectrometry. Sometimes, not only may the values of k_0 and k'_0 differ but also the values of the damping constant, a, for the absorption and fluorescence lines may differ significantly. Nevertheless, in this monograph, only results for the case where the damping constants for both lines are identical will be considered, i.e. resonance fluorescence.

If the equations above are combined, the fluorescence radiant flux, in W, due to analyte atoms within the elementary cell i is given by

$$\Delta\Phi_{\text{FiL}} = \left(\frac{2}{\sqrt{\pi}}\right) E_0' \left(\frac{\Delta\Omega}{4\pi}\right) \Delta x \Delta y \Delta z \left(\frac{\Delta\alpha_{\text{L}}}{\Delta x}\right) \times$$

$$\left\{1 - \frac{2}{\sqrt{\pi}} \int_{-\infty}^{+\infty} \delta_{v'}[1 - \exp(-b'k_0'\delta_{v'})]\, dv'\right\} Y\left(\frac{\Delta\Omega'}{4\pi}\right) \tag{2.14}$$

if the atomic vapour is excited by means of a line (L) source and is given by

$$\Delta\Phi_{\text{FiC}} = E_0 \left(\frac{\Delta\Omega}{4\pi}\right) \Delta x \Delta y \Delta z \left(\frac{\Delta\alpha_{\text{C}}}{\Delta x}\right) \times$$

$$\left\{1 - \frac{2}{\sqrt{\pi}} \int_{-\infty}^{+\infty} \delta_{v'}[1 - \exp(-b'k_0'\delta_{v'})]\, dv'\right\} Y\left(\frac{\Delta\Omega'}{4\pi}\right) \tag{2.15}$$

if the atomic vapour is excited by means of a continuum (C) source. The term in curly brackets is simply the fraction of fluorescence radiation excited in cell i which is emitted from the cell and is more simply $(1 - 2\alpha_{\text{L}}'/\sqrt{\pi})$ where $2\alpha_{\text{L}}'/\sqrt{\pi}$ is the fraction of fluorescence due to cell i which is absorbed within the path length b'.

It should be noted that pre-absorption of exciting radiation is accounted for in the $\Delta\alpha/\Delta x$ term and post-absorption of luminescence (re-absorption) is accounted for as above. The equations given here, which are for an elementary cell i, are very similar to the expressions given by Hooymayers [200] for the complete cell.

The total observed fluorescence radiant flux Φ_{F} emitted by analyte atoms in the entire atomizer cell is given by summation of all individual cells

$$\Phi_{\text{F}} = \sum_i \Delta\Phi_{\text{Fi}} \tag{2.16}$$

This expression must be solved numerically by the use of a computer for any particular case.

The spectral fluorescence radiant flux of the elementary cell i, $\Delta\Phi_{\text{Fiv}}$, is given by

$$\Delta\Phi_{\text{Fiv}} = \Delta\Phi_{\text{Fi}}\delta_{\text{v}} \tag{2.17}$$

where $\Delta\Phi_{\text{Fi}}$ is generally defined by Eq. (2.1) and is precisely represented by Eq. (2.14), for line-source excitation and by Eq. (2.15) for continuum-source excitation, and δ_{v} is the Voigt profile factor for the fluorescence radiation. The total spectral radiant flux of fluorescence Φ_{Fv} for the entire atomizer cell is simply a summation (over the entire atomizer cell) of individual elementary cell spectral fluorescence radiant flux given by

Eq. (2.17); the expression is similar to Eq. (2.16) except that Φ_{Fv} replaces Φ_F and $\Delta\Phi_{Fiv}$ replaces $\Delta\Phi_{Fi}$, i.e.

$$\Phi_{Fv} = \sum_i \Delta\Phi_{Fiv} \tag{2.18}$$

2.2.1.3. Computer Calculations

All programmes were written in FORTRAN IV. With the exception of the Voigt profile computations, where double precision arithmetic was used, all numerical operations were carried out with single precision (approximately 7 decimal digits). Numerical integration of Eq. (2.3) was performed by Simpson's rule using transformation involving the trigonometric function arctan. For each damping constant value, 153 values of δ_v were computed. The α_C and α_L functions and their first derivatives were computed for k_0t values (t = absorption path length of atomizer cell) ranging from 0.01 to 10,000 in steps increasing by a factor of $\sqrt[20]{10}$. The computed values of α_C were compared with those published by van Trigt, Hollander and Alkemade [409] and no differences greater than 0.1% were found. It was not possible to calculate growth curves and fluorescence line shapes of atoms in round flame cells in which the atomizer cell was totally illuminated and the fluorescence was totally measured at the same angle. The atomizer cell was arbitrarily divided into 10^4 equal rectangular elementary cells, and computations were carried out by means of the equations derived above. When all elementary cells were of the same size, large integration errors resulted at high values of k_0 (high concentration). On the other hand, changing the size and shape of the elementary cell according to its position in the atomizer cell to take into account the shape and direction of the exciting and observed light beams made the programme unduly complicated in the region of high optical densities where the greatest deviations occurred.

Fortunately, it is possible to represent almost all configurations of exciting beam, atomizer (flame) cell and observed beam used in experimental atomic fluorescence spectrometry by one of three limiting cases. These cases are specified by two characteristic parameters: k_0t, where k_0 is the peak absorption coefficient for the excitation process, assuming pure Doppler broadening, and t is the total absorption path length of the exciting beam, and by a factor $\varkappa$ which is the ratio of the optical densities for the excitation and the fluorescence processes (see $\varkappa$ factors for the limiting cases described below).

Limiting Case I. Single beam illumination and integral front surface observation (see Fig. 2.5b). The plane-parallel cell is illuminated with a single exciting beam entering the front boundary of the atomizer (flame) cell at an angle of Θ with respect to the front surface. The entire array of fluorescence beams leaving the front cell surface at an angle of $\Theta + \gamma$ is observed. It is apparent that identical fluorescence fluxes are obtained for

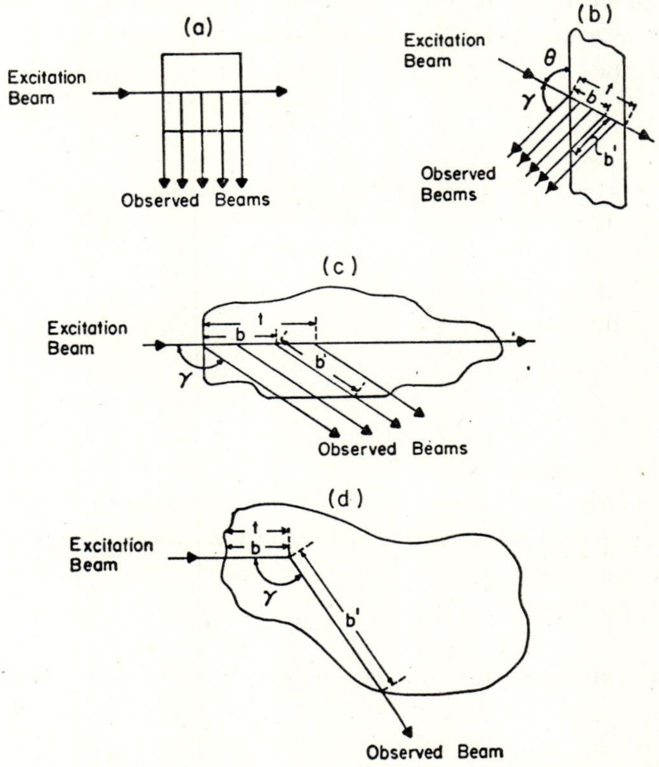

FIG. 2.5. Arrangements of excitation beam-flame cell-measurement beam for atomic fluorescence flame spectrometry: (a) Hooymayers system; (b) Limiting case I. Illumination by a single beam—integral front surface observation of fluorescence; (c) Limiting case II. Illumination by a single beam—integral observation of fluorescence; (d) Limiting case III. Illumination by a single beam—single beam observation.
(Reprinted from Ref. [539] by permission of The Society for Applied Spectroscopy)

all configurations with the same total path length of the primary beam t and the same ratio $\varkappa_1$ of distances from the same arbitrary elementary cell. For this case, the characteristic ratio of distances $\varkappa_1$ is defined by

$$\varkappa_1 = \frac{k_0 b}{k_0' b'} = \frac{k_0 \sin(\Theta + \gamma)}{k_0' \sin \Theta} \tag{2.19}$$

Limiting Case II. Single beam illumination and integral side observation (see Fig. 2.5c). The irregularly shaped atomizer (flame) cell is illuminated with a single excitation beam, and *all* fluorescence radiation resulting within the absorption path length t and leaving the atomizer cell at an angle of γ to the exciting beam is observed. It is also assumed here that the region of the cell where the exciting beam enters is perpendicular to the atomizer (flame) border where the fluorescence beam is observed. For this case, the characteristic ratio of distance $\varkappa_2$ is defined by

$$\varkappa_2 = \frac{k_0 t}{k_0' b'} \tag{2.20}$$

Limiting Case III. Single beam illumination – single beam observation (see Fig. 2.5d). In this case, only fluorescence at a single point of the exciting beam within the irregularly shaped atomizer (flame) cell is observed. The characteristic ratio of distances $\varkappa_3$ is again defined by Eq. (2.20), i.e.

$$\varkappa_3 = \frac{k_0 t}{k_0' b'} \tag{2.21}$$

It should be stressed that Case I (also Case III) exhibits some degree of symmetry: for Case I, if the observed and exciting beams are interchanged, then the resulting growth curves are identical to those corresponding to the original arrangement except that $\varkappa_1$ is now $k_0 b'/k_0 b$ rather than $k_0 b/k_0' b'$.

From the discussion above, it is obvious that *none of the three limiting cases is limited to some particular angle between the exciting and observation beams.* In other words, the angle between the exciting and observation beams in Case I may be greater than 90° to suppress the noise in the flame due to scattering of incident radiation by particles in the flame gases.

It is possible to compute atomic fluorescence growth curves and fluorescence line shapes for all of the limiting cases, using relatively simple computer programmes based upon the equations above. With the exception of some of the results obtained for $a = 0.1$ and $k_0 t = 1000$, the error in computation of Φ_F and $\Phi_{F\nu}$ values is believed to be less than 0.5%.

2.2.1.4. Computed Atomic Fluorescence Growth Curves

Growth curves in atomic fluorescence spectrometry are log-log plots of fluorescence radiant flux (or some linear function of radiant flux) *vs* optical density (or some linear function of optical density). In the growth curve plots in this section, the *ordinate* is a normalized radiant flux, i.e.

$$\Phi_F \bigg/ \left[\left(\frac{\Omega'}{4\pi}\right) Y A_c E_0 \left(\frac{\Omega}{4\pi}\right)\right]$$

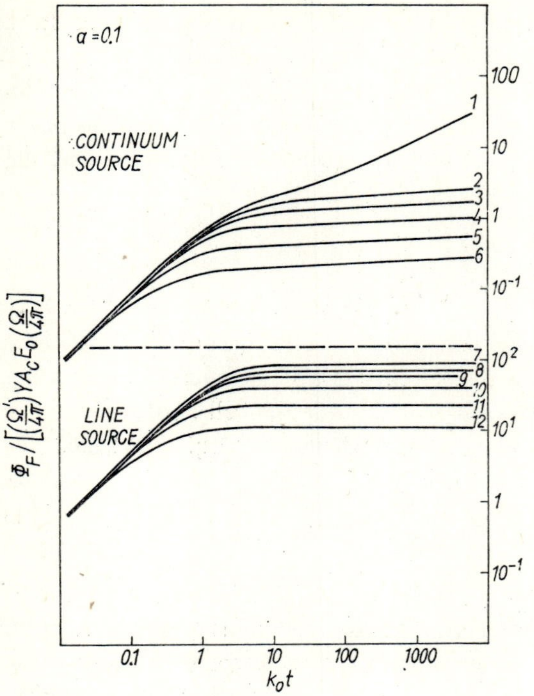

FIG. 2.6a. Computed atomic fluorescence growth curves for Limiting case I with continuum and line sources of excitation: *a*-parameter 0.1. Key to curves: (1) α_C; (2) $\varkappa_1 = 10.0$; (3) $\varkappa_1 = 3.16$; (4) $\varkappa_1 = 1.00$; (5) $\varkappa_1 = 0.316$; (6) $\varkappa_1 = 0.100$ (growth curves 2—6 are with continuum source); (7) α_L; (8) $\varkappa_1 = 10.0$; (9) $\varkappa_1 = 3.16$; (10) $\varkappa_1 = 1.00$; (11) $\varkappa_1 = 0.316$; (12) $\varkappa_1 = 0.100$ (growth curves 8—12 are with line source). (Reprinted from Ref. [539] by permission of The Society for Applied Spectroscopy)

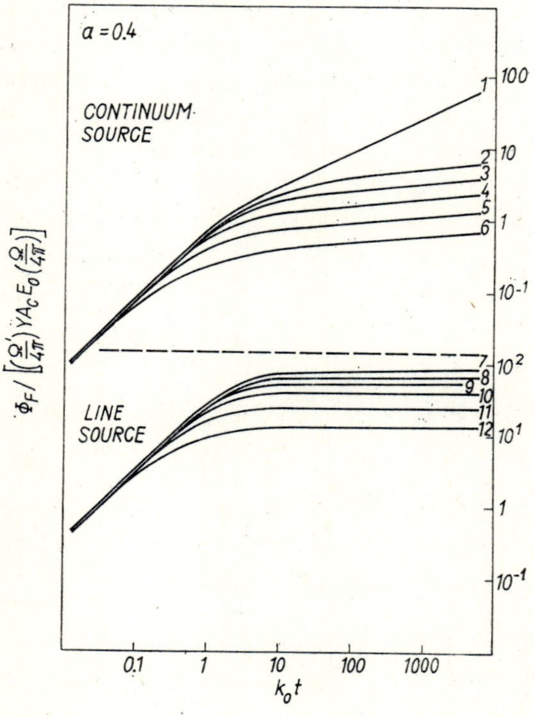

FIG. 2.6b. Computed atomic fluorescence growth curves for Limiting case I with continuum and line sources of excitation: *a*-parameter 0.4. (Key to curves see Fig. 2.6a.) (Reprinted from Ref. [539] by permission of The Society for Applied Spectroscopy)

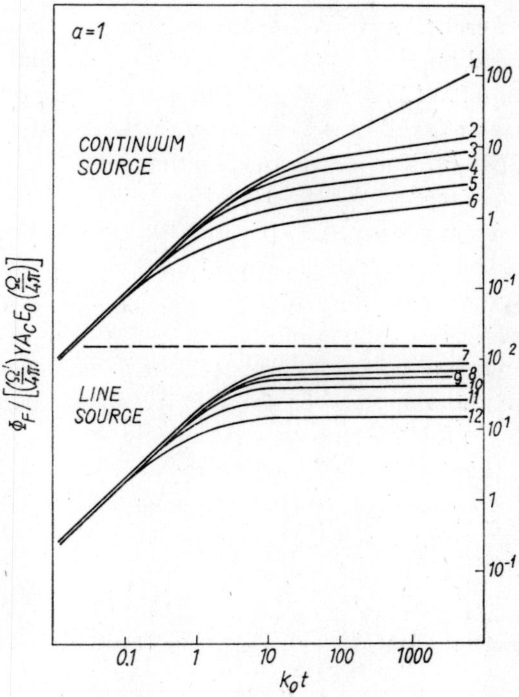

FIG. 2.6c. Computed atomic fluorescence growth curves for Limiting case I with continuum and line sources of excitation: *a*-parameter 1.0. (Key to curves see Fig. 2.6a.)
(Reprinted from Ref. [539] by permission of The Society for Applied Spectroscopy)

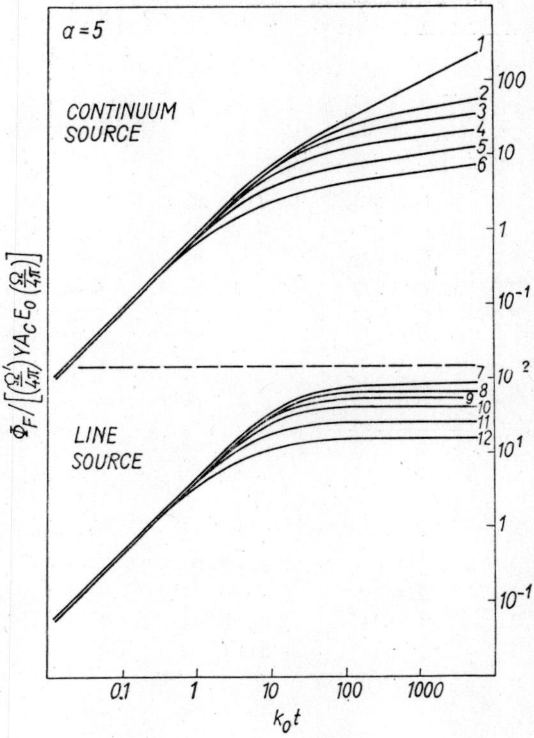

FIG. 2.6d. Computed atomic fluorescence growth curves for Limiting case I with continuum and line sources of excitation: *a*-parameter 5.0. (Key to curves see Fig. 2.6a.)
(Reprinted from Ref. [539] by permission of The Society for Applied Spectroscopy)

where A_c is the cross-sectional area of the exciting beam in cm^2, Ω is the solid angle of excitation in sr, and E_0 and Y are as defined as in section 2.2.1.2, and the *abscissa* is the optical density, $k_0 t$. The ordinate parameter is valid for a continuum source. For a line source, E_0 is replaced by $2E_0'/\sqrt{\pi}$. For the various limiting cases, four graphs—each with a different a-parameter (the a-parameters used are 0.1, 0.4, 1, and 5)—are given. On each graph, a set of six curves is given for both the continuum and the line source case. The uppermost curve for the set of curves for the continuum source is a log-log plot of α_C *vs* $k_0 t$, and for the line source, the uppermost curve is a log-log plot of α_L *vs* $k_0 t$. The five remaining curves for either source are growth curves for the designated a-parameter which differ *only* in the value chosen for $\varkappa$: This ranges from 10.0 for the upper curve to 0.100 for the bottom curve of any set of five curves.

Limiting Case I—continuum source. Growth curves for this case are shown in Figs. 2.6a–d (upper set of curves). It should be noted that all curves approach the same curve with slope of unity at low optical densities

TABLE 2.2. Slope of fluorescence growth curves at high optical densities for continuum and line sources (case I—single beam illumination and integral front surface observation of fluorescence)

Reprinted from Ref. [539] by permission of The Society for Applied Spectroscopy

	Continuum Source — $\varkappa_1$				
a	10.0	3.16	1.00	0.316	0.100
0.1	0.048 01	0.040 68	0.035 90	0.034 16	0.031 97
0.4	0.091 80	0.081 14	0.074 04	0.070 18	0.066 36
1.0	0.113 27	0.100 52	0.092 19	0.087 40	0.083 80
2.0	0.127 32	0.112 34	0.102 82	0.097 50	0.094 10
5.0	0.148 17	0.128 11	0.116 18	0.110 21	0.105 46

	Line Source — $\varkappa_1$				
a	10.0	3.16	1.00	0.316	0.110
0.1	0.000 06	0.000 03	0.000 00	0.000 00	0.000 00
0.4	0.000 14	0.000 09	0.000 05	0.000 03	0.000 00
1.0	0.000 39	0.000 25	0.000 17	0.000 13	0.000 08
2.0	0.000 84	0.000 54	0.000 39	0.000 31	0.000 27
5.0	0.002 17	0.001 43	0.001 06	0.000 88	0.000 79

and a *plateau* (near zero slope) at high optical densities. In Table 2.2, computed slopes of growth curves are given for the high optical density region (k_0t between 10^4 and 10^5). In all cases, the *slope* of the high optical density region was either *slightly positive* or *essentially zero*. With increasing values of $\varkappa$ and/or the a-parameter, the positive slope increases slightly for high optical densities. This high optical density region differs greatly from that for the right-angle case (see Fig. 2.5a) of Hooymayers [200].

Limiting Case I – line source. Growth curves for this case are also shown in Figs. 2.6a – d (lower sets of curves). It should again be noted that all curves approach the same curve with slope of unity at low optical densities and a plateau (near zero slope) at high optical densities. It is extremely interesting that all of these curves at high optical densities have a *slightly positive or zero slope and never have a negative slope as obtained in the Hooymayer's right-angle arrangement* [200] (see Table 2.2).

Limiting Case II – continuum source. Growth curves for this case are shown in Figs. 2.7a – e (solid lines). In the low optical density region, all growth curves approach the same curve with slope of unity, whereas in the high optical density region, the growth curves approach an asymptotic value (slope zero) as predicted by Hooymayers [200]. It is apparent that the

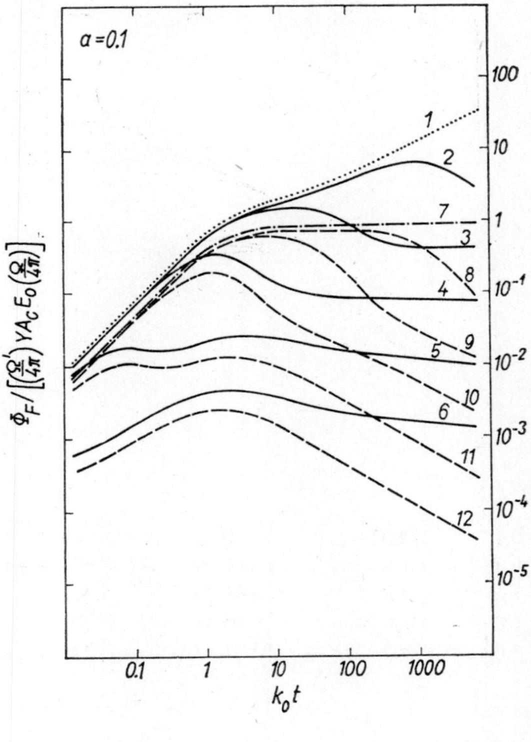

FIG. 2.7a. Computed atomic fluorescence growth curves for Limiting case II with continuum and line sources of excitation: a-parameter 0.1. (Key to curves is the same as in Fig. 2.6a except that $\varkappa_1$ is replaced by $\varkappa_2$). (Reprinted from Ref. [539] by permisison of The Society for Applied Spectroscopy)

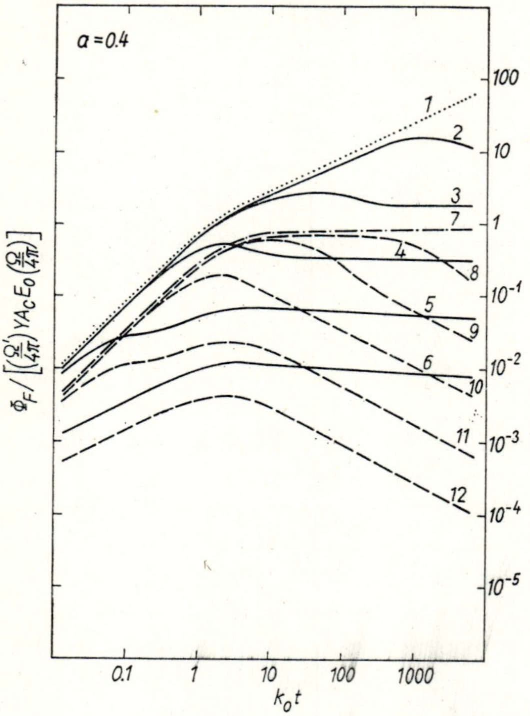

FIG. 2.7b. Computed atomic fluorescence growth curves for Limiting case II with continuum and line sources of excitation: a-parameter 0.4. (Key to curves is the same as in Fig. 2.6a except that $\varkappa_1$ is replaced by $\varkappa_2$.) (Reprinted from Ref. [539] by permission of The Society for Applied Spectroscopy)

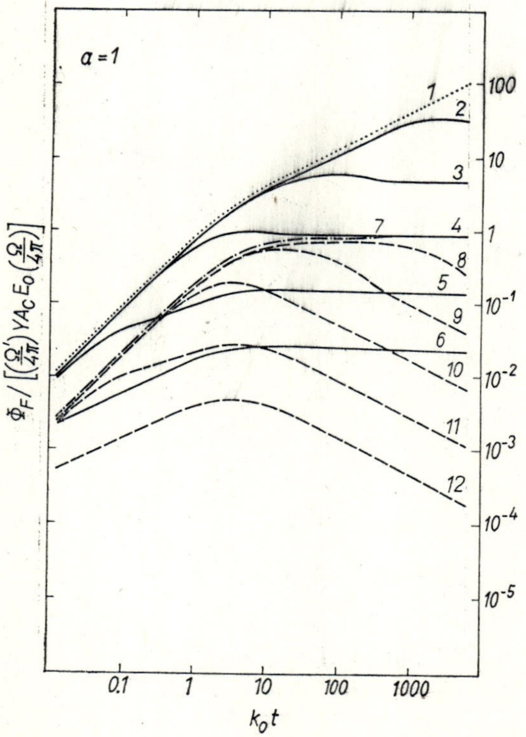

FIG. 2.7c. Computed atomic fluorescence growth curves for Limiting case II with continuum and line sources of excitation: a-parameter 1.0. (Key to curves is the same as in Fig. 2.6a except that $\varkappa_1$ is replaced by $\varkappa_2$.) (Reprinted from Ref. [539] by permission of The Society for Applied Spectroscopy)

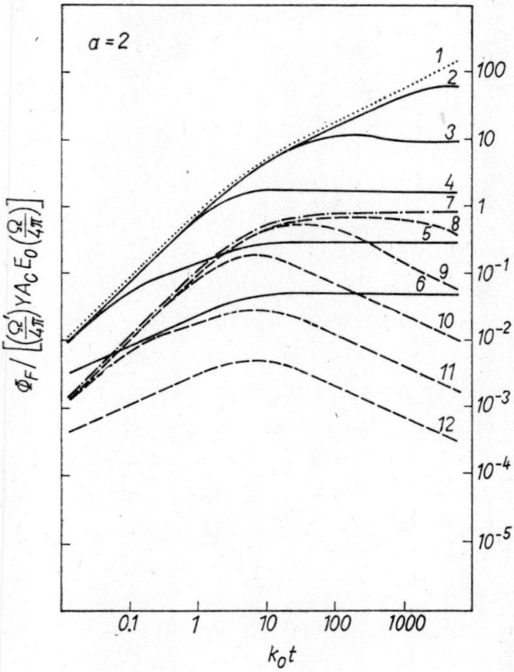

FIG. 2.7d. Computed atomic fluorescence growth curves for Limiting case II with continuum and line sources of excitation: *a*-parameter 2.0. (Key to curves is the same as in Fig. 2.6a except that $\varkappa_1$ is replaced by $\varkappa_2$.) (Reprinted fi om Ref. (539) by permission of The Society for Applied Spectroscopy)

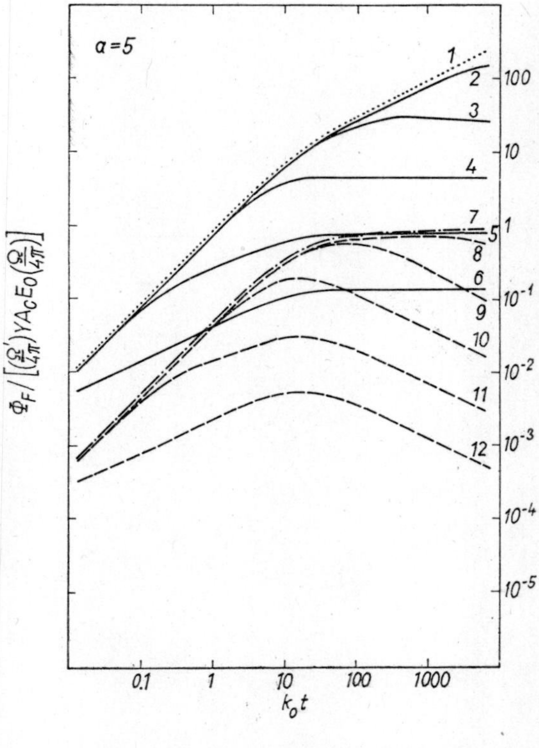

FIG. 2.7e. Computed atomic fluorescence growth curves for Limiting case II with continuum and line sources of excitation: *a*-parameter 5.0. (Key to curves is the same as in Fig. 2.6a except that $\varkappa_1$ is replaced by $\varkappa_2$.) (Reprinted from Ref. [539] by permission of The Society for Applied Spectroscopy)

length of the linear (slope unity) portion of the curves depends upon the value of $\varkappa_2$. The higher its value, the *smaller* is the fluorescence self-absorption effect and the *larger* the optical density at the onset of curvature. When $\varkappa_2 = 1$, these curves are identical with those computed by Hooymayers [200].

Limiting Case II – line source. Growth curves for this case are also shown in Figs. 2.7a – e (dashed lines). In the low optical density region, all growth curves approach the same curve with slope of unity, whereas at high concentrations, the growth curves exhibit either one sharp maximum (e.g., $\varkappa_2 = 0.100$ or 1.00), two maxima (e.g., $\varkappa_2 = 0.316$), or one broad maximum ($\varkappa_2 = 3.16$ or 10.0) and ultimately a negative slope (less steep than the positive slope at low concentrations). Again when $\varkappa_2 = 1$, these curves are identical with those computed by Hooymayers [200].

Limiting Case III – continuum source. In this case, only one small region of the flame is observed, and the atomizer (flame) cell is illuminated by a narrow beam of radiation. The resulting growth curves in Figs. 2.8a – e (solid lines) are independent of the atomizer (flame) cell shape. In principle, the total fluorescence radiant flux of a completely illuminated irregularly (or regularly) shaped atomizer (flame) cell in which *all* of the fluorescence towards the detection system is measured could be calculated by summation

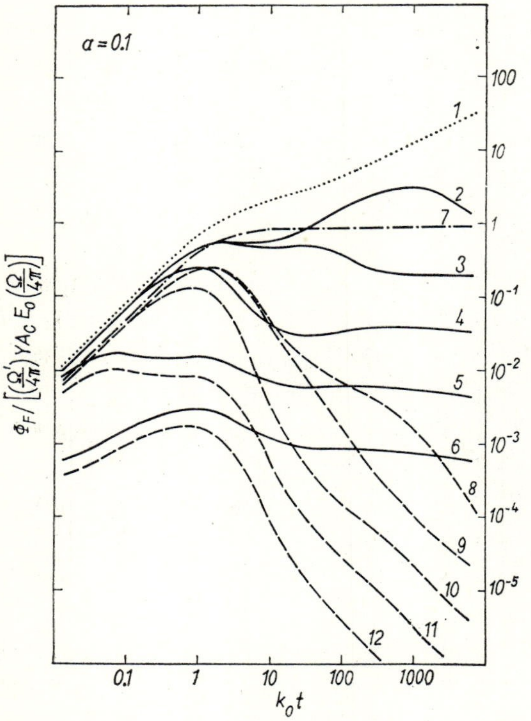

FIG. 2.8a. Computed atomic fluorescence growth curves for Limiting case III with continuum and line source of excitation: *a*-parameter 0.1. (Key to curves is the same as in Fig. 2.6a except that $\varkappa_1$ is replaced by $\varkappa_3$.)
(Reprinted form Ref. [539] by permission of The Society for Applied Spectroscopy)

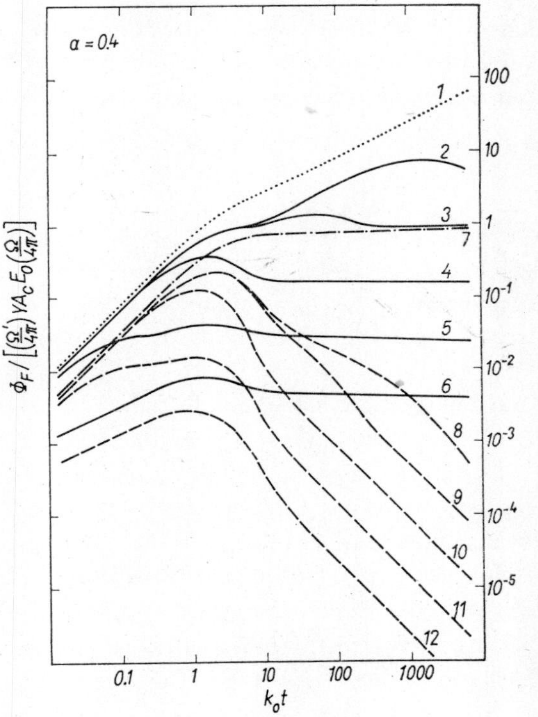

FIG. 2.8b. Computed atomic fluorescence growth curves for Limiting case III with continuum and line sources of excitation: *a*-parameter 0.4. (Key to curves is the same as in Fig. 2.6a except that $\varkappa_1$ is replaced $\varkappa_3$.) (Reprinted from Ref. [539] by permission of The Society for Applied Spectroscopy)

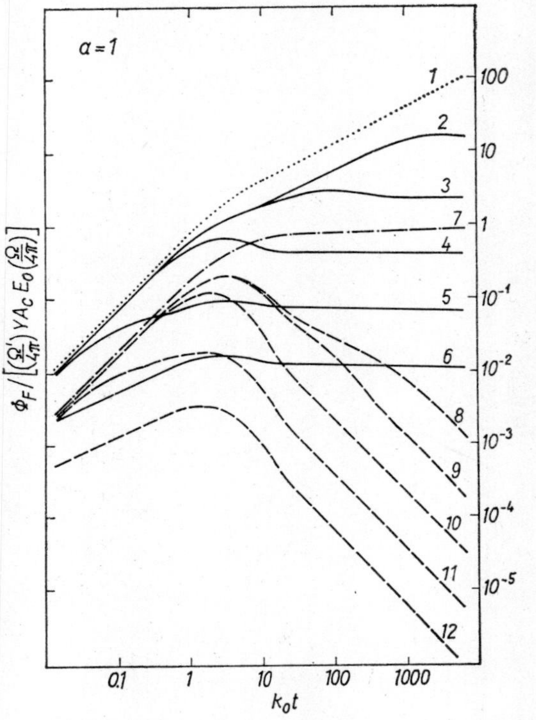

FIG. 2.8c. Computed atomic fluorescence growth curves for Limiting case III with continuum and line sources of excitation: *a*-parameter 1.0. (Key to curves is the same as in Fig. 2.6a except that $\varkappa_1$ is replaced by $\varkappa_3$.)
(Reprinted from Ref. [539] by permission of The Society for Applied Spectroscopy)

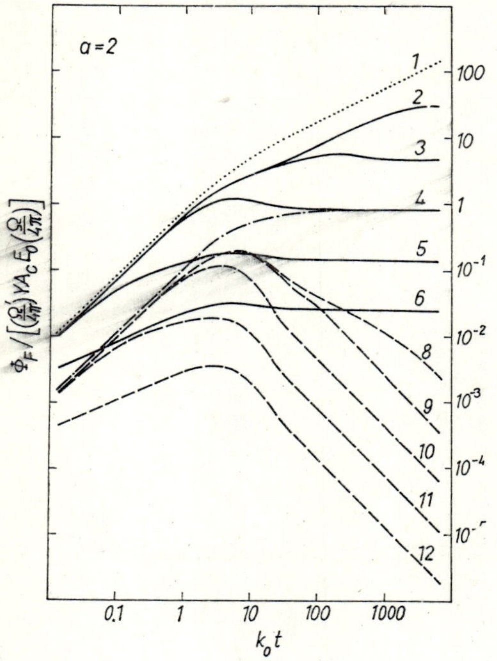

FIG. 2.8d. Computed atomic fluorescence growth curves for Limiting case III with continuum and line sources of excitation: a-parameter 2.0. (Key to curves is the same as in Fig. 2.6a except that $\varkappa_1$ is replaced by $\varkappa_3$.)
(Reprinted from Ref. [539] by permission of The Society for Applied Spectroscopy)

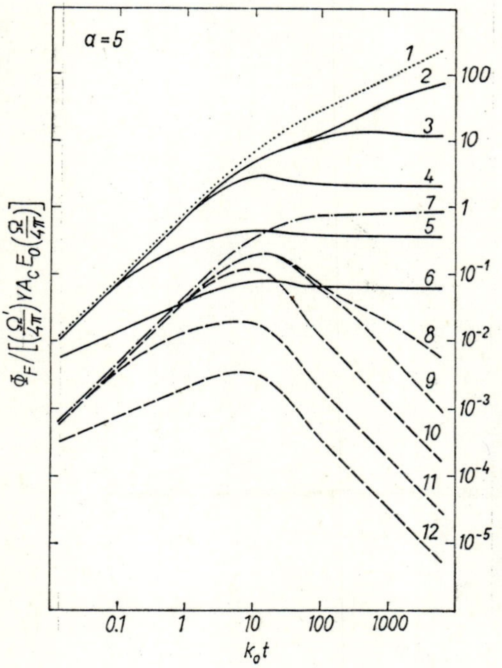

FIG. 2.8e. Computed atomic fluorescence growth curves for Limiting case III with continuum and line sources of excitation: a-parameter 5.0. (Key to curves is the same as in Fig. 2.6a except that $\varkappa_1$ is replaced by $\varkappa_3$.)
(Reprinted from Ref. [539] by permission of The Society for Applied Spectroscopy)

of curves such as those shown in Figs. 2.8a–e, assuming sufficient values of $\varkappa_3$ were used to cover the region of the atomizer cell observed. Again at low optical densities, the growth curves approach the same curve with slope of unity. At high optical densities, the growth curves contain several or just

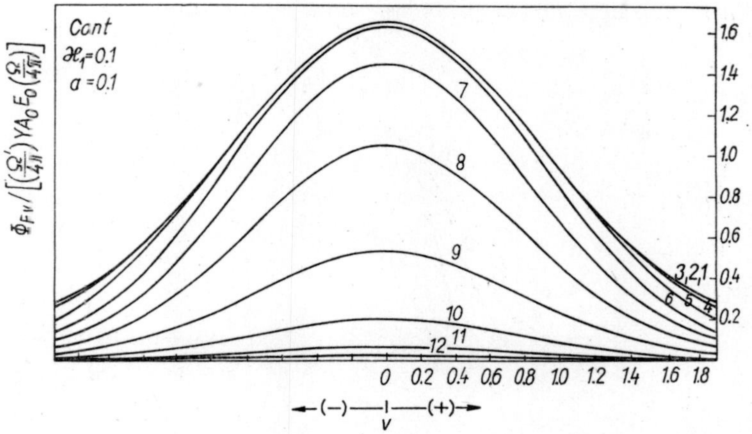

FIG. 2.9a. Computed atomic fluorescence line profiles for Limiting case I with continuum source: a-parameter 0.1; $\varkappa_1 = 0.1$. Key to curves: (1) $k_0 t = 10{,}000$; (2) $k_0 t = 3{,}160$; (3) $k_0 t = 1{,}000$; (4) $k_0 t = 316$; (5) $k_0 t = 100$; (6) $k_0 t = 31.6$; (7) $k_0 t = 10.0$; (8) $k_0 t = 3.16$; (9) $k_0 t = 1.00$; (10) $k_0 t = 0.316$; (11) $k_0 t = 0.100$; (12) $k_0 t = 0.0316$. Note that scales of ordinate and abscissa may vary between figures.

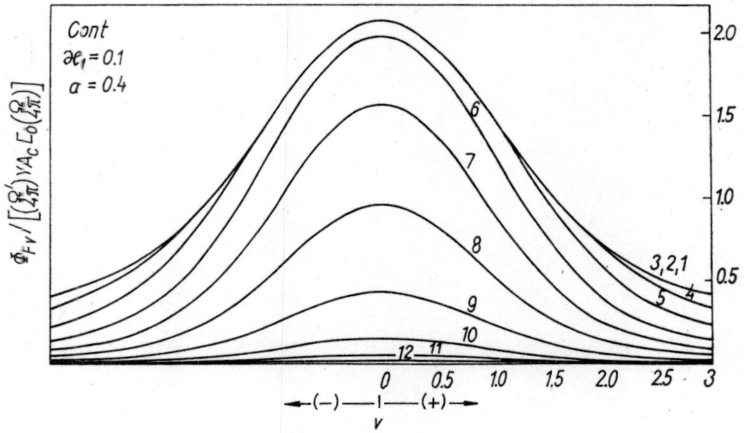

FIG. 2.9b. Computed atomic fluorescence line profiles for Limiting case I with continuum source: a-parameter 0.4; $\varkappa_1 = 0.1$. (Key to curves see Fig. 2.9a.)

one maxima, and at still higher optical densities approach a region of nearly zero slope.

Limiting Case III—line source. Growth curves for this case are also shown in Fig. 2.8a—e. In the low optical density region, the curves again approach a slope of unity. In the high optical density region, the growth curves reach a maximum or may contain several maxima before exhibiting *a rather steep negative slope followed by a less steep negative slope.*

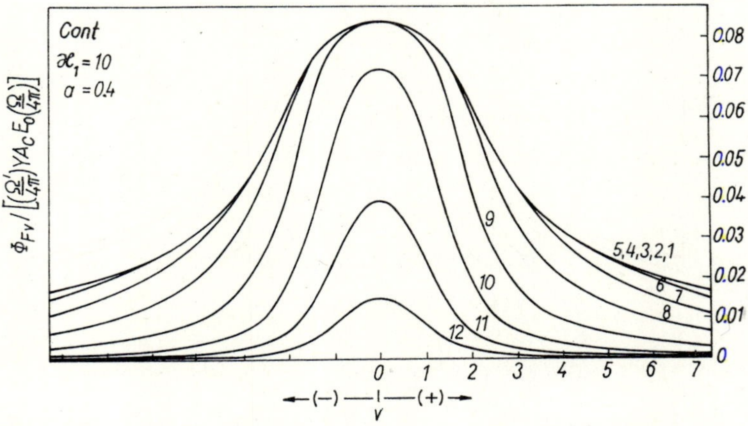

FIG. 2.9c. Computed atomic fluorescence line profiles for Limiting case I with continuum source: *a*-parameter 0.4; $\varkappa_1 = 10$. (Key to curves see Fig. 2.9a.)

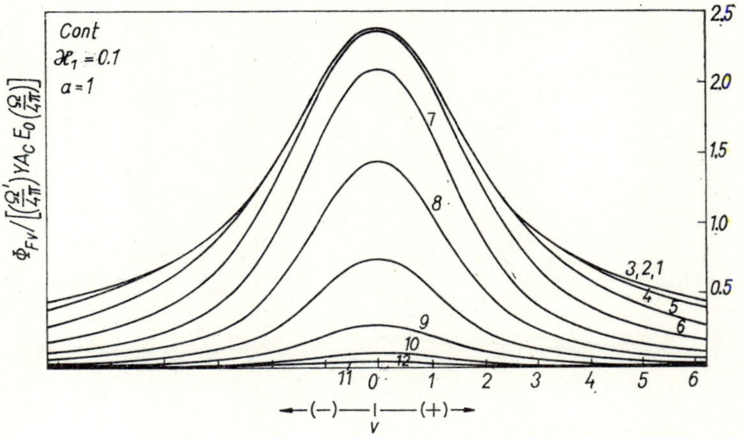

FIG. 2.9d. Computed atomic fluorescence line profiles for Limiting case I with continuum source: *a*-parameter 1.0; $\varkappa_1 = 0.1$. (Key to curves see Fig. 2.9a.)

2.2.1.5. Fluorescence Line Profiles

Fluorescence line profiles are plots of normalized fluorescence spectral radiant flux

$$\Phi_{\mathrm{F}\nu}\left[\left(\frac{\Omega}{4\pi}\right) YA_{\mathrm{c}}E_0\left(\frac{\Omega'}{4\pi}\right)\right]$$

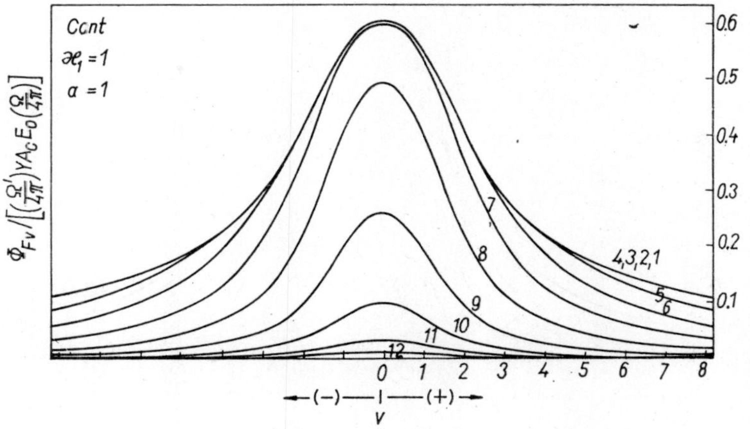

FIG. 2.9e. Computed atomic fluorescence line profiles for Limiting case I with continuum source: a-parameter 1.0; $\varkappa_1 = 1$. (Key to curves see Fig. 2.9a.)

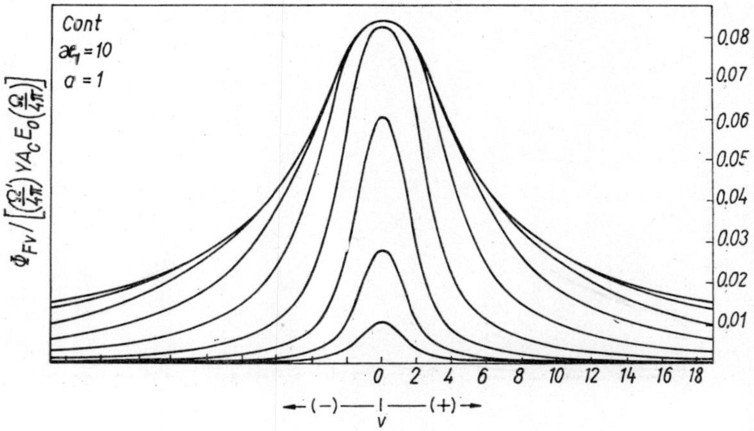

FIG. 2.9f. Computed atomic fluorescence line profiles for Limiting case I with continuum source: a-parameter 1.0; $\varkappa_1 = 10$. (Key to curves see Fig. 2.9a.)

vs relative frequency, *v*. For the various limiting cases, several profiles – each with a different combination of *a*-parameter and $\varkappa$ are given. Each group contains a set of twelve profiles of fluorescence lines corresponding to k_0t values of 0.0316 (bottom profile) 0.100, 0.316, 1.00, 3.16, 10, 31.6, 100, 316, 1,000, 3,160, and 10,000 (top profile). In some figures, the lowest and(or the uppermost curves cannot be distinguished from adjacent curves.

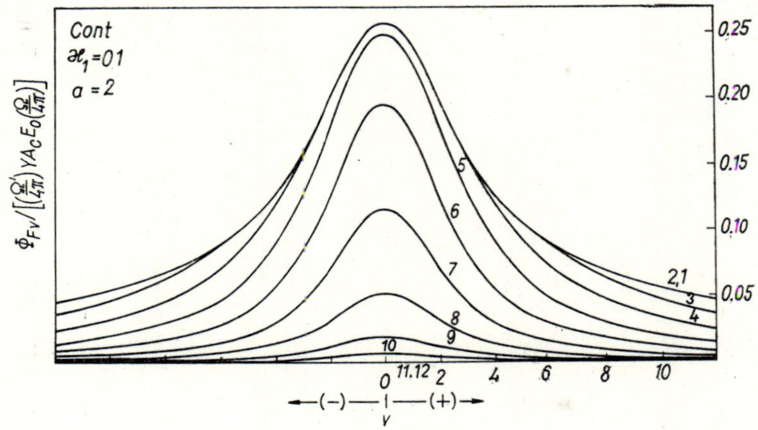

FIG. 2.9g. Computed atomic fluorescence line profiles for Limiting case I with continuum source: *a*-parameter 2.0; $\varkappa_1$ = 0.1. (Key to curves see Fig. 2.9a.)

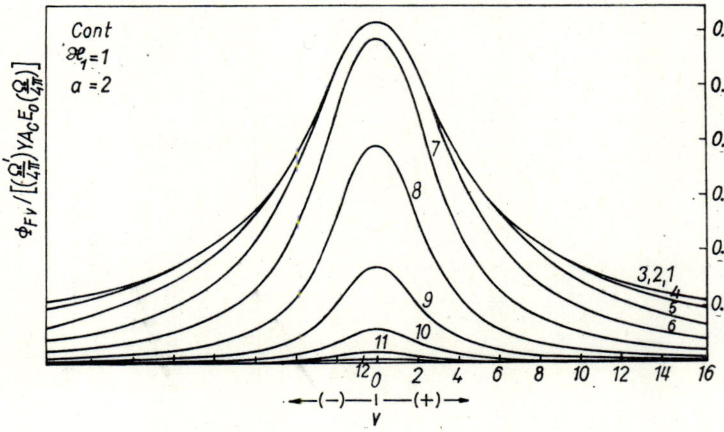

FIG. 2.9h. Computed atomic fluorescence line profiles for Limiting case I with continuum source; *a*-parameter 2.0; $\varkappa_1$ = 1. (Key to curves see Fig. 2.9a.)

Limiting Case I—continuum source. Profiles of fluorescence lines excited by a continuum source and representative of a number of experimental conditions are shown in Figs. 2.9a–k. It should be noted that as $\varkappa_1$ and/or the a-parameter become smaller, the profiles become narrower and the spectral radiant flux at the peak increases for any given optical density. Table 2.3 shows the variation in fluorescence line half-intensity widths for an a-parameter of 5 as a function of $\varkappa_1$ and $k_0 t$. These results indicate that

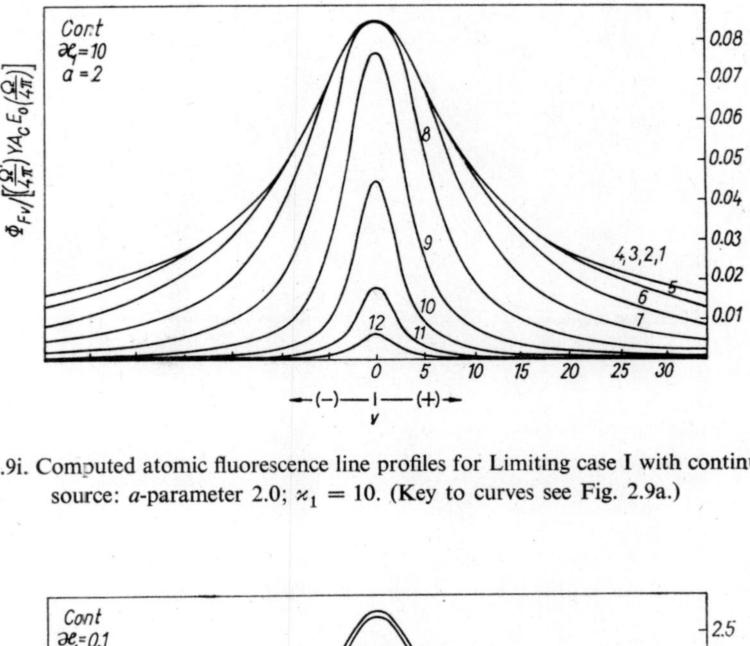

FIG. 2.9i. Computed atomic fluorescence line profiles for Limiting case I with continuum source: a-parameter 2.0; $\varkappa_1$ = 10. (Key to curves see Fig. 2.9a.)

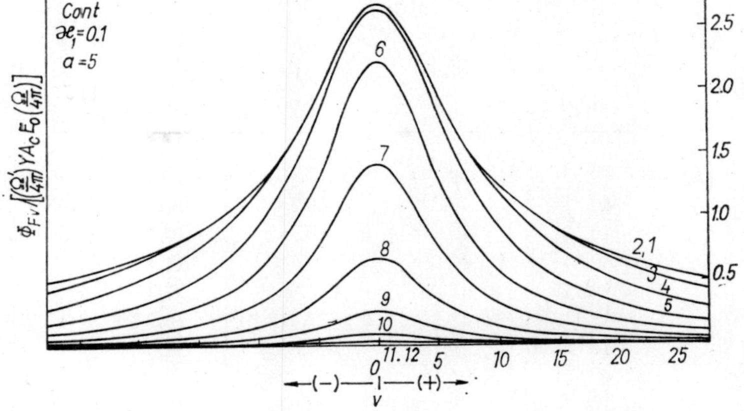

FIG. 2.9j. Computed atomic fluorescence line profiles for Limiting case I with continuum source: a-parameter 5.0; $\varkappa_1$ = 0.1. (Key to curves see Fig. 2.9a.)

substantial changes in optical density ($k_0 t$) or in the method of illumination ($\varkappa_1$) exert little influence over half-width of a fluorescence line. In all cases, the maximum spectral radiance occurs at $v = 0$, (i.e. at the line centre).

Limiting case I—line source. Profiles of fluorescence lines excited by a line source and representative of a number of experimental conditions are shown in Figs. 2.10a–f. Once again the line profiles become narrower

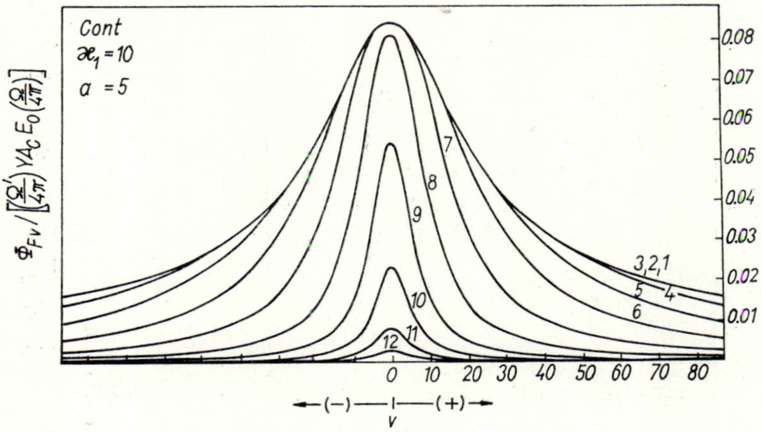

FIG. 2.9k. Computed atomic fluorescence line profiles for Limiting case I with continuum source: *a*-parameter 5.0; $\varkappa_1 = 10$. (Key to curves see Fig. 2.9a.)

TABLE 2.3. Theoretical fluorescence line half-widths (in units of v) for an *a*-parameter of 5 and for limiting case I

$k_0 t$	$\varkappa_1$				
	10.0	3.16	1.00	0.316	0.100
0.0316	10.314	10.314	10.320	10.376	10.388
0.100	10.314	10.320	10.336	10.388	10.552
0.316	10.322	10.336	10.388	10.552	11.134
1.00	10.336	10.388	10.550	11.128	13.046
31.6	10.972	12.442	17.074	27.828	47.646
100.	12.118	15.570	22.412	34.252	56.604
316.	14.664	19.166	24.672	35.680	57.910
1 000.	17.808	20.166	24.776	35.692	57.914
3 160.	18.520	20.176	24.776	35.692	57.914
10 000.	18.524	20.176	24.776	35.692	57.914

as $\varkappa_1$ and/or the a-parameter become smaller and the spectral radiant flux at the line peak increases for any given optical density. In all cases, the maximum spectral radiance occurs at $v = 0$.

Limiting Cases II or III – continuum or line source. Profiles of fluorescence lines excited by continuum or line sources are shown in Figs. 2.11a – d for representative a-parameters. It can be seen that both the half-width and self-reversal increase as the optical density ($k_0 t$) increases. As the a-parameter

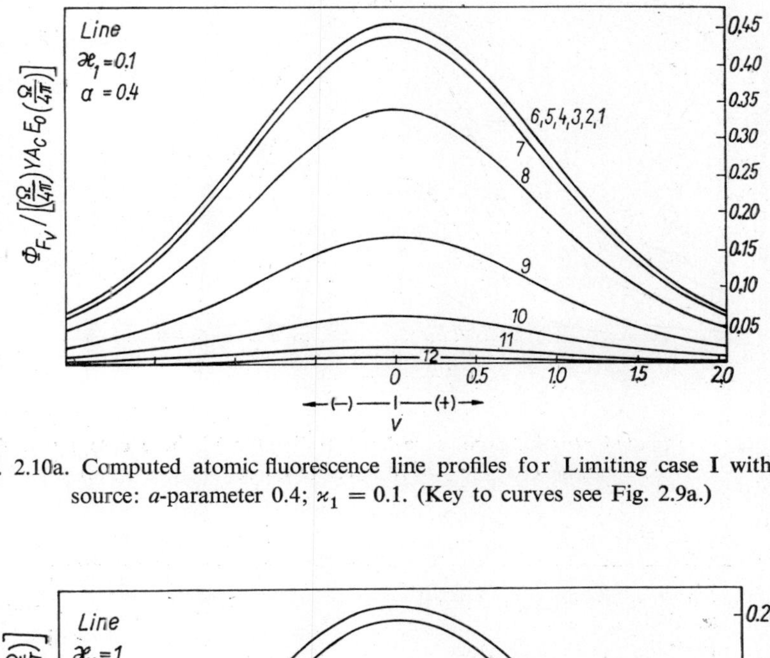

FIG. 2.10a. Computed atomic fluorescence line profiles for Limiting case I with line source: a-parameter 0.4; $\varkappa_1 = 0.1$. (Key to curves see Fig. 2.9a.)

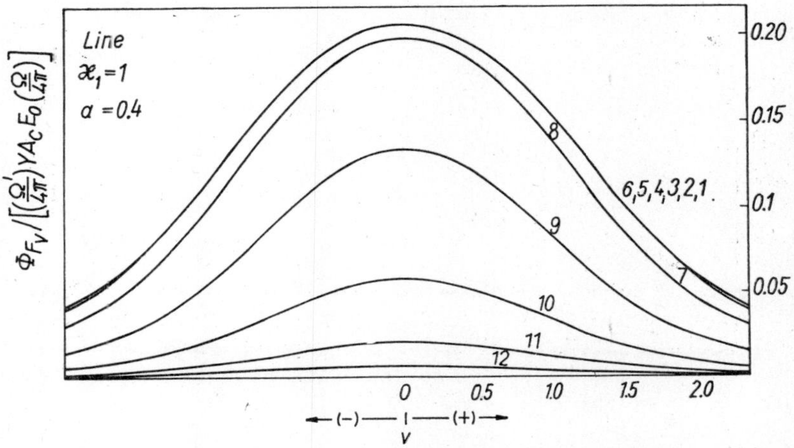

FIG. 2.10b. Computed atomic fluorescence line profiles for Limiting case I with line source: a-parameter 0.4; $\varkappa_1 = 1$. (Key to curves see Fig. 2.9a.)

becomes smaller, self-reversal for any given optical density increases. For Cases II and III, the curves are normalized so that the total area under each curve in every figure is constant. It should be noted that the relative frequency interval v is also given approximately by

$$v \sim \frac{2(\lambda_0 - \lambda)\sqrt{\ln 2}}{\Delta\lambda_D}$$

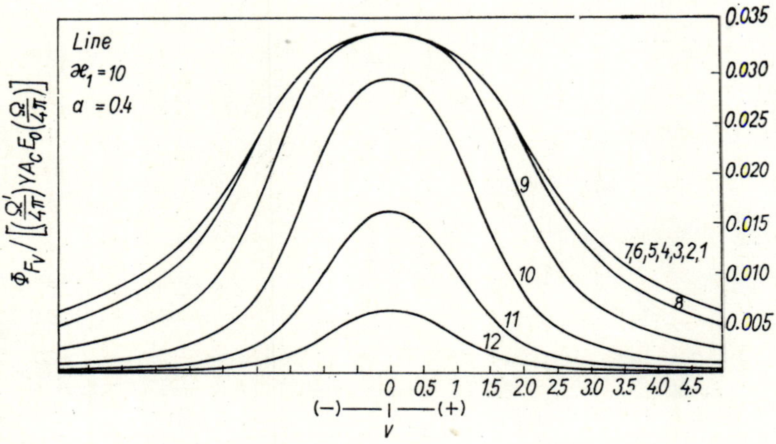

FIG. 2.10c. Computed atomic fluorescence line profiles for Limiting case I with line source: a-parameter 0.4; $\varkappa_1 = 10$. (Key to curves see Fig. 2.9a.)

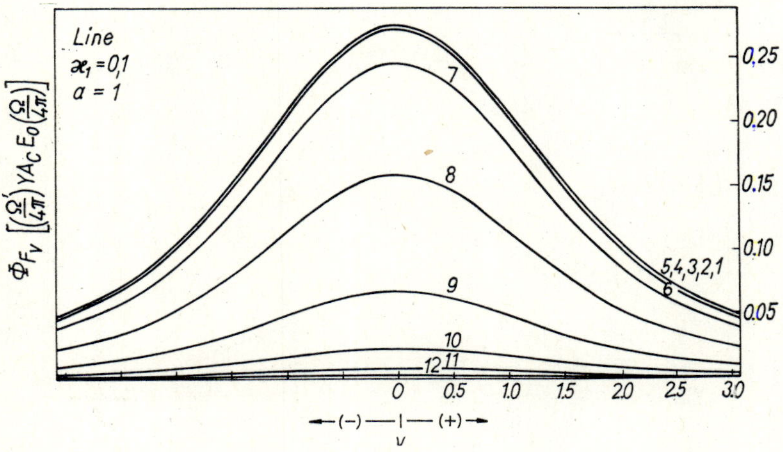

FIG. 2.10d. Computed atomic fluorescence line profiles for Limiting case I with line source: a-parameter 1.0; $\varkappa_1 = 0.1$. (Key to curves see Fig. 2.9a.)

where λ_0 is the peak fluorescence wavelength, λ is any fluorescence wavelength, and $\Delta\lambda_D$ is the Doppler half-width in wavelength units. For values of v up to about 100, λ_0 and λ will differ by only about 1%, and so v is given by the more convenient expression above.

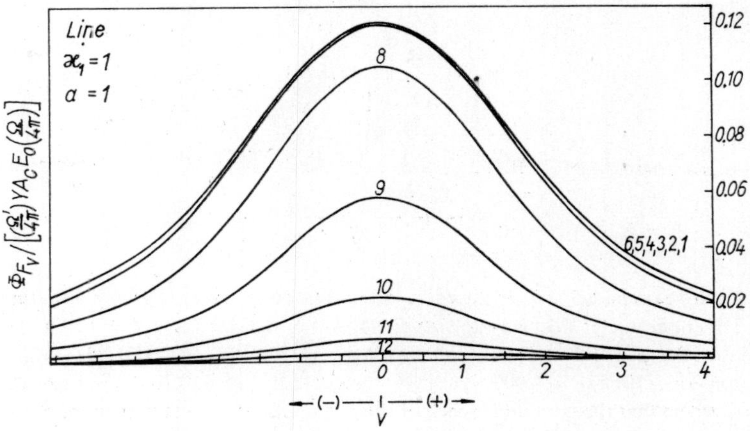

FIG. 2.10e. Computed atomic fluorescence line profiles for Limiting case I with line source: a-parameter 1.0; $\varkappa_1 = 1$. (Key to curves see Fig. 2.9a.)

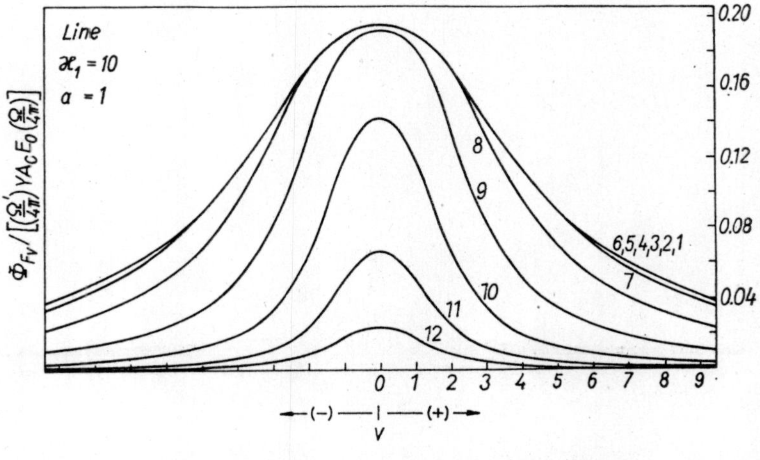

FIG. 2.10f. Computed atomic fluorescence line profiles for Limiting case I with line source: a-parameter 1.0; $\varkappa_1 = 10$. (Key to curves see Fig. 2.9a.)

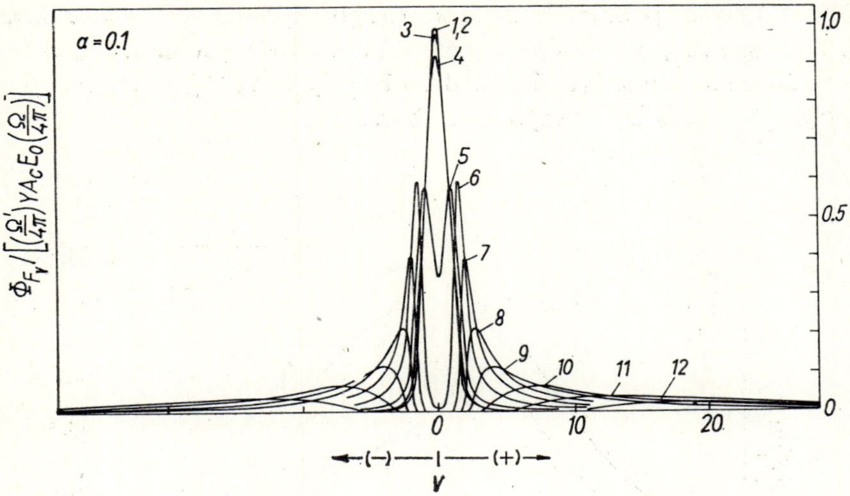

FIG. 2.11a. Computed atomic fluorescence normalized line profiles for Limiting cases II or III: *a*-parameter 0.1. Key to curves: (1) $k_0t = 0.0316$; (2) $k_0t = 0.100$; (3) $k_0t = 0.316$; (4) $k_0t = 1.00$; (5) $k_0t = 3.16$; (6) $k_0t = 10.0$; (7) $k_0t = 31.6$; (8) $k_0t = 100$; (9) $k_0t = 316$; (10) $k_0t = 1{,}000$; (11) $k_0t = 3{,}160$; (12) $k_0t = 10{,}000$. All these curves are normalized so that the area under each of them is constant. To the maximum of the curve with $k_0t = 0.0316$ (practically not self-absorbed) an arbitrary value of Φ_F equal to 1.0 was assigned.

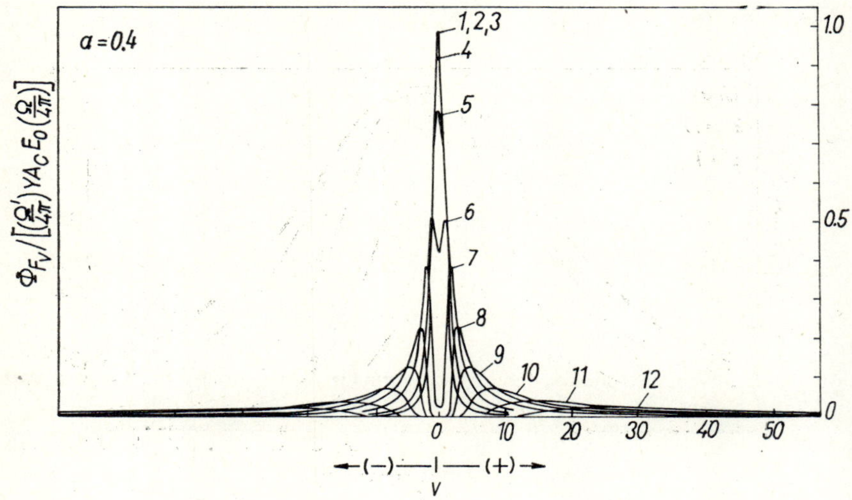

FIG. 2.11b. Computed atomic fluorescence normalized line profiles for Limiting cases II or III: *a*-parameter 0.4. (Key to curves see Fig. 2.11a.)

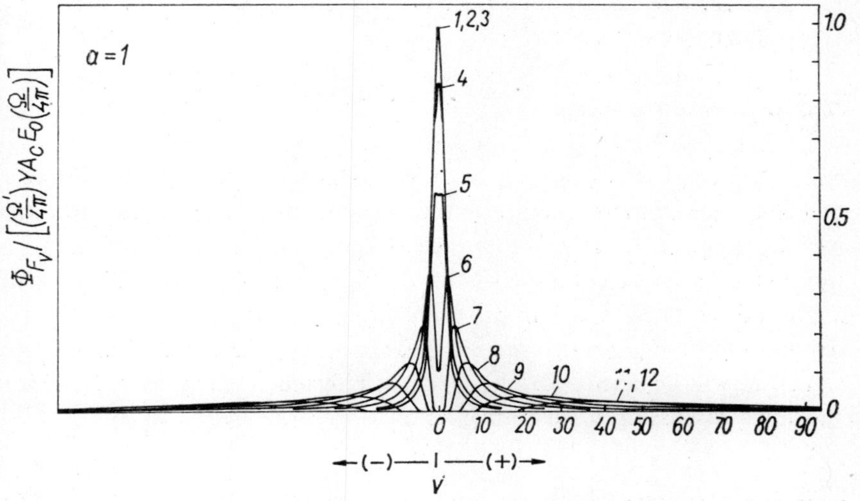

FIG. 2.11c. Computed atomic fluorescence normalized line profiles for Limiting cases II or III: *a*-parameter 1.0. (Key to curves see Fig. 2.11a.)

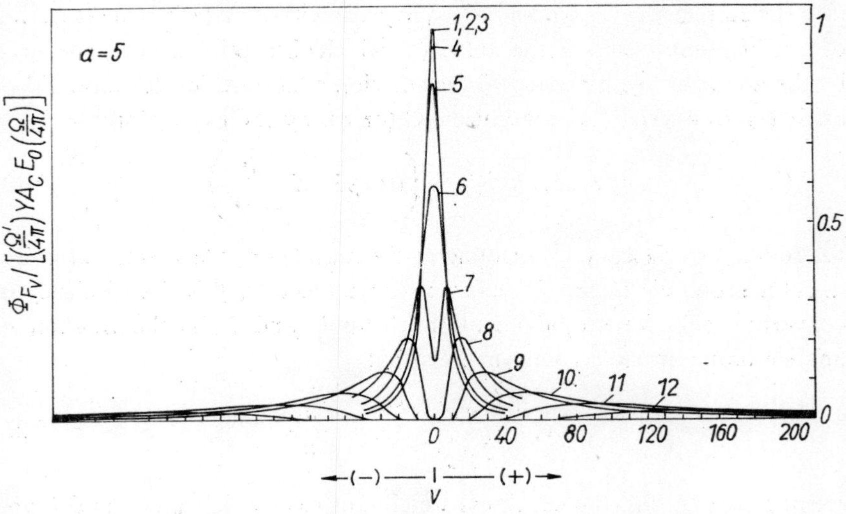

FIG. 2.11d. Computed atomic fluorescence normalized line profiles for Limiting cases II or III: *a*-parameter 5.0. (Key to curves see Fig. 2.11a.)

2.2.2. COMPARISON OF THEORETICAL AND EXPERIMENTAL GROWTH CURVES

2.2.2.1. General Comments

Experimental growth curves are plots (which may be log-log) of photodetector signal, S, *vs* concentration, C, or amount, q, of analyte introduced into an atomizer. An atomic fluorescence spectrometer consists of an excitation source (line or continuum), an atomizer, a monochromator or filter, a photodetector, and a current (or charge) measurement device. In this section, only the general principles governing the measured photodetector signal, S, will be discussed. The reader is referred to Chapter 3 for a more thorough discussion of instrumentation.

2.2.2.2. The Atomizer Cell

The atomizer, e.g., a flame, a furnace, etc., converts the sample containing the analyte into an atomic vapour. Samples in solution can either be continuously introduced into the atomizer *via* a nebulizer or discretely introduced into the atomizer *via* a hypodermic syringe. In either case, the concentration of analyte atoms in state i (the state from which absorption occurs) produced by the atomizer, n_i, is related to the solution concentration (or amount) of analyte introduced to the atomizer, C (or q), by the expression:

$$n_i = \varepsilon\beta K_1 X_i C \qquad \left(\text{or } \varepsilon\beta K_2 X_i \frac{q}{A_i V}\right) \tag{2.22}$$

where ε is the efficiency of introducing the sample analyte into the atomizer in submicroscopic forms, i.e. as atoms or molecules, β is the efficiency of converting submicroscopic forms into atoms, and X_i is the fraction of analyte atoms in state i, and is given by:

$$X_i = \frac{g_i e^{-E_i/kT}}{\Sigma_j g_j e^{-E_j/kT}} = \frac{n_i}{n} \tag{2.23}$$

where g_i and g_j are the statistical weights of electronic states i and j, and E_i and E_j are the energies of states i and j, k is the Boltzmann constant, T is the atomizer temperature (in K), and the summation is over all electronic states of the analyte atom.

[If i is the ground electronic state 0 (the most common case), then

$$X_0 = \frac{g_0}{\Sigma_j g_j e^{-E_j/kT}} \tag{2.24}$$

For many atoms the first excited electronic state is of the order of 0.5 eV or more above the ground state, and in this instance $X_0 = 1$.] V in Eq. (2.22) is the volume of solution containing q grams of analyte atoms of atomic weight A_i, and K_1 (or K_2) is a factor characteristic of the atomizer. For example, for a flame, $K_1 = F_s N_A / F_g$, where F_s is the flow-rate of solution into the flame, F_g is the flow-rate of burnt gases in the flame, and N_A is Avogadro's number. If discrete sampling is utilized, K_2 depends critically upon the exact method of atomization, gas flow-rates, etc., and no general expression will be given here. Fortunately, over a large range of concentrations, C (or q), $\varepsilon\beta K_1$ (or $\varepsilon\beta K_2$) will be independent of C (or q). However, for very small values of C (or q), β will usually decrease owing to increased ionization of analyte atoms; for large values of C (or q), $\varepsilon\beta K_1$ (or $\varepsilon\beta K_2$) will decrease owing to a decrease in the efficiency of production of submicroscopic forms of the analyte. For a thorough discussion of atomization in flames and in non-flame furnaces, the reader should consult the monograph by L'vov [262].

2.2.2.3. Quenching in the Atomizer

The physical and chemical environments within an atomizer cell have a profound effect upon the magnitude of the fluorescence flux emitted from the cell in addition to their influence on atomization efficiency.

In practice, the excited atom is always subject to radiationless deactivation due to bimolecular collisions with the other species present in the cell. Collisions of this type (quenching collisions) are characterized by a quantum yield, Y, i.e. the number of quanta emitted as fluorescence to the number of quanta absorbed capable of producing fluorescence.

The observed fluorescence radiant flux is directly proportional to the quantum yield (see Eq. 2.1), and so Y appears, as a normalization factor, in the ordinates of both the theoretical growth curve plots (Figs. 2.6–2.8) and the atomic fluorescence line profiles (Figs. 2.9–2.11). In the most general instance, the quantum yield for a fluorescence transition from an upper excited state of an atom, u, to a state of lower electronic energy, l, (which need not be the ground state of the atom), Y_{ul}, is given by:

$$Y_{ul} = \frac{A_{ul}}{\sum_i A_{ui} + r_u} \tag{2.25}$$

A_{ul} is the Einstein coefficient of spontaneous emission for the radiative transition u → l, in Hz, $\sum_i A_{ui}$ is the summation of the Einstein coefficients of spontaneous emission, in Hz, for all the possible radiative transitions from the upper state states i (including the transition u → 1), r_u is the total pseudo first-order rate constant for quenching from state u, in Hz, and is given by:

$$r_u = \sum_x n_{Qx} r_{Qx} \tag{2.26}$$

where n_Q is the concentration of a foreign species, in particles cm^{-3}, and r_Q is the second-order rate constant for the quenching process, in $cm^3\ s^{-1}$: The summation is for all quenching species of type x, and can be related to the quenching cross-section* σ_Q^2, in cm^2 (the product of the kinetic cross-section of the two colliding species and the probability that a collision will result in quenching of the excited atom), to the atomic weight of the atom M_a in amu, and to the molecular weight of the colliding species M_x, by the expression:

$$\sum_x n_{Qx} r_{Qx} = \sqrt{8\pi RT} \sum_x \sigma_{Qx}^2 n_{Qx} \left(\frac{1}{M_a} + \frac{1}{M_x} \right)^{1/2} \tag{2.27}$$

where R is the universal gas contant, J K^{-1} mol^{-1}.

Although quite general, Eq. (2.25) is only valid when photon excitation is the predominant pathway to the excited state, u, of the atom (i.e. collisional excitation is assumed to contribute negligibly to the excited state population). Also, only one absorption process is considered. If there is the possibility of an appreciable thermally or chemically excited population of an energy level from which a separate process of absorption can occur to give rise to fluorescence emission of frequency ν_{ul}, this must also be considered in the expression for the quantum efficiency.

In order to evaluate Eq. (2.25) *exactly* for an atomic fluorescence line emitted from an atomizer cell, it is necessary to know the Einstein coefficients for spontaneous emission for *all* transitions originating in the upper excited state, u. Also, the quenching rate constant, r_u, must be known for the combination of the specific excited state of the atom, u, and the gases present in the cell. Values of Y_{ul} have been calculated for a few elements in the flames most commonly used in atomic fluorescence spectrometry ($H_2/O_2/Ar$, $H_2/O_2/N_2$ and $C_2H_2/O_2/N_2$) and the values are given in

---

* The quenching cross-section, σ_Q^2, must be clearly differentiated from the collisional broadening cross-section for spectral lines, σ_C^2. The *quenching* process results in shortening of the mean lifetime of an electronic state of an atom. Collisional broadening *is* influenced by quenching collisions, but the major contribution to collisional broadening results from adiabatic perturbations of the electronic state, which occur without any permanent change in the electronic state of the atom.

Table 2.4. The high quantum yields (up to 0.75) for the argon-diluted flame are of particular interest.

TABLE 2.4. Calculated quantum yields (Y_{ul}) for various atoms in several flames and hot argon[1]

Reprinted from Ref. [539] by permission of The Society for Applied Spectroscopy

Flame composition	Temperature (K)	Fluorescence quantum yield				
		Na	K	Li	Tl	Pb
$2\,H_2/O_2/4\,N_2$	2100	0.066	0.047	0.021	0.070	0.079
$6\,H_2/O_2/4\,N_2$	1800	0.049	0.049	0.014	0.099	0.10
$0.4C_2H_2/O_2/4N_2$	2200	0.042	0.028	0.017	0.042	0.067
$2\,H_2/O_2/10\,Ar$	1800	0.75	0.37	0.15	0.33	0.22
$H_2/O_2/4\,N_2$	1600	0.044	0.03	—[3]	0.051	0.069
Hot Ar [2]	3000	≥ 0.98	≥ 0.99	≥ 0.95	≥ 0.99	≥1

[1] Values from [442].
[2] Values calculated from [211, 214, 215, 216].
[3] Value not reported in [442].

When the atomizer cell environment differs from the examples considered here, the quantum yield must be calculated by using A-values [287, 436] and quenching cross-sections [10, 197, 201, 211, 214, 215, 216] available in the literature.

In order to utilize the atomic fluorescence growth curves (Figs. 2.6–2.8) and the atomic fluorescence line profiles (Figs. 2.9–2.11), it is necessary to be able to estimate the a-parameter(s) for the atomic line(s) (absorption and/or fluorescence) of concern and to estimate the value of $k_0 t$, the absorption optical density.

Estimates of the a-parameter require a knowledge of line half-widths: These may be calculated by the method of Parsons, McCarthy and Winefordner [319]. Details of the calculation are not considered here, but the equations involved have been applied to calculating line-widths for two hypothetical atoms of mass (m_a) 10 and 100 in a variety of flames, and also in a furnace containing an atmosphere of argon at 3000 K. In the latter instance the effective cross-section for collisional broadening, σ_C^2, was assumed to have a value of 2 Å^2, and the average molecular weight (M_f) of the colliding species was taken as 40. Other assumptions made were identical to those of Parson *et al.* [319]. For hypothetical wavelengths of 2000 and 4000 Å, the results are shown in Table 2.5. The data in Table 2.5

TABLE 2.5. Typical half-widths of spectral lines
Reprinted from Ref. [539] by permission

Atom	Atomizer		$\Delta\lambda_C$ (Å)
$m_a = 10, \lambda_0 = 2000$ Å	O_2/H_2 ($T = 2700, m_f = 10$)	$\alpha_C^2 = 20$	0.002 5
		$\alpha_C^2 = 100$	0.012
	O_2/C_2H_2 ($T = 3000, m_f = 22$)	$\alpha_C^2 = 20$	0.002 0
		$\alpha_C^2 = 100$	0.008 0
	Air/C_2H_2 ($T = 2500, m_f = 30$)	$\alpha_C^2 = 20$	0.002 1
		$\alpha_C^2 = 100$	0.010
	Hot Ar ($T = 3000, m_f = 40$)	$\alpha_C^2 = 2$	0.000 18
$m_a = 100, \lambda = 2000$ Å	O_2/H_2 ($T = 2700, m_f = 10$)	$\alpha_C^2 = 20$	0.001 8
		$\alpha_C^2 = 100$	0.009 2
	O_2/C_2H_2 ($T = 3000, m_f = 22$)	$\alpha_C^2 = 20$	0.001 2
		$\alpha_C^2 = 100$	0.005 8
	Air/C_2H_2 ($T = 2500, m_f = 30$)	$\alpha_C^2 = 20$	0.000 12
		$\alpha_C^2 = 100$	0.005 0
	Hot Ar ($T = 3000, m_f = 40$)	$\alpha_C^2 = 2$	0.000 10
$m_a = 10, \lambda_0 = 4000$ Å	O_2H_2 ($T = 2700, m_f = 10$)	$\alpha_C^2 = 20$	0.010
		$\alpha_C^2 = 100$	0.048
	O_2/C_2H_2 ($T = 3000, m_f = 22$)	$\alpha_C^2 = 20$	0.008 0
		$\alpha_C^2 = 100$	0.032
	Air/C_2H_2 ($T = 2500, m_f = 30$)	$\alpha_C^2 = 20$	0.008 4
		$\alpha_C^2 = 100$	0.040
	Hot Ar ($T = 3000, m_f = 40$)	$\alpha_C^2 = 2$	0.000 72
$m_a = 100, \lambda = 4000$ Å	O_2/H_2 ($T = 2700, m_f = 10$)	$\alpha_C^2 = 20$	0.007 2
		$\alpha_C^2 = 100$	0.037
	O_2/C_2H_2 ($T = 3000, m_f = 22$)	$\alpha_C^2 = 20$	0.004 8
		$\alpha_C^2 = 100$	0.023
	Air/C_2H_2 ($T = 2500, m_f = 30$)	$\alpha_C^2 = 20$	0.004 8
		$\alpha_C^2 = 100$	0.023
	Hot Ar ($T = 3000, m_f = 40$)	$\alpha_C^2 = 2$	0.000 40

[a] The Einstein coefficient of spontaneous emission for the spectral line is taken as 10^8 sec^{-1}; temperatures are given in K.

of several hypothetical atoms in several atomizers[a]
of The Society for Applied Spectroscopy

$\Delta\nu_C$ (Hz)[b]	$\Delta\lambda_N$ (Å)	$\Delta\nu_N$ (Hz)[b]	$\Delta\lambda_D$ (Å)[b]	$\Delta\nu_N$ (Hz)[b]	$\Delta\lambda_T$ (Å)	$\Delta\nu_T$ (Hz)[b]
1.9	0.000 02	0.015	0.024	18	0.024	18
9.0	0.000 02	0.015	0.024	18	0.027	20
1.5	0.000 02	0.015	0.025	19	0.025	19
6.0	0.000 02	0.015	0.025	19	0.026	19
1.6	0.000 02	0.015	0.023	17	0.023	17
8.0	0.000 02	0.015	0.023	17	0.025	19
0.14	0.000 02	0.015	0.025	19	0.025	19
1.4	0.000 02	0.015	0.0075	5.6	0.0077	5.8
7.0	0.000 02	0.015	0.0075	5.6	0.012	9.0
0.9	0.000 02	0.015	0.0079	5.9	0.0080	6.0
4.4	0.000 02	0.015	0.0079	5.9	0.0098	7.4
0.9	0.000 02	0.015	0.0072	5.5	0.0073	5.5
4.4	0.000 02	0.015	0.0072	5.5	0.0093	7.0
0.08	0.000 02	0.015	0.0079	5.9	0.0074	5.6
1.9	0.000 08	0.015	0.048	9.0	0.0049	0.9
9.0	0.000 08	0.015	0.048	9.0	0.068	13.0
1.5	0.000 08	0.015	0.050	9.5	0.051	9.7
6.0	0.000 08	0.015	0.050	9.5	0.060	11.4
1.6	0.000 08	0.015	0.046	8.5	0.047	8.9
8.0	0.000 08	0.015	0.046	8.5	0.061	11.5
0.14	0.000 08	0.015	0.050	9.5	0.050	9.5
1.4	0.000 08	0.015	0.015	2.8	0.017	1.3
7.0	0.000 08	0.015	0.015	2.8	0.040	7.6
0.9	0.000 08	0.015	0.016	2.9	0.017	3.2
4.4	0.000 08	0.015	0.016	2.9	0.028	5.3
0.9	0.000 08	0.015	0.014	2.6	0.015	2.9
4.4	0.000 08	0.015	0.014	2.6	0.027	5.1
0.08	0.000 08	0.015	0.015	2.9	0.015	2.9

[b] Frequency intervals have been multiplied by 10^{-9}, i.e., 2.0×10^9 Hz.

should be useful in estimating the factors (and the sensitivity to change in factors) which are important in influencing the broadening of spectral lines in analytical atomizers. Also the values in Table 2.5 should allow estimation of the various half-widths for a specific spectral line of a specific atom in an analytical atomizer. Of course, if more accurate values are desired, then it is necessary to use the half-width expressions.

Knowledge of atomic line half-widths of absorption and fluorescence lines of analyte atoms also allows an estimate to be made of the a-parameter, which is defined by Eq. 2.28, i.e.

$$a = \sqrt{\ln 2}\left(\frac{\Delta\nu_L}{\Delta\nu_D}\right) = \sqrt{\ln 2}\left(\frac{\Delta\lambda_L}{\Delta\lambda_D}\right) = 0.83\left(\frac{\Delta\lambda_L}{\Delta\lambda_D}\right) \quad (2.28)$$

For flames, a-parameters are in the range of $0.1-5$ (see the data in Table 2.5); therefore the atomic fluorescence growth curves plotted in Figs. 2.6–2.8 are for this range of values. For the one non-flame cell (furnace), the a-parameter is of the order of 0.01; since no atomic fluorescence growth curves are plotted for atomic lines with a-parameters less than 0.1, a reasonable estimate of the shape and approximate position of the growth curves can be obtained by reference to curves corresponding to $a = 0.1$.

2.2.2.4. Estimates of Optical Density

The total absorption path length t can be simply estimated by illuminating the atomizer with a source of light and measuring the width of the cell image. The atomic absorption coefficient for the absorption process k_0 is given by Eq. (2.7), i.e.

$$k_0 = \frac{2\sqrt{\ln 2}Xn_0 f}{\sqrt{\pi}\Delta\nu_D} = \frac{2\sqrt{\ln 2}\,Xn_0\lambda_0^2}{c\sqrt{\pi}\Delta\lambda_D} = 8.1\times 10^{-13}\left(\frac{n_0\lambda_0^2}{\Delta\lambda_D}\right) \quad (2.29)$$

The parameters in parentheses can be evaluated as follows: the oscillator strength f and the peak wavelength λ_0, in cm, by reference to suitable tables [287, 436]; the Doppler half-width, $\Delta\lambda_D$, in cm, can be estimated as in [319]; and the concentration, n_0, of atoms in the lower state involved in the absorption process can be estimated if the atomizer is calibrated. For example, for a flame produced with a premixed burner and a chamber type nebulizer to introduce analyte into the flame, n_0, in atoms cm^{-3}, is given by Eq. (2.30).

$$n_0 = \frac{\varepsilon\beta F_s N_A C}{F_g} \quad (2.30)$$

where ε is the efficiency of producing submicroscopic species from the analyte in solution (no units), β is the efficiency of producing atoms from the submicroscopic species (no units), F_s is the solution transport rate, in $cm^3\ s^{-1}$, C is the concentration of analyte in the sample solution, in mole cm^{-3}, N_A is Avogadro's number (no units), and F_g is the flow-rate of burnt gases into the flame, in $cm^3\ sec^{-1}$. Typical solution flow-rates are 0.03 – 0.2 $cm^3\ sec^{-1}$. Equation (2.30) is also valid for direct injection nebulizer burners. Typical burnt gas flow-rates, F_g, for analytical premixed flames are: $3 \times 10^3\ cm^3\ sec^{-1}$ for nitrous oxide-acetylene ($T \sim 2950$ K, $e_f \sim 16.4$); $3.4 \times 10^3\ cm^3\ sec^{-1}$ for nitrous oxide-hydrogen ($T \sim 2900$ K, $e_f \sim 9{,}6$); $1.5 \times 10^3\ cm^3$ for air-acetylene ($T \sim 2450\ K$, $e_f \sim 8.5$); $\sim 2 \times 10^3\ cm^3\ sec^{-1}$ for air-hydrogen ($T \sim 2000$ K, e_f 6.1); and $3 \times 10^3\ cm^3\ sec^{-1}$ for argon-oxygen-hydrogen ($T \sim 2200$ K, $e_f \sim 7$). The burnt gas flow-rate F_g is given by $F_g = F_u e_f$, where F_u is the flow-rate of unburnt gases into the flame and e_f is the flame gas expansion factor and is given by

$$e_f = \frac{n_T T}{n_R T_R}$$

where T and T_R are the flame temperature and room temperature respectively, and n_T and n_R are the number of moles of flame gas products at T and T_R, respectively.

The first four values are taken from an article by de Galan and Samey [136] and the value for the argon-oxygen-hydrogen flame from an article by Smyly *et al.* [371]. Typical ε values for air-acetylene and nitrous oxide-acetylene flames produced with premixed nebulizer burners are 0.08 and 0.04, respectively [136, 442]. Typical β values for a variety of atoms are given in Table 2.6.

Example of estimation of n_0 and k. Assuming the use of an air–acetylene flame, a solution flow-rate of 0.1 $cm^3\ sec^{-1}$, and the measurement of Mn atoms ($\beta \sim 0.62$ and $f = 0.057$ for the 4030 Å line) we find

$$n_0 \sim \frac{0.07 \times 0.62 \times 0.1 \times 6 \times 10^{23} C}{1.5 \times 10^3} = 1.7 \times 10^{18} C\ \text{cm}^{-3}$$

Thus, if the concentration is 1 ppm (1 μg/cm^3), i.e. $C = 1.8 \times 10^{-8}$ mole cm^{-3}, then $n_0 \sim 3 \times 10^{10}$ atoms cm^{-3}. Therefore, k_0 is

$$k_0 \sim \frac{8.1 \times 10^{-13} \times 3 \times 10^{10} \times 0.057 \times (4.03 \times 10^{-5})^2}{2 \times 10^{-10}} = 0.011\ \text{cm}^{-1}$$

Thus, if the flame thickness is taken as 1 cm, $k_0 T \sim 0.01$ which is certainly on the linear (slope of unity) portion of any of the analytical curves. Actually

TABLE 2.6. Measured free atom fractions (β-values) of various atoms in several flames

Reprinted from Ref. [539] by permission of The Society for Applied Spectroscopy

Element	Ref. [136]				Ref. [134]	Ref. [487]	Ref. [242]	Ref. [242]	Ref. [442]	Ref. [442]	Ref. [371]
	$C_2H_2 - N_2O$	$H_2 - N_2O$	C_2H_2 – air	H_2 – air	C_2H_2 – air			$C_2H_2 - N_2O$	C_2H_2 – air	$C_2H_2 - N_2O$	$H_2 - O_2 - Ar$
Ag	0.57	0.72	0.70	0.85	0.66	—	—	—	—	—	1.00*
Al	0.13	$<10^{-4}$	$<5\times10^{-5}$	$<8\times10^{-5}$	$<10^{-5}$	—	10^{-4}	0.29	—	0.28	—
Au	0.27	0.43	0.40	0.54	$<10^{-3}$	—	0.21	0.16	0.76	0.48	—
B	0.0035	$<10^{-3}$	$<6\times10^{-4}$	$<10^{-3}$		—	—	—	—	—	—
Ba	0.17	0.0046	0.0018	0.005	0.0011	0.0009§	0.0034	0.074	0.0031	0.20	—
Be	0.095	4×10^{-4}	4×10^{-5}	2×10^{-5}	—	—	—	—	—	—	—
Bi	0.35	0.26	0.17	0.63	—	—	—	—	—	—	—
Ca	0.52	0.036	0.07	0.15	0.14	0.066§	0.052	0.34	0.083	0.04	—
Cd	0.56	0.62	0.38	0.37	0.50	—	0.80	0.60	—	—	—
Co	0.25	0.28	0.28**	0.21	0.41†	—	0.023	0.11	—	—	—
Cr	0.63	0.042	0.071**	0.31	0.064	0.065	0.55	1.00	—	—	—
Cu	0.66	0.92	0.88	0.96	0.98	1.00*	0.40	0.49	1.19	0.68	0.97
Fe	0.83	0.91	0.84	0.82	0.66	0.38	—	—	—	—	—
Ga	0.73	0.16	0.16	0.45	0.16	—	—	—	—	—	—
In	0.93	0.61	0.67**	1.07	0.67	—	0.10	0.37	—	—	—
K	0.12	0.20	0.32	0.40	0.25	—	—	—	0.34	—	—
Li	0.34	0.094	0.12	0.14	0.20	—	—	—	0.31	0.30	—
Mg	0.88	0.97	1.06	0.87	0.59	0.64	1.05	1.07	1.00*	1.00*	0.29
Mn	0.77	0.54	0.62	0.75	0.45	0.59	0.33	0.39	0.83	—	0.26
Na	0.97	0.90	1.04	1.06	1.00	—	0.63	0.32	0.63	0.23	—
Pb	0.84	0.93	0.77	1.03	0.44	—	—	—	—	—	—
Rb	—	—	—	—	—	—	—	—	0.19	—	—
Si	0.055	$<10^{-3}$	$<10^{-3}$	$<2\times10^{-3}$	—	—	—	—	—	—	—
Sn	0.82	0.059	0.043	0.38	$<10^{-4}$	—	0.061	0.35	0.092	0.52	—
Sr	0.26	0.039	0.063	0.17	0.13	0.075§	0.068	0.57	—	—	0.094
Ti	0.11	$<3\times10^{-3}$	$<10^{-3}$	$<2\times10^{-3}$	—	—	—	—	—	0.21	—
Tl	0.55	0.61	0.52	0.68	0.36	—	—	—	—	—	—
V	0.32	0.0081	0.015	0.018	—	—	4×10^{-4}	—	—	0.61	—
Zn	—	—	—	—	0.45	—	—	—	1.31	0.61	—

† Recalculated from Ref. [218]; here the non-resonance character of the transition was not taken into account.
* Assumed reference value.
§ Theoretically calculated values [134]: Ca – 0.05, Sr – 0.06, Ba – 0.0007.
** From absolute emission intensities Co – 0.15, Cr – 0.078, In – 0.43.

most analysts are interested in analysing dilute solutions of analyte, i.e. $C < 10$ ppm, and so optical densities, $k_0 t$, of most analytical flames containing analyte atomic vapour will be on the linear portion of the analytical curve. For atoms, e.g. zinc, cadmium, copper, etc., with large absorption oscillator strengths, the non-linear region of the growth curves (slope $\neq$ unity) may begin at concentrations near 1 ppm. For atoms, e.g. manganese and other atoms with absorption oscillator strengths of the order of 0.1, the linear region of the growth curves will extend to about 10–1000 ppm. For atoms, e.g. mercury, with absorption oscillator strengths of the order of 0.01 or less, the linear region of the growth curves will extend to 1000 ppm or higher.

Chapter 3

INSTRUMENTATION

3.1. EXPERIMENTAL ARRANGEMENTS FOR MEASURING ATOMIC FLUORESCENCE

The instrumental requirements of atomic fluorescence spectrometry are analogous to those of molecular fluorimetry and, apart from the need for external excitation sources, to those of flame emission spectrometry. A block diagram of an atomic fluorescence spectrometer is shown in Fig. 3.1. The

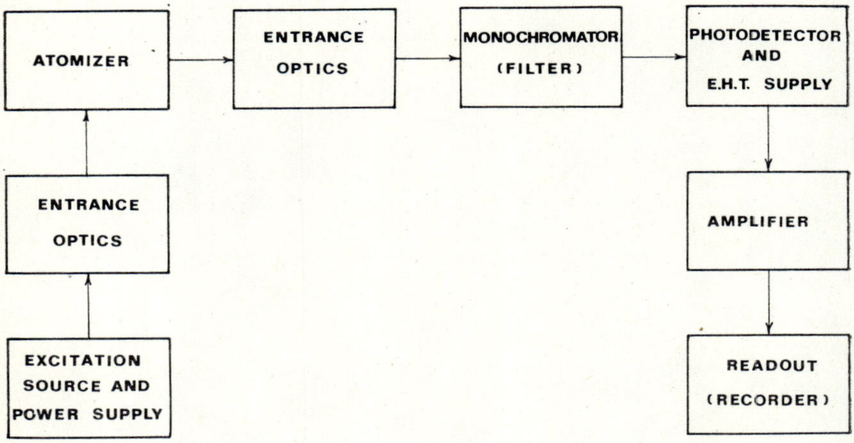

FIG. 3.1. Block diagram of an atomic fluorescence spectrometer.

basic instrument consists of an excitation source with an appropriate power supply, an atomizer, a method of isolating the wavelength region of interest, a photodetector, and an electronic amplifier/read-out system. Illumination at right angles to the detection axis is preferred to suppress detection of radiation from the excitation source. With various minor modifications, this instrumental arrangement has been the basis of all atomic fluorescence systems built to date.

The excitation source can be either a spectral line source (e.g., metal vapour-discharge lamp, microwave-excited electrodeless discharge lamp, one of the various types of hollow-cathode lamp) or a continuum source (e.g. high-pressure xenon arc lamp). The atomizer is usually a nebulizer-burner system producing either a turbulent or a premixed flame of the type commonly used in atomic absorption and flame emission spectrometry, or a non-flame atom reservoir (e.g. electrically heated graphite furnace or metal filament operated in an inert gas atmosphere).

Lenses are normally used to focus the exciting radiation on the atomic vapour and the emitted fluorescence onto the entrance slit of a monochromator or a detector window; mirrors may be used to increase the effective intensity of the excitation source and the fraction of the fluorescence radiation reaching the detector; baffles and slits minimize the effects of scattered and stray radiation. The fluorescence wavelengths can be isolated by a monochromator, a narrow-band, interference filter, or broad-band optical filter, depending on the source and atomizer type, the element being determined, the sample matrix, etc., but in some cases it is advantageous to omit the monochromator.

The photodetectors employed are without exception photomultiplier tubes either of the conventional type or especially sensitized or "solar blind" photomultipliers which are sensitive only in the ultraviolet (1800–3200 Å) region. The electronic measurement devices used range from a simple d.c. amplifier and direct read-out system to a more complex tuned a.c. phase-sensitive amplifier and recorder read-out system or a photon-counting detection system. When a.c. systems are used, the excitation source output is either electronically modulated or mechanically modulated (chopped).

There is no commercial instrumentation available at present designed specifically for atomic fluorescence spectrometry, except for a prototype multichannel atomic fluorescence/flame emission spectrometer introduced by the Technicon Instruments Corporation, Tarrytown, N. Y., U.S.A. Very good results, however, have been obtained with modified atomic absorption/flame emission spectrometers and particularly with purpose-built instruments assembled from selected commercial and/or laboratory made units.

In the following sections, the individual components of an atomic fluorescence instrument are discussed in more detail.

3.2. EXCITATION SOURCES

3.2.1. GENERAL CONSIDERATIONS

In Section 2.2 it was pointed out that the fluorescence intensity is proportional to the intensity of the exciting radiation. Consequently, sources of high radiance are required for good limits of detection. The general requirements for an excitation source are as follows.

1. High radiance over the centre of the absorption line of the analyte atoms (at least $10^{-4}-10^{-3}$ W cm^{-2} sr^{-1} for the most sensitive ultraviolet lines which usually correspond to a total output radiant flux of *ca.* $10^{-2}-10^{-1}$ W)
2. Good short-term and long-term stability
3. Availability for a large number of elements
4. Simplicity of operation
5. Long lifetime
6. Short warm-up time
7. Low cost of source and associated power supplies

Purity of the line spectrum does not matter so much in atomic fluorescence spectrometry as in atomic absorption because only those wavelengths that give rise to the fluorescence are important i.e. the fluorescence process functions as a "self-monochromating" device). This is in contrast to scatter, where any spectral lines which are not resolved by the monochromator result in erroneous fluorescence signals.

Similarly, the width of the source line is also not as critical as in atomic absorption flame spectrometry (see Sections 2.2 and 2.4). Consequently, both line and continuum sources can be used with success. Narrow line sources are preferable where elements with very closely spaced lines occur together in the atomic vapour.

The development of suitable excitation sources constitutes one of the major fields of research in atomic fluorescence. Many types of source have been prepared and tested with differing degrees of success. At present none of them completely fulfils the requirements postulated above. Line sources generally yield higher radiance and are usually less expensive than suitable continuum sources, but the latter offer the advantage of multielement analysis, are generally more stable, simpler to focus and operate, and have longer lifetimes. Electrodeless discharge lamps (particularly the multielement type) still show most promise for the future despite some unsolved problems regarding their stability, lifetime, and combination of more than just a few elements. More work is required to develop satisfactory excitation sources that fulfil all the requirements mentioned.

Line and continuum excitation sources may be powered either by a conventional d.c. supply or some form of electronically modulated supply, so that the exciting radiation is modulated either mechanically (chopped) or electronically. Modulation of the exciting radiation coupled with a.c. detection of the resulting modulated photodetector signals eliminates measurement of the unwanted emission signals resulting from both the analyte species and the background of a flame or non-flame atomizer, as well as of other d.c. signals. The sinusoidal or square-wave modulation of exciting radiation does not usually result in any substantial increase in the atomic fluorescence signal-to-noise ratio and can even result in a decrease [451].

Pulse techniques may also be applied to the operation of excitation sources [131, 145, 168, 333]. For a source such as a hollow-cathode lamp, the maximum signal-to-noise ratio in atomic fluorescence spectrometry is obtained by operating the lamp at the maximum possible current, provided the lamp does not exhibit strong self-reversal. For d.c. or continuous wave operation, the maximum intensity obtainable from a given lamp is limited by the maximum current which may be used without overheating the cathode and damaging the lamp. Operation of lamps by using short-duration pulses of current with a sufficient off-period between pulses to maintain the mean lamp current at a low level results in an increase in peak intensity and a reduction of the total signal observation time. For example, if a hollow-cathode lamp capable of continuously emitting spectral radiant flux $\Phi_{\lambda O}$ at a steady-state current i_0 (which corresponds to a fluorescence radiation of spectral radiant flux $\Phi_{\lambda F}$) is pulsed at an on-to-off time ratio of 1 : 10, and if the mean discharge current is constant, then the peak lamp current i_{pk} can be increased to ca. 10 i_0, since

$$i_{pk} = \frac{i_0}{w_p f_p} \tag{3.1}$$

where w_p = pulse width in sec, and f_p = pulse repetition frequency in sec^{-1}. The peak lamp spectral radiant flux $\Phi_{\lambda pk}$ increases to at least 10 $\Phi_{\lambda O}$ according to

$$\Phi_{\lambda pk} \sim i_{pk}^n \tag{3.2}$$

where n for a given lamp usually lies within the range 1–3 [223, 333]. Consequently, spectral radiant flux $\Phi_{\lambda pk}$ gives rise to a fluorescence intensity of at least 10 $\Phi_{\lambda F}$ assuming that there is no significant increase in width or self-reversal of the excitation source line.

When the signal processing system is gated between pulses, 90 % of the noise arising from the analyte emission and background emission from the atomizer, etc. is rejected, which results in a significant increase in the mean fluorescence signal-to-noise ratio. Mitchell [145] derived the following expression for the ratio of signal-to-noise ratio values for pulsed and continuously operated hollow-cathode lamp sources operating at the same mean current under optimized conditions for the same measurement time:

$$\frac{[S/N]_{\text{pulsed}}}{[S/N]_{\text{coutinuous}}} = \frac{1}{[w_p f_p]^{n-1/2}} \tag{3.3}$$

For pulsed operation, by design $w_p f_p < 1$, and the signal-to-noise ratio increases if $n > 1/2$. For example, if the lamp is pulsed for 40 msec at 1 Hz, pulsing improves the signal-to-noise ratio by a factor of 5 if $n = 1$, 25 if $n = 1.5$, and 125 if $n = 2$.

According to the discussion above, an ideal excitation source would be a stable, high-power repetitively-pulsed source with a small ratio of on-to-off time. Very good results have been obtained with pulsed hollow-cathode lamps for a large number of elements [32, 143, 144, 146, 291, 294] and with pulsed, tunable dye lasers [168]. At the time of writing this book, limited information on the use of pulsed excitation sources in atomic fluorescence spectrometry was available.

For practical purposes, in the choice of the excitation source account should be taken of the analytical requirements, i.e. number of elements to be determined, type of matrix, interferences expected, and radiance required to obtain adequate precision at the analyte levels under investigation. The common types of source used in atomic fluorescence studies are discussed in the following sections.

3.2.2. SPECTRAL LINE SOURCES

3.2.2.1. Metal Vapour-Discharge Lamps

Metal vapour-discharge lamps (usually known by the name of the manufacturer, e.g., Wotan, Osram, or Philips lamps) were the first sources employed in the early analytical studies on atomic fluorescence spectrometry [102, 266, 424, 445, 446]. These sources are available only for some volatile elements, such as cadmium, zinc, mercury, gallium, indium, thallium, and the alkali metals (except for lithium). The inner discharge tube is usually made from silica and contains the metal, an inert gas at low pressure (usually 1 – 10 torr) and two tungsten (oxide-coated) electrodes. The inner tube is mechanically and thermally protected by an evacuated glass jacket. If ultraviolet radiation is to be transmitted an opening must be cut into this protecting jacket: this may lead to excessive cooling of the discharge tube, which results in unstable radiation output. To prevent this, the outer envelope should be provided with a sealed quartz window and re-evacuated, or the jacket made of quartz. Figure 3.2 shows a schematic diagram of a vapour-discharge lamp modified to transmit in the ultraviolet region.

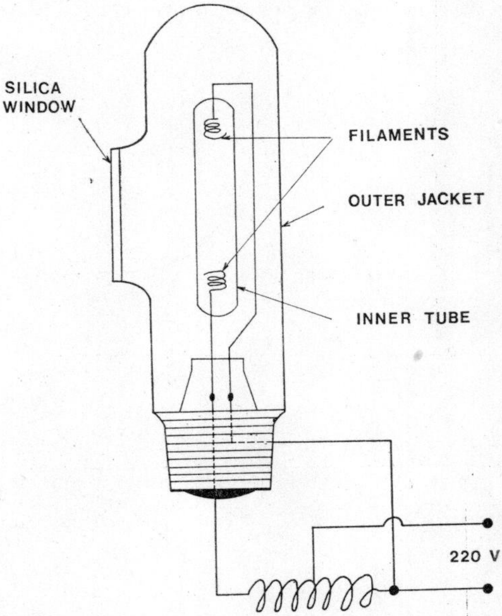

FIG. 3.2. Sketch of a metal vapour-discharge lamp modified to transmit ultraviolet lines.

Vapour-discharge lamps are usually powered either from a.c. mains with a choke switched in series, or by d.c. with a series resistance. Commercial lamps have a relatively high radiation output but suffer from serious self-reversal when operated at the currents recommended by the manufacturers [231, 446]. To minimize the degree of self-reversal the lamps must be run at currents far below the recommended values (usually at 0.4 – 0.6 A). Cooling of the lamps by a stream of air is less efficient and can result in condensation of metal on the lamp walls and a serious decrease in the emission intensity.

Only cadmium, zinc and perhaps thallium vapour-discharge lamps have been found to be satisfactory excitation sources for analytical atomic fluorescence purposes [102, 103, 186, 266, 272, 312, 420, 445, 446]. Radiances of about $10^{-3} - 10^{-2}$ W cm^2 sr^{-1} were reported [141] for cadmium and zinc lamps operated at carefully optimized currents. These are capable of exciting detectable atomic fluorescence from concentrations ranging from *ca.* $10^{-3} - 10^{-5}$ μg/ml, depending on the instrumental system employed (see Sections 5.7 and 5.3.9).

The effect of lamp current on the fluorescence intensity of cadmium is shown in Fig. 3.3. The maximum fluorescence signal corresponds to an optimum lamp current of *ca.* 0.5 A. Calibration curves obtained with the lamp operated at low and high currents are compared in Fig. 3.4. At low currents, when the emitted line is sharp, the shape of the curves is normal

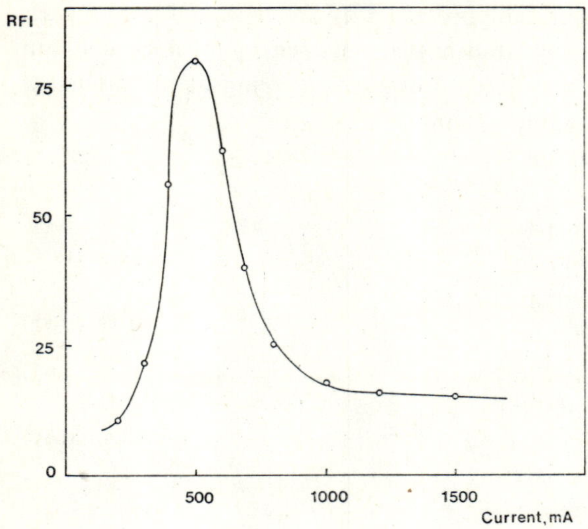

FIG. 3.3. Variation in fluorescence intensity with current of a cadmium Wotan vapour-discharge lamp [46].

(Reprinted from Ref. [46] by permission of Varian-Techtron Pty. Ltd.)

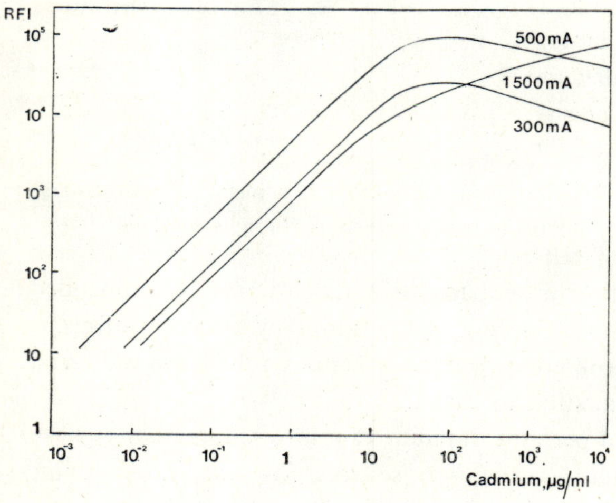

FIG. 3.4. Calibration curves for cadmium using a Wotan vapour-discharge lamp at several operating currents [46].

(Reprinted from Ref. [46] by permission of Varian-Techtron Pty. Ltd.)

and inversion occurs, while at high current, when the line is considerably broadened by self-absorption, the curve maintains a positive gradient (i.e. the lamp behaves as a continuum source).

Even at low currents, lamps for gallium and indium and, to a much greater extent, lamps for mercury and the alkali metals, still exhibit appreciable self-absorption and consequently low radiation output. This results in rather poor detection limits [375, 445, 446]. However, these lamps were used successfully for studying some atomic fluorescence excitation mechanisms [45, 313, 316] and excitation and deactivation processes in flames [10, 11, 197, 198, 217].

The popular mercury sources known as "Hg-pen lamps" or germicidal discharge lamps are of similar construction. These lamps consist of double bore quartz tubing with two electrodes at one end sealed into a rugged plastic handle. The quartz discharge tube can be screened by removable shields. Very low argon or neon pressure and low operating currents (10 – 20 mA) result in relatively cool lamps and in the emission of narrow spectral lines. No air-cooling is normally required. Approximately 90 % of the total output is normally concentrated at the 2536.5 Å mercury line, which makes the lamps useful as intense sources of short-wave ultraviolet radiation. The lamps are powered from a high-leakage reactance type a.c. power supply which provides a high starting voltage which automatically falls to the normal operating voltage as the lamp lights.

The C-13-61 mercury-argon lamp manufactured by Oriel Optics Corp., Stamford, Conn. has often been used in mercury atomic fluorescence studies [165, 300, 301, 361, 414, 415, 416]: the lamp has the following parameters: o.d. 6.5 mm, overall length 110 mm, spectral radiance ca. 0.15 W cm^{-2} sr^{-1} at the 2536.5 Å line when operating at an optimum current of 17 mA, and a 30-min warm-up time. The lifetime is rated from 1000 to 5000 hr.

The Philips 90-W mercury high-pressure discharge lamp was found to be effective in exciting the fluorescence of several elements as a result of spectral overlap of either source and atomic spectral lines or the analyte lines and strong source background continuum [312]. However, even when the line coincidence was good, e.g. for iron (Fe 2483.27 – Hg 2482.72 Å), thallium (Tl 3775.72 – Hg 3776.26 Å), and magnesium (Mg 2852.13 – Hg 2852.42 Å), fairly high detection limits were obtained, which are of little practical importance.

In conclusion, conventional metal vapour-discharge lamps cannot be generally recommended as excitation sources for atomic fluorescence spectrometry, although those for cadmium and zinc proved quite good. However particularly for these two elements microwave-excited electrodeless discharge lamps with excellent characteristics have been prepared [268, 483].

3.2.2.2. Hollow-Cathode Lamps

Detailed description of the construction and performance of these devices has been presented elsewhere; readers are referred particularly to references [76] and [79]. The following sections include only essential general information, together with an outline of the more recent trends in the development of these devices which are of particular relevance in atomic fluorescence spectrometry.

3.2.2.2.1. *Standard Hollow-Cathode Lamps*

This type of hollow-cathode lamp (which includes all kinds of lamps without a boosted output) was designed primarily to meet the needs of atomic absorption spectrometry, i.e. the emission at moderate intensity of very narrow spectral lines free from self-reversal. It is, therefore, not surprising that early attempts to excite atomic fluorescence with ordinary types of unshielded hollow-cathode lamps were completely unsuccessful owing to the very low output intensities [342].

Shielded small-bore hollow-cathode lamps which are now offered by most manufacturers exhibit much higher radiative output than their predecessors. Prugger [332] made absolute radiation measurements on 40 different shielded hollow-cathode lamps supplied by Carl Zeiss, Oberkochen, W. Germany, and found the radiance of most of the resonance lines to lie between 10^{-4} and 10^{-5} W cm^{-2} sr^{-1}. The output of some lamps, e.g. those for silver, magnesium, copper, and manganese, was found to be even higher by one order of magnitude. Manning and Heneage [265] and Larkins [258] tested a large number of Perkin-Elmer "Intensitron" and A.S.L. shielded lamps as excitation sources for atomic fluorescence. These lamps have been shown to provide detection limits of 1 μg/ml or less for about 15 elements (see Table 3.7) in air-hydrogen or sheathed air-acetylene flames. Unfortunately, the intensity of shielded lamps is still too low for detection at very low concentrations.

Some shielded hollow-cathode lamps, however, can prove useful when operated in the pulsed mode, for reasons already explained (Section 3.2.1). In addition to giving rise to a considerable improvement in the signal-to-noise ratio, pulsed lamps are found to exhibit rapid warm-up and longer lifetime, and to be applicable to multielement atomic fluorescence analysis [40, 143, 144, 291, 294, 333]. Excellent detection limits have been obtained for a large number of elements as can be seen in Table 3.7 (p. 110). It should be stressed, however, that such low limits of detection were attributable not only to the pulse technique but also to significant improvements in the optical and signal processing systems used as can be seen when values from references [143, 145, 191] are compared with those from reference [32]. It may be expected that pulsed operation of hollow-cathode lamps will receive much greater attention in the near future.

3.2.2.2.2. *Boosted-Output ("High-Intensity") Hollow-Cathode Lamps with an Auxiliary Discharge*

This type of hollow-cathode lamp was developed to improve the atomic absorption determination of elements having complex atomic spectra or for which satisfactory standard hollow-cathode lamps were not available. Their principal feature is the use of two separate discharges. One of these is a hollow-cathode discharge which produces an atomic vapour by sputtering from the internal walls of the hollow cathode and excites some of the vapour in the negative glow, and the other is a low-voltage high-current arc discharge which causes increased excitation of the sputtered atomic vapour in the anodic column. In the original (Sullivan and Walsh) design [380] the secondary discharge is directed across the open end of a normal hollow cathode so that the positive column of the secondary discharge causes increased excitation of the cathode vapour which is in front of and

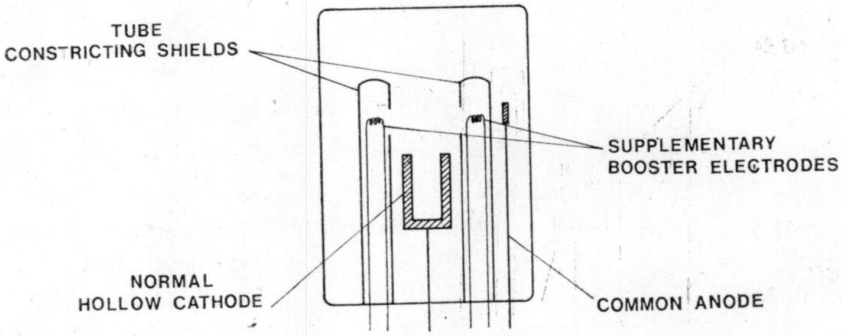

FIG. 3.5. Diagram of electrode assembly of a boosted-output hollow-cathode lamp of the Sullivan–Walsh design.
(Reprinted from Ref. [380] by permission of Pergamon Press Ltd.)

just inside the hollow cathode (Fig. 3.5). This results in higher total radiation output and more favourable emission-line profiles than from standard lamps [129]. Hollow-cathode lamps of the Sullivan-Walsh pattern may give resonance-line intensities up to 100 times that from standard "open" hollow-cathode lamps and 20 times that from shielded hollow-cathode lamps [79, 260, 388]. Radiances reported for these lamps by Prugger [332] are a little misleading, probably because the real profile of the emitted lines was not taken into account.

Hollow-cathode lamps of the Sullivan-Walsh design have been used to excite the atomic fluorescence of many elements [123, 258, 259, 264, 345, 368, 428, 429]. They have received a great deal of attention, particularly in our laboratory [245, 276, 278, 387, 390, 391, 394]. They have proved to be excellent excitation sources, and in most cases yield detection limits at least as good as those obtained with electrodeless discharge lamps at their present stage of development (see Table 3.7). They exhibit high output intensity and good stability, and are substantially simpler to operate than other high-intensity sources. The disadvantages are the limited range of elements for which the sources are available and the shorter lifetime. The latter is due to the necessity (in most instances) to run the lamps at maximum primary and secondary operating currents, which often results in rapid loss of the cathode material, especially for volatile elements, and in lamp overheating and/or damage to the supplementary oxide-coated filaments. Higher cost and the need for two power supplies are not considered by the authors to be a serious disadvantage, since some other excitation sources need more complex and expensive power supplies. It is a great pity that lamps of the ordinary Sullivan-Walsh design are not extensively manufactured.

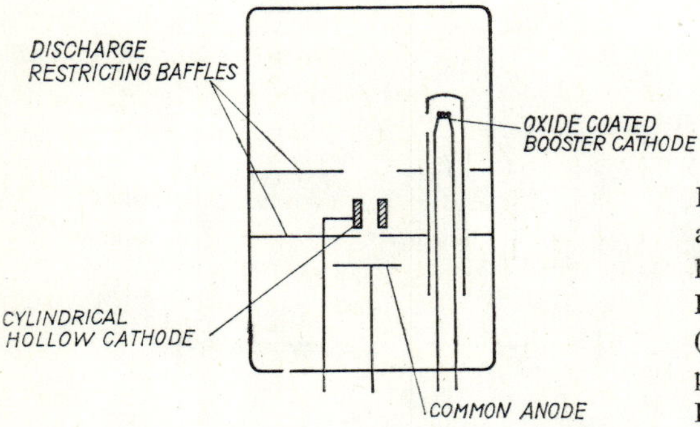

FIG. 3.6. Diagram of electrode assembly of a boosted-output hollow-cathode lamp of the Löwe design [260].
(Reprinted from Ref. [260] by permission of Pergamon Press Ltd)

The newer (Löwe) design of boosted-output hollow-cathode lamp was developed primarily as a source for atomic fluorescence spectrometry [260]. The construction of the lamp is shown in Fig. 3.6. An open-ended cylinder is used for the cathode and the secondary discharge is directed through its centre [408] and then through a thin baffle with a central hole. As the lamp is designed for maximum intensity, the size of the hole is chosen so that the maximum amount of a radiative output from inside the hollow cathode can be collected. With this electrode geometry the positive column of the secondary discharge is mainly confined to the region of the negative glow of the hollow-cathode discharge, which gives this lamp two advantages over the Sullivan-Walsh design.

(1) Confining the positive column by the use of a small diameter cylindrical cathode greatly increases the electron density in the positive column, causing an increased efficiency in the excitation of sputtered vapour.

(2) The high electron-density positive column is directed through the region where the atom concentration produced by the primary discharge is greatest.

The power supplies for the lamps are similar to those used for Sullivan-Walsh type lamps, but the intensity of the Löwe-type lamp is more critically dependent on the secondary discharge current. The secondary current for maximum lamp intensity is different for each lamp but is usually in the range 100 – 300 mA.

The performance of lamps of the Löwe design in atomic fluorescence has been evaluated and compared for several elements with the performance of shielded hollow-cathode lamps and hollow-cathode lamps of the Sullivan-Walsh type supplied by A.S.L. Pty. Ltd., Melbourne. Intensity comparisons obtained by measuring the fluorescence signals produced by 1-μg/ml solutions of various elements are shown in Table 3.1. These results indicate that the Löwe-type hollow-cathode lamps emit resonance lines with considerably greater intensity than those from the Sullivan-Walsh type. Despite some broadening of the emitted resonance lines at high primary currents, they seem to be very useful excitation sources for atomic fluorescence especially where low limits of detection are required.

TABLE 3.1. Comparison of the intensity of fluorescence excited by various hollow-cathode lamps sources (arbitrary units) [260]

Element	Standard shielded hollow-cathode lamp	Sullivan-Walsh hollow-cathode lamp	Löwe hollow-cathode lamp
Ag	1	17	119
Au	1	12	50
Co	1	0.75	15
Cr	1	—	12
Cu	1	11	44
Fe	1	4	50
Ni	1	4	50

No information is at present available on the use of these lamps in atomic-fluorescence studies apart from several detection limits listed by Larkins and Willis [258, 259]. Some of these are given in Table 3.7, p. 110.

3.2.2.2.3. *Hot Hollow-Cathode Lamps*

The design and use of hot hollow cathodes as excitation sources in emission spectrography is well established [207, 219]. In the search for intense light-sources for excitation purposes in atomic fluorescence spectrometry, attention was also turned to this type of source. The possibility of operating hollow cathodes at high currents, and the resulting gain in radiant power was expected to lead to high fluorescence intensities.

Massmann used compact hot hollow-cathode lamps to excite atomic fluorescence in a furnace-type atomizer [274, 275]. These lamps were run at very high currents for short periods and were turned off before excessive cathode sputtering and appreciable self-reversal of emitted lines occurred. Except for the lamps of some volatile elements, the emitted intensities were too low to give rise to satisfactory fluorescence signals.

Dinnin and Helz [151, 152] described the design and performance of a demountable water-cooled hot hollow-cathode lamp and examined the atomic fluorescence characteristics of eighteen elements with this source. The replaceable cathode consisted of a graphite rod with a cavity at one end to hold a small amount of the element (about 50 mg of metal turnings or powder). Power (d.c.) was supplied to the cathode holder by a high-voltage supply capable of providing as much as 1.2 A at voltages as high as 2 kV. A ballast resistance of 1250 Ω was used to control the current. The lamp was operated at low pressures (2–8 torr) of helium or argon. Source currents of 400–600 mA were applied to get the best fluorescence signals. The dependence of fluorescence intensity on the source current for a gold hot hollow-cathode lamp operated at 4 torr of helium is shown in Fig. 3.7. The application of higher currents is limited by the melting point of the metal in the cathode. Despite incomplete optimization of experimental

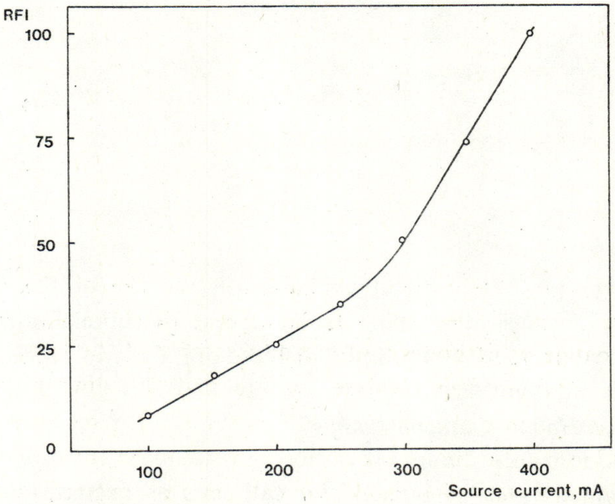

FIG. 3.7. Fluorescence intensity of gold as a function of hot hollow-cathode lamp current [151].

Reprinted from Ref. [151] by permission of The American Chemical Society parameters good results were obtained with this lamp for copper, silver, and gold as can be seen in Table 3.7 (p. 110).

The fluorescence excited by the hot hollow-cathode lamp and two commercially-available boosted hollow-cathode lamps (gold and silver, A.S.L. Pty., Ltd.) has been found to be about the same for the same instrumental settings and flame conditions.

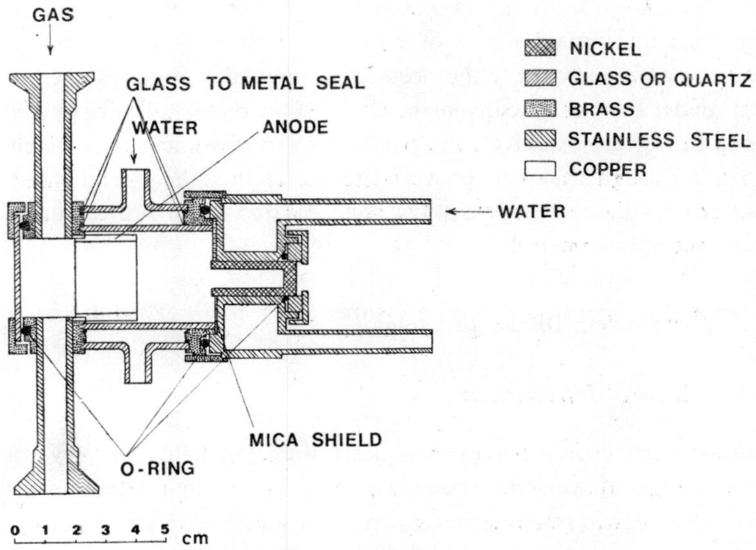

FIG. 3.8. Sketch of a demountable hot hollow-cathode lamp [349]. (Reprinted from Ref. [349] by permission of Pergamon Press Ltd.)

A sketch of the demountable water-cooled hot hollow-cathode lamp used by Rossi and Omenetto in the study of the fluorescence of twelve elements [349] is shown in Fig. 3.8. The lamp is supplied with currents ranging from 200 to 500 mA. The filler-gas is argon at a pressure of 0.4 – 4 torr, depending on the physical properties of the element under investigation. The unique design of the lamp, which incorporates conical nickel cathodes, allows rapid interchange of cathodes without disassembly of the lamp. Moreover, reproducible and uniform cathode cooling and shielding permits steady discharge conditions to be reached within about 10 min. Most of the elements introduced into the nickel cathodes were in the form of metal discs, 6 mm in diameter and 2 mm thick; for gallium and thallium, gallium oxide and thallium sulphate were pressed into the bottom of the cathode.

The authors compared the performance of the lamp with that of metal vapour-discharge lamps for gallium, indium, and thallium. They obtained better limits of detection for these elements with the hot hollow-cathode lamp as source despite the greater emission intensity (by a factor of 10–40) of the vapour-discharge lamps. They concluded that this is mainly due to the more favourable emission line-profile; efficient cooling and continuous flushing of the hot hollow-cathode lamp results in the emission of very narrow spectral lines which do not exhibit self-reversal.

Demountable hot hollow-cathode lamps can be considered as medium-intensity excitation sources. Their use is often restricted by their construction to a somewhat limited range of elements. Relatively complex handling procedures and particularly the need for a large and cumbersome vacuum system render their use inconvenient for routine analysis. Their applicability depends to a great extent on the particular circumstances in a given laboratory. Although they can prove quite useful in some special cases, they cannot be considered as a generally applicable excitation source for atomic fluorescence spectrometry.

3.2.2.3. Electrodeless Discharge Lamps

3.2.2.3.1. General Introduction

Radio- and microwave-frequency electromagnetic fields are very efficient for generating and accelerating electrons and thereby maintaining a gaseous glow discharge without internal electrode contact with the plasma. This electrodeless system represents the ultimate in simplicity among the other types of gaseous discharges and is not affected by some of the problems and limitations associated with systems requiring internal electrodes. The radio frequency or microwave-frequency powered electrodeless discharge in sealed-off tubes containing an inert gas at low pressure and some volatile pure metal or metal salt (usually a halide) thus offers a means of obtaining a simple spectral line source with very good spectral characteristics when operated under proper conditions.

The electrodeless discharge yields, owing to its non-isothermal nature [196], a so-called "cold plasma". The temperature of high mass particles (atoms, ions, and molecules), which are not able to follow the high frequency of the electromagnetic field to the same extent as electrons approaches 10^3 K, while the temperature of electrons is of the order of 10^4 K [35]. This means that the particles effective for excitation are almost exclusively electrons [273, 283]. Such a plasma exhibits a "skin effect", i.e. the confinement of the discharge to the thin surface layers ("skin layers") of the plasma, near the tube walls [68, 180, 283]. The radiation emitted, therefore, does not pass through a layer of cold atomic vapour, and the tendency toward self-reversal is decreased. Since many independent variable parameters are involved in the operation of low-pressure electrodeless discharges, it is very difficult to give a precise and detailed kinetic and thermodynamic description of the plasma. For some general information, the reader is referred to the the publications by Hollahan [196], Reed [338], Steers [376], and Worden *et al.* [478].

Sealed-off electrodeless discharge lamps have been effectively operated over a wide range of radio and microwave frequencies (10 – 3000 MHz). On reviewing the literature, it becomes apparent that no thorough critical evaluation of the various types of power system has been made with regard to intensity of output, stability, and other factors which are appropriate to atomic fluorescence. Most of the applications of radiofrequency-operated electrodeless discharge lamps (usually at 10 – 100 MHz) have been concerned with physical studies of atomic and fluorescence spectra [24, 77, 180, 285, etc.] and more recently, with use in analytical atomic absorption [e.g., 29, 30, 34, 36, 37, 208, 209]. Only in two cases [40, 188] have these lamps been used in analytical atomic fluorescence spectrometry.

The use of microwave frequencies is reported to result in more efficient conversion of electrical power into radiative power [174, 210] and longer lamp lifetime [210, 285] than that of radiofrequencies. Because of these two advantages and the ready availability of inexpensive microwave power generators, the operation of electrodeless discharge lamps for analytical atomic fluorescence studies has been confined to microwave excitation at 2450 MHz.

Most of the pioneer work in the preparation and use of microwave-excited electrodeless discharge lamps as excitation sources for atomic fluorescence has been done by Winefordner and his colleagues at the University of Florida [267, 268, 483] and by West and his group at Imperial College, London [e.g. 8, 71, 91, 105, 115, 433].

3.2.2.3.2. Microwave Power-Sources and Coupling Devices

Most of the experimental work involving the use of electrodeless discharge lamps for analytical atomic fluorescence studies has been carried out with commercially available microwave generators. Medical diathermy units operating at a frequency of 2450 MHz and at powers of up to 150 W have been found to be useful as microwave sources. The heart of the microwave generator is the "magnetron". This device is basically a vacuum tube operated in a magnetic field to produce microwave energy directly.

Because most commercial microwave generators are designed specifically as medical diathermy units, their power output stability is frequently inadequate for analytical usage. Such generators may exhibit considerable mains-frequency modulation, and may have unstabilized power supplies to the magnetron oscillator. The resulting mains ripple superimposed on the microwave output can often cause trouble when a.c. amplifiers with mains-derived synchronous demodulation are used. For this reason it is advisable to use a current-stabilized power supply for the magnetron.

Modern microwave generators (e.g., Microtron 200 Mk. II, Electro Medical Supplies Ltd., London), which are designed specifically for spectroscopic purposes, have a low ripple d.c. supply to the magnetron. Tests performed with the magnetron JP 2−0.2 in the "Microtron 200" microwave generator indicated that voltage stabilization to $\pm 0.1\,\%$ could give a 3-W fluctuation in the power output, and current stabilization to $\pm 0.1\,\%$ would give only a 0.1-W fluctuation [8]. It should be stressed that the stability of spectral output depends not only on the stability of the power supply, but also on the efficiency with which the power is converted into radiation. Also modulation of the microwave energy may improve both the short-term and long-term stability of the spectral output [cf. Section 3.2.2.3.3].

Suitable microwave generators are now available from a number of suppliers, together with the necessary coupling units: a reflected-power meter is either built-in or supplied as accessory. In order to protect the generator it is essential that the microwave energy is dissipated in the gas discharge or terminating load and not reflected back down the cable to the magnetron, as this can cause excessive overheating and possible damage to the magnetron cathode. A reflected-power meter gives a direct reading in watts of the power not being absorbed at the gas discharge cavity or other terminating load, and its use greatly facilitates tuning for maximum efficiency and safety.

Microwave energy from the generator is led through the coaxial output cable and coupled to the discharge lamp by means of either a resonant cavity or a director-antenna. The purpose of these devices is to transfer as much of the output power into the discharge lamp as possible. Sufficient energy must be absorbed to effect vaporization, excitation, and maintenance of a suitable discharge. The efficiency with which the electrodeless discharge lamps operate depends to a large extent on the proper design and choice of these coupling devices.

Director-antennae were primarily developed for medical therapeutic use in conjunction with diathermy units and are nothing more than antennae with metal reflectors for directing the microwave beam. An A-type antenna was found to be quite adequate for operating some discharge lamps [267, 268]. However, its major limitation for operation of a wide variety of lamps (particularly those with very involatile compounds) is a lack of thermal energy for maintaining an adequate vapour pressure of the metallic species in the discharge; this occurs because the microwave energy is less efficiently coupled to the lamp by the A-antenna, than by a resonant cavity. Since the microwave discharge itself generates heat, this problem was solved by inserting the lamp into a quartz insulating jacket containing quartz wool, mounted directly above the antenna, or by enclosing the lamp in a special quartz vacuum jacket [483]. These arrangements, shown in Fig. 3.9, simultaneously exclude extraneous draughts, which also contribute to the erratic operation of the lamps. A C-type antenna was found to couple less efficiently and is unsuitable for use in atomic fluorescence spectrometry [267, 268].

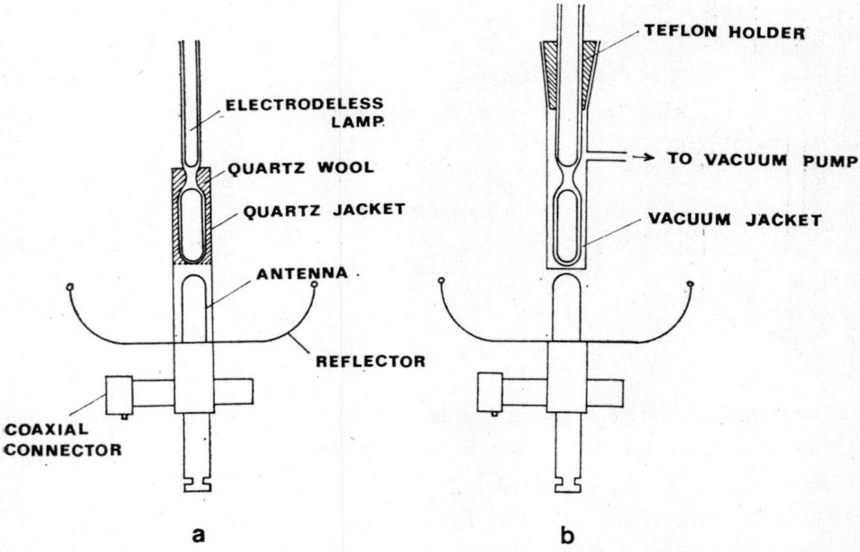

FIG. 3.9. Operating of A-antenna with (a) quartz jacket and with (b) evacuated jacket [483].

Reprinted from Ref. [483] by permission of The American Chemical Society

A resonant cavity transfers microwave energy to a discharge by matching its own resonance frequency to that of the microwave source. When this frequency and the impedance are properly matched, the power reflected from the cavity is at a minimum value. The more efficient the cavity, the lower the reflected power. By silver plating of the cavity walls, a high Q-factor can be realized. The conditions for resonance and the loaded Q-factor usually change after initiation of the discharge and during any change of the discharge conditions; they are also unfavourably affected by excessive generation of heat. It is therefore advantageous (and even essential for some types of cavities) to provide facilities for tuning and matching and for direct air-cooling of the cavity to obtain efficient operation over a wide range of discharge conditions. Cavities which are relatively independent of the discharge conditions and the material in the discharge lamp, and therefore require only simple tuning are more suitable.

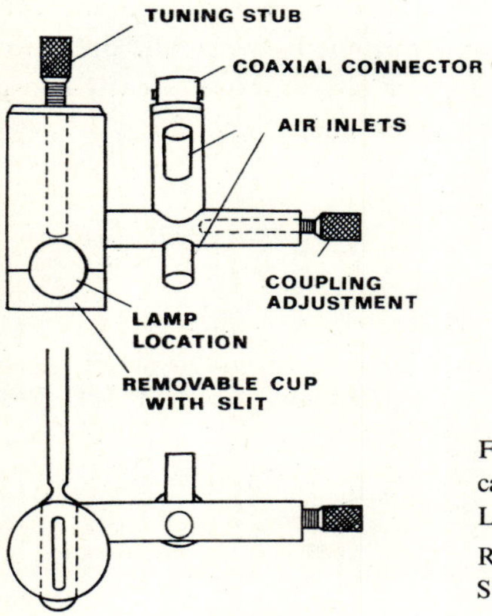

FIG. 3.10. Schematic diagram of the 1/4-wave cavity (No. 214 L, Electro Medical Supplies Ltd., London).

Reprinted by permission of Electro Medical Supplies (Greenham) Ltd.

Various types of microwave resonant cavities have been developed [168]. Of these, the 1/4-wave coaxial ("Evenson") cavity and the 3/4-wave coaxial ("Broida") cavity are the most popular. Various modifications of the two basic types exist; (e.g., Opthos Instrument Co., Rockville, Md., U.S.A., Evans Electroselenium Ltd., U.K., Beckman-RIIC, London, U.K., Electro Medical Supplies, London, U.K., etc.): they differ with respect to the location of tuning stubs, coupling adjustments and provisions for air cooling [8, 91, 172, 190, 280].

In most atomic fluorescence studies with electrodeless discharge lamp sources, 1/4-wave and 3/4-wave cavities supplied by Electro Medical Supplies, Ltd. (designated No. 214 L and No. 210 L, respectively) have been employed. The 214 L cavity (Fig. 3.10) is more efficient and has external dimension of approximately 2.5 cm diameter and 6 cm length with variable coupling at right angles to the longitudinal axis of the cavity. The coaxial input connector is integral with the cooling air inlet, and the main tuning stub is in line with the axis of the cavity. The end cap with a slit may be completely removed to facilitate viewing the tube. The cavity has two cooling air inlets. If it is required to keep the discharge cool, the inlet that is coaxial to the coaxial connector is used. This directs most of the air into the cavity, and cools the discharge. If the temperature of the discharge lamp is to be allowed to rise, the cooling-air inlet at right angles to the coaxial connector should be used. This keeps the connector and its Teflon insulation at a reasonable temperature without unduly affecting the discharge lamp.

Tuning of this cavity is rather tedious, and as the two tuning stubs are mutually dependent in operation, successive readjustment of each must be used to attain maximum efficiency. It is, therefore, strongly recommended that a reflected-power meter is used to optimize tuning of the cavity.

The 214 L cavity proved to be quite satisfactory for some elements [72, 73, 106, 107, 108, 112, 113, 118, 191] but suffers from several disadvantages. It is frequently thrown out of tune as the vapour and condensed phases shift within the lamp during operation, but new tuning during operation is very inconvenient. Forced air cooling is necessary in most cases, although this does not entirely eliminate thermal instability, nor does it prevent localized heating ("hot spots") from developing. Moreover, it is possible to sustain a discharge outside the discharge tube which can cause substantial deterioration of the interior finish of the cavity, and its subsequent coupling efficiency.

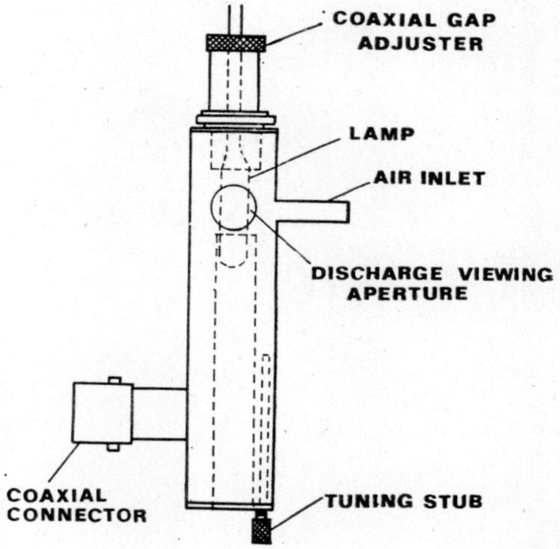

FIG. 3.11. Schematic diagram of the 3/4-wave cavity (No. 210 L, Electro Medical Supplies Ltd., London). Reprinted by permission of Electro Medical Supplies (Greenham) Ltd.

The 210 L cavity (Fig. 3.11) is cylindrical in shape, about 5 cm in diameter and 13 cm long, has an adjustable re-entrant gap, coaxial with the discharge lamp, and a tuning stub offset from and parallel to the discharge lamp. The cavity may be air-cooled and possesses adequate apertures either at right angles to, or in line with, the discharge lamp. Only simple tuning is generally required. This is in marked contrast to the critical tuning of the 214 L cavity. Because the lamp is almost totally "immersed" in the microwave field, the lamps prepared from relatively involatile or very volatile materials are able to reach equilibrium conditions because, no matter how much movement of material takes place within the lamp, it is always within the microwave field. In addition, the cylindrical cavity functions to some extent as a thermostat for the lamp by excluding extraneous draughts.

It is evident that the thermal stability offered by the 3/4-wave cavity and the uniformity of the microwave field obtained has been achieved at the expense of high coupling efficiency. Moreover, because of the lower Q-factor and larger thermal capacity of the cavity, the warm-up time has been found to be considerably longer. In general, the 210 L cavity must be operated at powers about 25 % higher than the 214 L cavity, and the lamps require higher filler-gas pressures. Very good atomic fluorescence results have been obtained with a number of electrodeless discharge lamp sources excited in the 210 L cavity [5, 71, 73, 112, 118, 119, 157, 158, 193]. This cavity also proved satisfactory when the lamps were electronically modulated [71, 118, 126, 399].

The addition of capacitative tuning stubs has been suggested to increase the coupling efficiency of the Broida 3/4-wave cavity [8, 91, 179]; the main advantage was that of improved warm-up time.

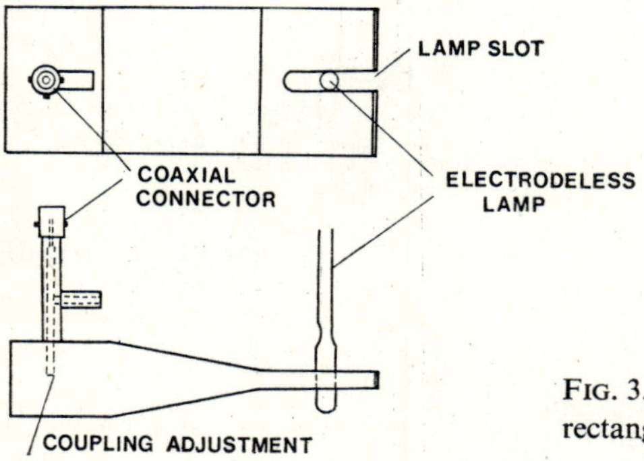

FIG. 3.12. Schematic diagram of the tapered rectangular cavity.

The tapered rectangular cavity, shown in Fig. 3.12, has also been used in several investigations [267, 268, 483]. Its coupling efficiency is less than that of the other two cavities and, therefore, it is very difficult to obtain a large forward-to-reflected power ratio for most lamps. However, it has other features which are advantageous in operation. It operates without forced air-cooling for many lamps, and, even in cases where the lamps require cooling, the cavity itself does not heat up. Furthermore, it does not require tuning during operation. Slight tuning can be effected by altering the position of the lamp in the cavity slot.

With all the coupling devices described, initiation of the discharge, which relies on the production of free electrons, is achieved with a simple Tesla-coil vacuum-tester.

In conclusion, it should be stressed that electrodeless discharge lamps prepared for any one of these coupling devices generally operate with the others but usually with significantly poorer performance. For this reason it is important that the descriptions of electrodeless discharge lamps in the literature are accompanied with the type of coupling device used, otherwise misleading statements on the performance of the lamps could result.

3.2.2.3.3. Modulation

In early atomic fluorescence studies, microwave-excited electrodeless discharge lamps were operated in the d.c. mode in conjunction with d.c. amplifiers [e.g. 106, 108] or mechanical choppers were used to provide the necessary modulation to match the a.c. amplification frequency of the spectrophotometers used [73, 114, 268, 483]. The disadvantages of such systems are obvious: measurement in the d.c. mode does not differentiate between atomic fluorescence and flame emission signals; mechanical choppers, on the other hand, physically prevent part of the radiation from reaching the atom reservoir, which has a deleterious effect upon the limits of detection attainable.

Electronic modulation, tuned to the frequency of the a.c. detecting system, obviates the need for a mechanical chopper and enables the distance between the source and atom reservoir to be reduced: this allows a larger solid angle of exciting radiation to reach the atomic vapour and thus the limits of detection can be improved.

Two basic ways have been suggested for amplitude modulation of microwave energy. Browner *et al.* [71] and later Wildy and Thompson [399, 438] described methods of electronically modulating electrodeless discharge lamps via transformer coupling of a low voltage (*ca.* 12 V peak-to-peak) a.c. signal into the anode circuit of the magnetron. This method suffers from certain disadvantages. Because the secondary windings of the transformer are in the anode circuit, the high d.c. current may cause saturation of the core material and the modulation wave form applied to the primary windings of the transformer may not be faithfully reproduced in the microwave output. Moreover, the frequencies and functions that can be used are limited by the use of a transformer which in general is most efficient for low frequency sinusoidal waveforms. Despite these limitations, this modulation system was found to be satisfactory with a number of electrodeless discharge lamps used in atomic fluorescence studies [71, 117, 118, 157, 399] and modulator units developed for this design (e.g., Microtron modulator unit, Electro-Medical Supplies, Ltd., London) have been used satisfactorily in conjunction with most atomic-absorption flame-emission spectrophotometers, at modulation frequencies of 50 – 400 Hz.

The second method of electronic modulation is a direct addition of the modulation waveform at the error-sensing or reference point of the stabilizing element of the stabilized magnetron power supply. In this case, the modulation can be easily effected with a low-power signal from a function generator, which may be of sine, square, triangle, or pulse waveform at frequencies up to *ca.* 100 kHz. Dagnall *et al.* [126] have investigated various modulation waveforms at different microwave power levels. The optimum frequency of operation with respect to intensity of the radiation output was found to be *ca* 20 kHz, with a square waveform. In addition, modulation of the stabilizing circuit was found to produce less noise and drift than transformer-coupled modulation (see Table 3.2).

The degree of modulation in both types of modulator is controlled by the amplitude of the modulating waveform. The degree of modulation that can be applied to an electrodeless discharge lamp without extinguishing the discharge depends on the type of lamp (i.e. element, and filler gas) and also the modulating frequency. At high frequencies (*ca.* 100 kHz) modulation can approach 100 %, but at low frequencies (*ca.* 100 Hz), because of the

TABLE 3.2. Optimum operating conditions, noise and drift for some microwave-excited electrodeless discharge lamps (EDL) used in a.c. and d.c. modes [126][a]

Reprinted from Ref. [126] by permission of Pergamon Press Ltd.

EDL Material	Wave-length Å	Optimum power, W		Average peak-to-peak noise measured over 30-min period			Drift, %/hour		
		d.c. mode	a.c. mode	d.c. mode	a.c. mode transformer modulation	a.c. mode stabilization modulation	d.c. mode	a.c. mode transformer modulation	a.c. mode stabilization modulation
Cd	2288.0	60	35	2.2	1.3	0.3	—3.6	—1.2	—1.0
Hg	2536.5	38	25	2.5	2.6	0.5	—4.6	—1.7	—1.2
Zn	2138.6	65	38	1.0	0.4	0.4	—3.5	—1.7	—1.2
Sb/I_2	2311.5	40	25	2.7	1.3	0.5	—4.3	—1.5	—1.2
Se	1960.3	38	25	3.1	1.6	0.7	—2.8	—1.3	—1.3
As/I_2	1937.0	40	28	1.8	0.7	0.4	—2.5	—0.9	+0.2
Pb/I_2	2833.1	50	30	3.0	1.7	0.9	—5.8	—2.3	—1.5

[a] With a Broida type three-quarter wave cavity (Evans Electroselenium Ltd.) fitted with two capacitative tuning stubs.

[b] Modulation frequency 20 kHz.

finite lifetime of the ions in the discharge, the useful modulation may only be 70 %. Because the voltage *vs* power output curve for the magnetron is not linear, the voltage required to give the same depth of modulation varies with power output. In practice this means that if the same percentage is to be maintained, then the modulation depth must be readjusted each time the power is varied.

In conclusion, electronic modulation of microwave-excited electrodeless discharge lamps results in better long-term and short-term stability, longer lifetime (as a result of lower effective microwave operating powers), and higher output intensity compared to unmodulated or mechanically modulated lamps: it also allows the use of phase-sensitive amplification systems which can increase the signal-to-noise ratio in analytical measurements.

3.2.2.3.4. Construction of Electrodeless Discharge Lamps

A microwave-excited discharge lamp consists of a sealed-off cylindrical reservoir of clean transparent quartz or glass of 3–12 mm bore, 30–40 mm long, with a wall thickness *ca.* 1 mm, containing a small amount of the element or one of its compounds and an inert gas at low pressure. Shapes of two typical electrodeless lamps are shown in Fig. 3.13. A vacuum-jacketed type ensures a high thermal stability of the discharge, resulting in a very stable spectral output. The microwave power required to give a stable, intense source for a single element is *ca.* 25 % less than that required for the more conventional, non-jacketed type. The warm-up time for the vacuum-jacketed lamp is also considerably less [8]. Because the inner tube can acquire a high running temperature it is possible to use comparatively involatile elements or compounds as a charge material, e.g., lead in place of lead iodide. Moreover, jacketed lamps can be used as multielement sources for elements of differing volatilities [97].

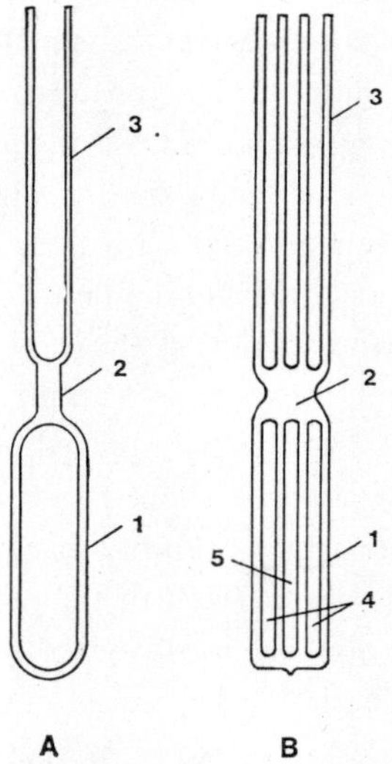

FIG. 3.13. Typical electrodeless discharge lamp: A-conventional 9 mm-bore lamp; B-vacuum jacketed (dual-cavity) lamp. (1) bulb; (2) seal; (3) stem; (4) outer cavity; (5) inner cavity.

In most cases, the lamps must be made of quartz in order to transmit ultraviolet radiation and to be resistant to the high temperatures that are usually generated in the resonant cavity. For elements which may react with the quartz envelope (e.g. alkali metals, magnesium, etc.) the quartz surface must be rendered inactive. This may be achieved by boration of the inner surface of the quartz tubing by first wetting it with a saturated aqueous solution of boric acid and then heating the lamp in a furnace for several hours at about 400 °C to form a protective layer of borate glass. When Pyrex (borosilicate) glass is used in place of quartz for the preparation of the lamps for alkali metals, the inner surface of the glass must be washed with 40 % hydrofluoric acid. This removes a layer of silicon, thereby leaving a boron-rich glass. Both protective methods rely on the fact that alkali metals will not attack borate glass and displace elementary boron, because of the stability of the B-O bond in comparison to that of the much weaker Si-O bond. However, both methods were found to be effective only for some metals and only for several hours of operation [6].

Aldous *et al.* [6] have prepared sufficiently intense and long-lived electrodeless lamps for alkali metals (except for lithium) and magnesium from two-ply glass (Glass Tubing and Components Ltd., Chesterfield, England; type X 93). The inner ply consists of a sodium-vapour resistant silica-free glass and the outer ply of normal soda-glass. Extreme care must be taken when handling this glass not to break the inner ply of sodium-resistant glass, otherwise attack then readily occurs. The Broida-type cavity (Electro Medical Supplies No. 210 L) is recommended for operation of these lamps.

The spectral transmission of X93 two-ply glass is limited in the ultraviolet region; little transmission occurs below 3000 Å. To prepare satisfactory magnesium lamps for use in atomic fluorescence spectrometry it is advisable first to seal a quartz window onto the open end of the two-ply glass and then proceed with the usual method of preparation. The lamp is operated in a horizontal position with the end of the lamp containing the quartz window out of the cavity area (1(4-wave Evenson type cavity), since only hot magnesium vapour attacks quartz [6].

The material placed in the conventional lamp blank must have a vapour pressure of about 1 torr at 200 – 400 °C to be vaporized easily and to achieve maximum stability and intensity. It is this property that determines whether the metal itself or its more volatile compounds are employed. Pure element lamps are mostly used for the softer metals such as mercury [73, 97, 483], zinc [97, 483], cadmium [97, 270, 483], selenium [97, 105, 106, 483], some alkali metals [6], magnesium [6], etc. Less volatile metals can also be used in special instances, but the lamps require higher operating powers [8, 97]. Dagnall *et al.* [108] used a pure iodine lamp for the excitation of bismuth.

Metal iodides and chlorides have been used almost exclusively in order to obtain higher vapour pressures than are attainable with some pure elements. Because these compounds are not always available or are too readily hydrolysed to be handled and stored, a number of specific preparations and methods for introduction of these compounds into the lamp blanks have been described [115, 268, 483]. In most cases, they are prepared *in situ* by a suitable reaction of the pure metal with a halogen, or are introduced into the lamp directly, most often by sublimation. Microgram amounts of the material may be accurately introduced into the lamp blank in the form of aliquots of a dilute aqueous standard solution of the element of interest. The water is removed under vacuum and the compound gently sublimed onto the walls of the lamp [91].

A large number of metal halide lamps have been prepared which are suitable for atomic fluorescence excitation purposes. Metal iodide lamps proved very satisfactory for aluminium, titanium, and zirconium [112, 118, 483], bismuth [108, 193, 483], tellurium [105, 106], tin [72], lead [71], antimony [107, 483], arsenic [111, 114, 115], germanium [113, 119, 483], silicon [117], beryllium [191], silver, gold, cobalt, chromium, copper, gallium, hafnium, indium, magnesium, manganese, nickel, scandium, and strontium [483]. The presence of metal iodides with a slight excess of metal is often recommended [115, 191, 483]. Metal chloride lamps are reported to give better spectral characteristics than their iodide counterparts for vanadium [112, 118], thallium [73], molybdenum [118], palladium, platinum and gold [5], silver [5, 114], iron [107], manganese [158], cobalt [114, 173], nickel and chromium [97, 114, 306], and lithium [6].

Sometimes the addition of a small amount of mercury has been found to be useful, particularly if the element in question tends to be adsorbed on the quartz. Mercury prevents this adsorption, presumably because of the greater tendency to produce an amalgam; however, the use of higher filler-gas pressure has a similar effect [115]. Mercury has been used both as a carrier of the initial discharge in the absence of argon and as a clean-up agent in cuprous chloride/amalgam lamps [22]. Mansfield *et al.* [268] have tested various metal amalgam and metal iodide/amalgam lamps, and concluded that the high noise associated with these lamps offsets any benefit of high intensity over metal/metal halide lamps.

It is not possible to give a completely uniform scheme for the choice of material used for the preparation of microwave-excited electrodeless discharge lamps. The requirement for a certain metal vapour pressure is not independent of other lamp parameters, such as the lamp diameter, filler-gas pressure, amount of the charge material, mode of the excitation, applied microwave power, type of coupling device, etc., and should always be correlated with them. The effect of some of these variables will be discussed later.

One of the most important features of electrodeless discharge lamps is simplicity of preparation in the laboratory. Although commercial lamps are now available (e.g., lamps with quite good spectral characteristics are manufactured by E.D.T. Supplies, Hampton Hill, Middlesex, U.K.), a considerable amount of research in atomic fluorescence spectrometry is still concentrated on the development of very intense electrodeless discharge lamps.

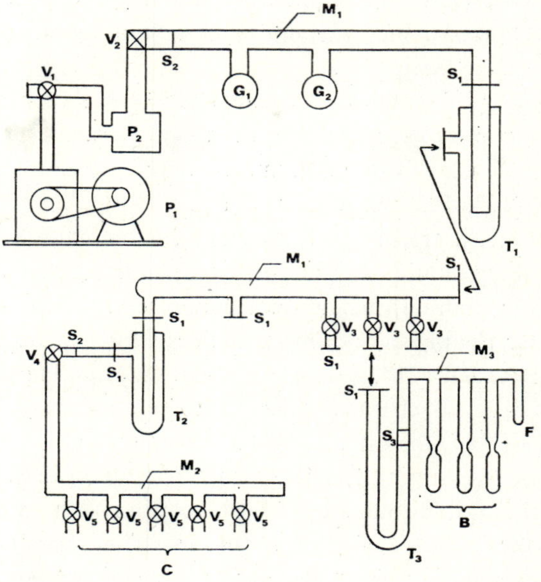

FIG. 3.14. Vacuum system used for preparation of electrodeless discharge lamps [267]: (P_1) mechanical pump; (P_2) oil diffusion pump; (V_1) 10 mm bore vacuum stopcock; (V_2) bellows-sealed vacuum valve; (S_2) Kovar-Pyrex seals; (G_1) cold cathode discharge (ionization) gauge; (G_2) oil differential manometer; (S_1) Viton O-ring connectors; (T_1) liquid-nitrogen cold-trap; (V_3) 2 mm bore high vacuum stopcock; (S_3) quartz-Pyrex seals; (V_4) diaphraghm-sealed metering valve; (V_5) diaphraghm-sealed open-shut valve; (T_2) molecular-sieve trap; (T_3) liquid-nitrogen cold-trap; (M_1) primary manifold (Pyrex glass); (M_2) gas manifold (Pyrex glass); (M_3) working manifold (quartz); (F) quartz finger; (B) lamp blanks; (C) connections to reservoirs with inert gases.

The vacuum line used for lamp preparation by Mansfield [267, 268] is reproduced in Fig. 3.14. This system was designed on the basis that the use of "O-ring" seals for construction would allow easy disassembly for cleaning purposes and operates at pressures of about 10^{-6} torr. The stopcocks, V_3, separate the working manifolds from the primary manifold and allow the primary system to be operated continuously. The working manifolds, M_3, are attached only during lamp preparation; because they have only a small volume, they are pumped down in a matter of minutes. The oil manometer, G_2, which may be temporarily disconnected by a stopcock at high vacuum, is used for measuring the gas pressures admitted to the lamps, and reads from 0.1 to 10 torr.

Although there are minor variations in procedure for making the various lamps, the general scheme is the same in most instances [267]. The cold trap, T_1, is charged with liquid nitrogen, and the ionization gauge, G_1, is checked to see that it reads below 10^{-6} torr. The gas manifold, M_2, may be evacuated at this time if necessary. After evacuation, V_4 is closed. The lamp blanks are sealed onto the working manifold, M_3, and the quartz finger, F, containing the metal or metal halide is then attached.* The working manifold is connected to the primary manifold at S_1, and a Dewar flask of liquid nitrogen is placed around T_3. By turning the stopcock V_3, the working manifold is evacuated. The lamp blanks are carefully heated with a fuel-rich oxygen-gas hand-torch until the quartz reaches a red glow. The blanks should be thoroughly out-gassed, especially at constrictions which are to be sealed. After the lamps have cooled to room temperature, the metal or metal halide is slowly sublimed into the lamp bulbs. The stopcock of the differential manometer, G_2, is opened and the filler-gas is admitted into the gas manifold, M_2, through one of the valves V_5. The metering valve, V_4, is opened slightly so that the pressure rises to the point where the reading at G_1 is nearly off-scale. This flushing procedure is carried out for several minutes, after which V_4 is closed and the system is pumped down to about 10^{-6} torr. Then V_2 is closed, G_1 is switched off, and V_4 is opened again, admitting more filler-gas to the system. The filler-gas pressure is read on G_2 and is controlled by manipulating V_2 and V_4. When the desired filler-gas pressure has been attained, the lamps are carefully sealed-off with a hand-torch with the bulbs partially immersed in cold water. In performing the sealing-off operation, it is important to produce an even seal, free of capillary inclusions, otherwise the lamp could be excessively noisy in use. After the lamps are sealed-off, V_3 is closed and V_2 is opened, allowing the system to pump down again. The working manifold assembly is removed from the system, finished lamps are broken off from the manifold M_3, and the lamp stem ends are flame-polished. The working manifold must be thoroughly cleaned before subsequent use.

* Some of the metals and metal halides decompose rapidly during sublimation and must be sublimed in through the bottom of the lamp blanks. For such elements, the lamp blanks are specially prepared with the quartz finger attached below the lamp bulb via a constriction. After the blank has been baked out under vacuum, the material is sublimed into the lamp, and the lower constriction is closed. Lamps containing iodides of nickel, cobalt, iron, calcium, and magnesium were prepared in this manner [267, 268, 483].

3.2.2.3.5. Factors Affecting the Spectral Characteristics of Lamps

It is known that the intensity and stability of microwave-excited electrodeless discharge lamps are subject to many variables, such as lamp diameter and volume, nature and pressure of the filler-gas, nature and mass of the material introduced into the lamp blank, and microwave power applied to the lamp by means of certain coupling devices, most of which are interdependent and cause changes in metal vapour pressure. Insufficient information is available at present to evaluate correctly the effect of each of these variables and particularly their mutual interactions. Improvements in performace of lamps via optimization of these variables have been largely attained by empirical methods and often without consideration of the correlation between individual parameters. It is, therefore, not surprising that in reviewing the literature, contradictory statements can be found. Moreover, few conclusions are valid for a wide range of elements; they are generally valid only for one element or, at best, a small group of elements or their compounds which yield a comparable vapour pressure at a given operating temperature.

Present knowledge of the effect of individual variables on the performance of microwave-excited electrodeless discharge lamps can be summarized as follows.

The lamp dimensions chosen (lamp diameter and the length of the lamp bulb) depend largely on the vapour pressure of the material used and the efficiency of the coupling device employed. For both the A-antenna and the resonant cavities, an internal diameter of 8–9 mm and a bulb length of 2.5–3 cm have been found to give the best compromise between noise, lifetime, and intensity in most cases [91, 268]. In general, smaller diameter lamps exhibit shorter lifetime owing to the higher frequency of atom-to-wall collisions, particularly at high power levels. The use of smaller dimensions may have some advantage when less volatile charge material is used because such lamps can be run at a much higher temperature and can offer a higher integrated line intensity. However, the analytical signal is frequently reduced under these conditions because of excessive line broadening and/or self-reversal. An upper limit on length is prescribed by the dimensions of the discharge; when the discharge plasma does not completely fill the lamp, intensity fluctuations occur. For a very volatile element or compound (e.g. mercury) some advantage may be gained by using lamps of larger dimensions because of the reduced vapour pressure for a given weight of charge material and applied microwave power. The use of large lamps, however, is somewhat limited by the construction of some coupling devices.

The nature and pressure of the inert filler-gas affect the efficiency of conversion of absorbed microwave energy into radiation. The concentration of free electrons at a given microwave power level depends on the ionization potential of the element; a higher filler-gas pressure is needed for a filler-gas with a higher ionization potential to produce the same concentration of electrons. Thus, the filler-gas pressure influences the distribution and energies of electrons in the discharge plasma. In general, at low filler-gas pressures (high electron temperature) an electrodeless discharge lamp operates at relatively high temperatures and the atomic vapour concentration is mainly produced thermally. Under these conditions, intense emission of lines with higher excitation potentials occurs. At high filler-gas pressures (low electron temperature) much cooler operation occurs and the emission of lines with lower excitation potentials prevails. In this case, sputtering also takes part in the production of the atomic vapour, particularly when an involatile element or compound is used as the charge material. Both mechanisms described are strongly influenced by the applied microwave power [360].

Filler-gas pressure also governs lamp lifetime. In general, the higher the filler-gas pressure, the longer will the lamp last. This phenomenon is related to the rate of "clean-up" of the filler-gas due to gettering action and/or diffusion into the material of the lamp walls. At very low pressures the clean-up process rapidly decreases the initial filler-gas pressure, resulting in very short lamp lifetime. The clean-up process also occurs at higher pressures of course, but the total fraction of filler-gas effectively removed from the lamp atmosphere is considerably less, resulting in longer lamp lifetime. However, it has been found that when a lamp with low filler-gas pressure has failed, the process may be reversed by simply heating the lamp to a bright orange in a flame for a few seconds; the lamp usually ignites again and operates for a period of time at its former intensity.

It should be stressed that lamp lifetime is affected not only by clean-up of the gas in the lamp, but also by reactions of the charge material with the material of the lamp walls, diffusion of metal species into the quartz, electron-wall collisions, etc. At higher filler-gas pressure, it is logical to assume that the mean free path of the charged particles is less, decreasing the probability of wall-collisions; thus, the lamp lifetime is also increased in this way.

The filler-gas most frequently used in previous studies has been argon at pressures of 1–10 torr [91, 105, 115, 268, 483]. The reasons for this choice have not usually been stated. The only thorough study on the effect of the nature and pressure of the filler-gas was carried out by Silvester and McCarthy [358, 360]. The effect of helium, neon, argon, and air as filler-gas over a wide pressure range ($10^{-4} - 10^{-2}$ torr) on the spectral characteristics of the lamps was investigated for three metals with widely different boiling points – mercury, cadmium, and manganese (as manganese chloride). The weight of the charge material was 5 mg, and the microwave energy was coupled to the lamp by means of an A-antenna. It was concluded that helium should be the filler-gas of choice for all the lamps studied because of the greater emission and fluorescence intensities obtained at higher pressures (where the lamp lifetime is the longest and vaporization is largely

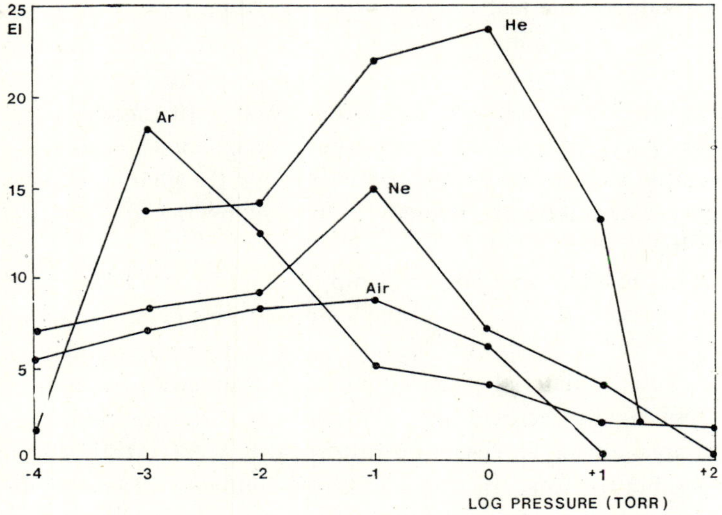

FIG. 3.15. Effect of nature of filler-gas on cadmium (2288.0 Å) electrodeless discharge lamp intensity (5 mg of cadmium metal, microwave power 200 W, A-antenna). Reprinted from Ref. [358] by permission of Marcel Dekker, Inc.

thermal). The results for cadmium are shown in Figs. 3.15 and 3.16. For individual inert gases distinct maxima (Fig. 3.15) appear in the order of their ionization potentials, the maximum for argon (lowest ionization potential) being at the lowest pressure. The most intense cadmium lamp is obtained with helium as filler-gas at about 1 torr. These maxima can be observed at various power levels, as shown for argon in Fig. 3.16. The rapid decrease in relative emission intensity at higher filler-gas pressures after the maximum is reached (which is more pronounced for high microwave power levels) is probably due to severe self-reversal caused by large temperature gradients from the central arc to the lamp walls.

Studies such as those described above for cadmium should accompany any investigation of the effect of varying the amount of charge material(s) (the reasons for this are explained later). Furthermore, it should be remembered that such results are only valid for the type of coupling device employed. The results of Cooke *et al.* [91], who found helium-filled lamps unsuccessful under the conditions used in their study (microgram amounts of less volatile charge material and coupling via a resonant cavity), confirm this.

The effect of various amounts of charge material on lamp intensity has been investigated in several papers, but a complete evaluation does not

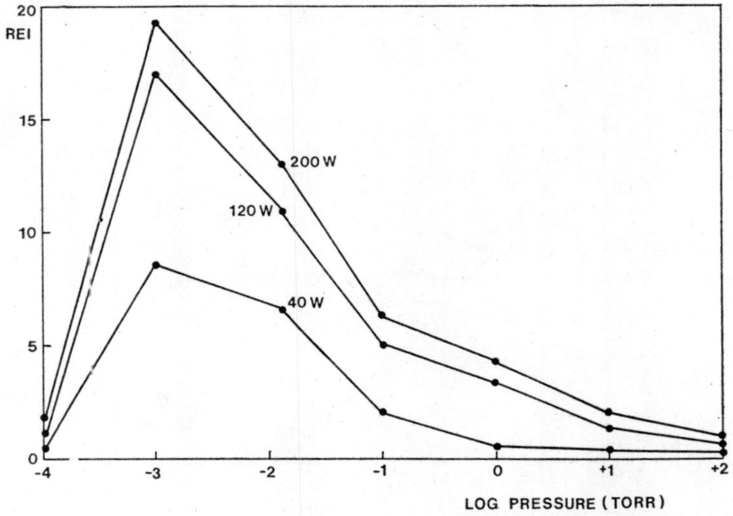

FIG. 3.16. Effect of pressure of argon filler-gas on cadmium (2288.0 Å) electrodeless discharge lamp intensity at various microwave power levels (5 mg of cadmium metal, A-antenna).

Reprinted from Ref. [358] by permission of Marcel Dekker, Inc.

appear to have been reported. Mansfield *et al.* [268] stated, on the basis of a statistical treatment, that there was no correlation between the weight of material and intensity within the range 0.23 – 10 mg of cadmium. In earlier studies of the preparation and operation of microwave-excited electrodeless discharge lamps, this conclusion was often assumed, and the effect of the amount of material introduced into the lamp bulb was mostly ignored. Silvester and McCarthy [358, 360] observed greater line intensities from lamps containing 5 mg of cadmium than from those containing larger amounts; lower quantities were not examined. The lower intensity and poorer stability of lamps with larger amounts of cadmium were shown to be a direct consequence of a metal film deposit on the interior of the lamp which effectively reduced the transmission of the lamp walls.

A systematic examination of the effect of introducing small (microgram) quantities of charge material into the lamp has been carried out by Cooke *et al.* [91] for zinc and cadmium lamps filled with argon (3 torr) and operated in the 3/4-wave Broida cavity. For all the zinc and cadmium lamps the integrated resonance line intensity increased with increasing amount of material used. The increase in intensity was generally linear with applied

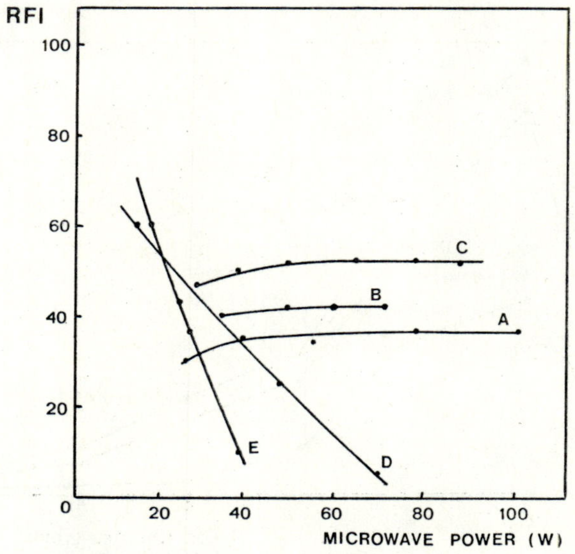

FIG. 3.17. Effect of microwave power on the atomic fluorescence signal of zinc for various zinc lamps: (A) 1 μg $ZnCl_2$; (B) 5 μg $ZnCl_2$; (C) 10 μg $ZnCl_2$; (D) 100 μg $ZnCl_2$; (E) 1000 μg $ZnCl_2$. (3/4-wave Broida cavity, 3 torr of argon) [91]. Reprinted from Ref. [91] by permission of Elsevier Publishing Company

microwave power up to about 75 W, and then began to level off. In the case of lamps containing 1 mg or more of material the intensity began to fall at very high values of applied power (> 100 W) owing to unduly high metal vapour pressure. The variation of the atomic fluorescence signal with the amount of material used in the zinc lamp at different applied microwave powers is shown in Figs. 3.17 and 3.18. From Fig. 3.17 it can be seen that the effect of power is much less critical for lamps containing low amounts of zinc chloride. This may be an advantage in that the analytical signal will be unaffected by any fluctuations in the microwave power supply.

The authors cited above have concluded that the optimum spectral performance of zinc and cadmium lamps is obtained when low quantities of the metal or metal halide are used to prepare the lamp. The effect of the amount of the charge material on the analytical atomic fluorescence signal at higher power levels is almost certainly due to the fact that the compound used is completely converted into atomic vapour at the operating temperature of the lamp. Under such conditions the partial pressure of the compound concerned within the lamp is directly proportional to the amount of that compound used. This means that for more than a certain quantity of material

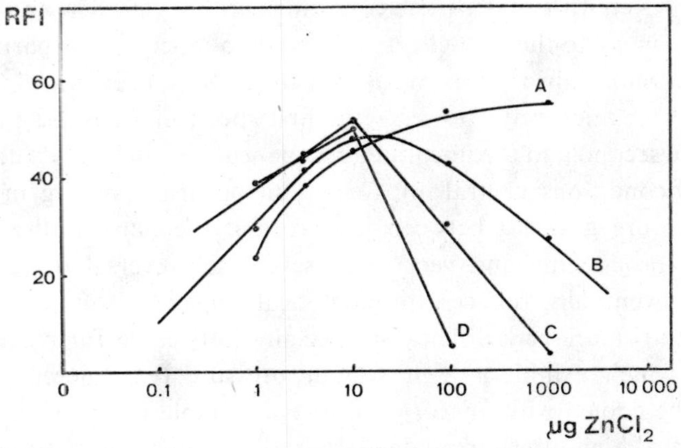

FIG. 3.18. Effect of various amount of zinc chloride in zinc lamps on atomic-fluorescence signal at various microwave power levels: (A) 20 W; (B) 30 W; (C) 50 W; (D) 70 W. (3/4-wave Broida cavity, 3 torr of argon) [91].
Reprinted from Ref. [91] by permission of Elsevier Publishing Company

the metal vapour pressure will be such that the additional material would not be volatilized, and so the amount present above this critical quantity would be without effect. It is to be expected that this would occur when milligram amounts of charge material are used, which undoubtedly explains the results obtained by other authors. The decrease in analytical signal with increasing amounts of material used is due to an increase in atomic vapour pressure, which results in a gradual transition for the light emission from lines free from self-reversal and line-broadening (skin-effect) to lines exhibiting line-broadening and self-reversal (formation of a central arc discharge) [8, 360].

It is to be expected that the performance of lamps prepared from less volatile elements or compounds would show little or no dependence on the amount of material used, once present in amounts above the maximum which may be converted into atomic vapour at the operating temperature of the lamp. The use of microgram amounts of material should be preferred because this largely obviates the problem of instability caused by movement of material within the lamp during operation.

The effect of applied microwave power is strongly dependent upon all the other parameters which affect the metal vapour pressure within the lamp, such as dimensions of the lamp, nature and weight of the charge material, nature and pressure of the filler-gas, and efficiency of the coupling device used. In general, at low metal vapour pressure (lower microwave powers, the so-called low-intensity mode of operation) the lamps emit predominantly first spectrum lines and the strongest lines of the second spectrum. Lines emitted under these conditions are very narrow and are not self-reversed, owing to the skin-effect. This mode of operation is particularly useful for atomic absorption measurements. At higher metal vapour pressures (higher microwave powers), the first spectrum lines are enhanced relative to the second and the integrated resonance line intensity is increased. Under these conditions central arc formation occurs, resulting in a very large temperature gradient between the arc and the lamp walls; the net result is line-broadening and very often severe self-reversal of resonance lines, which eventually reduces the analytical sensitivity of both atomic absorption and fluorescence. This is especially noticeable for elements or compounds which exhibit a high vapour pressure (e.g. mercury); such lamps must be cooled with an air-jet to give a suitable line profile.

This so-called high-intensity mode of operation is sometimes recommended as particularly useful for atomic fluorescence excitation purposes [192, 190]: however, only when the resonance line is broadened and not self-reversed, does a positive correlation exist between applied microwave power (integrated emission intensity) and fluorescence intensity. When severe self-reversal occurs, both analytical atomic absorption and atomic fluorescence signals are adversely affected to a similar extent. It is important to be aware of these different operational modes and to use them properly. The microwave power input should be set very carefully, because any transition between modes may result in considerable lamp instability.

Another source of instability that can be partly controlled by proper choice of applied microwave power is the efficiency of the transition of the charge compound from the solid to the vapour phase, and the reversibility of its dissociation. For example, when metal halides are used, the following equilibria occur:

$$MX_{solid} \rightleftharpoons MX_{vapour} \rightleftharpoons M + X$$

In these instances there can be a competition for the discharge between MX (giving rise to molecular bands) and M and X. This is especially noticeable if the second equilibrium is only partly reversible, and excess of halogen is generated in the lamp. Also at low microwave power levels, when the vapour pressure of MX is likely to be low, metal atoms can be sputtered directly from the solid phase as in the hollow-cathode type of discharge. Provided the charge compound exhibits a high vapour pressure at relatively low temperatures and dissociates reversibly, then stable and reliable electrodeless discharge lamps can be prepared easily. For more involatile compounds slight variations in operating temperature may result in considerable instability which can be partly overcome by the use of a vacuum-jacketed lamp [8].

Despite incomplete knowledge of the effect of all the parameters discussed above on the performance of electrodeless discharge lamps, lamps exhibiting very good spectral characteristics and capabilities for exciting atomic fluorescence have been prepared. However, there can be little doubt that this is not the ultimate performance that it is possible to achieve.

Good electrodeless discharge lamps properly operated usually yield line intensities which are from 100 to 1000 times those of the corresponding conventional unshielded hollow-cathode lamps, 10 – 100 times those of the shielded small bore hollow-cathode lamps and up to 10 times those of boosted-output hollow-cathode lamps of the Sullivan – Walsh type. Mansfield *et al.* [268] report radiances of about $10^{-3}-10^{-4}$ W cm^{-2} sr^{-1} for most of their lamps, which corresponds to a total output of radiant power

TABLE 3.3. Comparison of intensities of some hollow cathode and electrodeless discharge lamps [433]

Reprinted from Ref. [433] by permission of Butterworth and Co. (Publishers) Ltd.

Element	Wavelength Å	Relative intensity[a, b]	
		HCL[c]	EDL[d]
Tin	2246	1	820
	2863	220	2.75×10^5
Lead	2170	1	80
	2833	3.20×10^3	1.20×10^5
Thallium	2768	1	60
	3775	4	380
Mercury[e]	2536	1	910
Silver	3281	1	1

[a] At optimum operating conditions.
[b] Intensities uncorrected for spectral response of the instrument.
[c] Perkin—Elmer Intensitron lamps.
[d] Alternating current mode (electronic transformer modulation).
[e] Westinghouse lamp.

of about 0.15–0.01 W. These results are in good agreement with comparative measurements of relative emission intensities of electrodeless and hollow-cathode lamps [433]. Some comparative data for intensity, warm-up time, stability, etc. are shown for a few elements in Tables 3.3 and 3.4.

TABLE 3.4. Source characteristics of some hollow-cathode and electrodeless discharge lamps [433][a]

Element	Wavelength, Å	Warm-up time, min		Noise level on signal, %		Drift %/hour	
		HCL	EDL	HCL	EDL	HCL	EDL
Tin	2246	15	3	0.5	1	8	1
Lead	2170	12	10	1	0.5	3	3
Thallium	2768	12	10	1	1	3	2
Mercury	2536	12	5	1	1	4	4
Silver	3281	6	6	0.5	0.5	3	1

[a] See footnotes to Table 3.3.

Atomic flame-fluorescence detection limits attained with electrodeless discharge lamp sources for individual elements are listed in Table 3.7 (p. 110).

3.2.2.3.6. Multielement Electrodeless Discharge Lamps

One of the important features of the electrodeless discharge lamps is the possibility of constructing multielement lamps. Provided that groups of elements or compounds are selected that have approximately the same vapour pressures (i.e. similar running conditions), and which do not have resonance lines too close together to be resolved, there is no reason why several elements should not be grouped together in the same lamp bulb. If the charge species and their amounts are chosen very carefully, so that all material within the lamp bulb may be maintained in the vapour phase during operation, there should be no selective volatilization problem as there is in multielement hollow-cathode lamps. However, it should be borne in mind that for larger combinations of elements, conditions under which all elements present in the lamp would emit as intensely as in corresponding single element sources can practically never be realized.

Typical groupings which have been used, so far, include In–Ga (iodides)

[270], Cd–Zn (metals) [270], As–Sb (iodides) [179], Ni–Co (chlorides) [306], and Se–Te–Hg [270]. The Se–Te–Hg source is prepared from the elements plus iodine. The elements Se and Hg both function perfectly in the lamp in the elemental state, but Te is somewhat involatile and is best used as the iodide. The presence of iodine reduces the volatility of the Se (as selenium iodide) and slightly impairs the emission of the selenium lines in comparison with the single-element selenium lamp. All tubes containing iodides emit the 2061.6 Å non-resonance iodine line which has sufficient coincidence with the 2061.7 Å bismuth line to permit bismuth to be determined by resonance fluorescence at the same wavelength or even more sensitively by direct-line fluorescence at 2696.7 and 3024.6 Å (cf. Section 5.6). Thus, e.g. the Se–Te–Hg–I source becomes a multielement source for

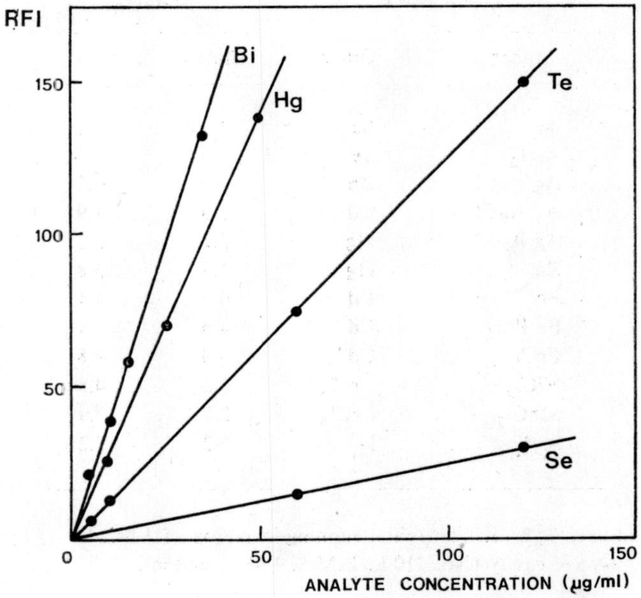

FIG. 3.19. Calibration curves for bismuth, mercury, tellurium, and selenium using a multielement electrodeless discharge lamp [270].
Reprinted from Ref. [270] by permission of Elsevier Publishing Company

Se–Te–Hg–Bi (Fig. 3.19). The jacketing method described for the preparation of single-element lamps (see Section 3.2.2.3.4 and Fig. 3.13) can be applied to the manufacture of dual-element electrodeless discharge lamps which contain elements or compounds of quite widely differing volatilities [97]. It is essential to place the element or compound of lower vapour pressure in the inner tube, which acquires the highest running temperature, and the volatile charge material in the outer tube. Because the separate sections of the source are sealed independently, it is possible to use different filler-gas at various pressures for each section. The charge material introduced into the inner tube should have a vapour pressure of about not more than 1 torr at 600 °C to prevent self-absorption [231, 235]. Alternatively, the amount of material introduced should be insufficient to provide a saturated vapour at the working temperature of the lamp. Some characteristics of dual-element lamps prepared in this way are summarized in

TABLE 3.5. Some characteristics of dual-element (dual-cavity) sources [97]
Reprinted from Ref. [97] by permission of Elsevier Publishing Company

Lamp	Components		Stability[a]		Power[b] [W]
	Inner	Outer	Inner	Outer	
Te/Se	Te	Se	1.7	3.8	60
As/Se	As/I_2	Se	1.9	2.5	50
Ag/Zn	Ag/AgCl	Zn	1.9	3.2	65
Ag/Cd	Ag/AgCl	Cd	1.8	1.9	65
Ag/Hg	Ag/AgCl	Hg	3.2	5.2	60
Zn/Hg	Zn	Hg	1.7	3.8	60
Pb/Cd	Pb	Cd	0.7	4.5	60
Pb/Cd	Pb/PbI_2	Cd	4.8	3.5	65
Cr/Cd	$CrCl_3$	Cd	3.0	4.8	50
Ni/Cr	$NiCl_2$	$CrCl_3$	3.2	4.8	60
Mn/Cr	$MnCl_2$	$CrCl_3$	2.7	7.9	60
Sb/Bi	Sb/I_2	I_2	2.2	2.2	50

[a] Expressed as ±%/h after an initial running-in period of 1 hr.
[b] Broida 3/4-wave cavity (No. 210 L, E.M.S. Ltd., London).

Table 3.5. It can be seen that the element in the inner tube tends to give a more stable discharge than the element in the outer tube. For certain elements, such as tellurium, lead, and thallium, stable and intense discharges can be obtained from the element, instead of the halide normally used, without running the lamps at high power.

Studies of multielement electrodeless discharge lamps have shown that the lamps can be operated over prolonged periods of time in exactly the same way as the individual lamps and that their intensity and stability are as good as those of the single-element lamps.

3.2.2.3.7. *Evaluation of Electrodeless Discharge Lamps*

The characteristics of microwave-excited electrodeless discharge lamps nearly approach the requirements of ideal line sources for atomic fluorescence spectrometry. They have been reported to exhibit the following properties.

(1) They are simple in design, construction and are inexpensive to prepare; only a small amount of charge material is required.

(2) They can be prepared for practically all elements.

(3) They emit spectra comprised almost exclusively of the principal atomic lines, with no carrier-gas background.

(4) They give high intensity at principal resonance lines and exhibit very favourable line profiles when operated under optimized conditions.

(5) They are very stable when prepared and operated under proper conditions.

(6) Multielement lamps can be prepared, although the single-element lamps are readily interchangeable.

(7) They appear to have a relatively short warm-up period and a very long running and shelf-life.

Improvements in electrodeless discharge lamps are being made daily. Very recently it was shown that stability and reproducibility of spectral radiant output of both single-element and multi-element electrodeless discharge lamps operated with an A-antenna can be greatly improved by precise temperature control [495, 496, 530]. A simple device for controlling the lamp temperature (rigidly fixed with respect to the antenna) consists of a small fan, nichrome heater coil and thermocouple connected to a temperature controller, which regulates the current to the heater coil. Heated air passing over the lamp envelope allows a precise adjustment of the optimum lamp temperature corresponding to maximum spectral radiant output of the particular element. The temperature controlled system was found to be relatively insensitive to changes in applied microwave power in marked contrast to other systems (both resonant cavities and antennas without temperature control).

For multielement electrodeless discharge lamps [530], the spectral radiant output from each individual element in a multielement lamp is very temperature sensitive, but largely uninfluenced by the presence of other elements (or compounds). Plots of the variation of spectral radiant output with temperature allow the rational choice of a compromise operating temperature for several elements in each lamp ("simultaneous mode"). Alternatively, the optimum temperature for each element present may be selected in turn ("sequential mode"). The ability to use a single operating temperature for each multielement lamp is attractive for the application of these sources to simultaneous multielement atomic fluorescence analysis.

Although some parameters involved in the preparation and operation of the electrodeless discharge lamps have not been fully examined, it can be presumed that they will probably become standard sources for atomic fluorescence spectrometry.

3.2.2.4. Radiofrequency Plasma Sources

The radiofrequency, induction-coupled plasma (Radyne Ltd. Model SC 15, U.K.) operating at a frequency of 36 MHz and 2.5 kW maximum power output was tested as an excitation source in studying the atomic fluorescence characteristics of several elements in conventional flames [206]. With an unsheathed air-propane flame as atomizer, and 1000–4000 μg/ml metal solutions as source material, relatively high detection limits were attained (Table 3.6). The main advantage of this type of line source compared to single-type sources is flexibility. Unfortunately, no other research on the use of plasma sources in atomic fluorescence spectrometry has been performed, so that it is difficult to make an objective evaluation of these excitation sources at present.

TABLE 3.6. Atomic fluorescence detection limits with the radiofrequency plasma source [206]

Element Wavelength (Å)	Solution concentration into the plasma (μg/ml)	Power (kW)	Plasma carrier, argon flow-rate (1/min)	Detection limit (μg/ml) air-propane flame
Cd 2288.0	2000	2	4	0.08
Ca 4226.7	4000	2	4—6	0.1
Cu 3247.5	3000	2	5	0.05
Mg 2852.1	1000	2	4—6	0.005
Mn 2798.3	4000	2	5	0.1
Zn 2138.6	2000	2	4—6	0.08

3.2.2.5. Lasers

Very recently, an investigation of the use of lasers as excitation sources for atomic fluorescence flame spectrometry has been reported. Tunable dye lasers [373] show the greatest promise, because of their high intensities and simple optical requirements. Denton and Malmstadt [148] demonstrated the analytical utility of a system consisting of pumped tunable dye laser (the dye cell was filled with a $5 \times 10^{-4} M$ solution of 2′, 7′-dichlorofluoresein adjusted to pH 11) and an ultrasonic nebulizer-burner for the measurement of the atomic fluorescence of barium at the 5535.5 Å resonance line. A narrow-band dye laser has been shown to be a possible light-source for atomic fluorescence analysis of subnanogram amounts of sodium [254].

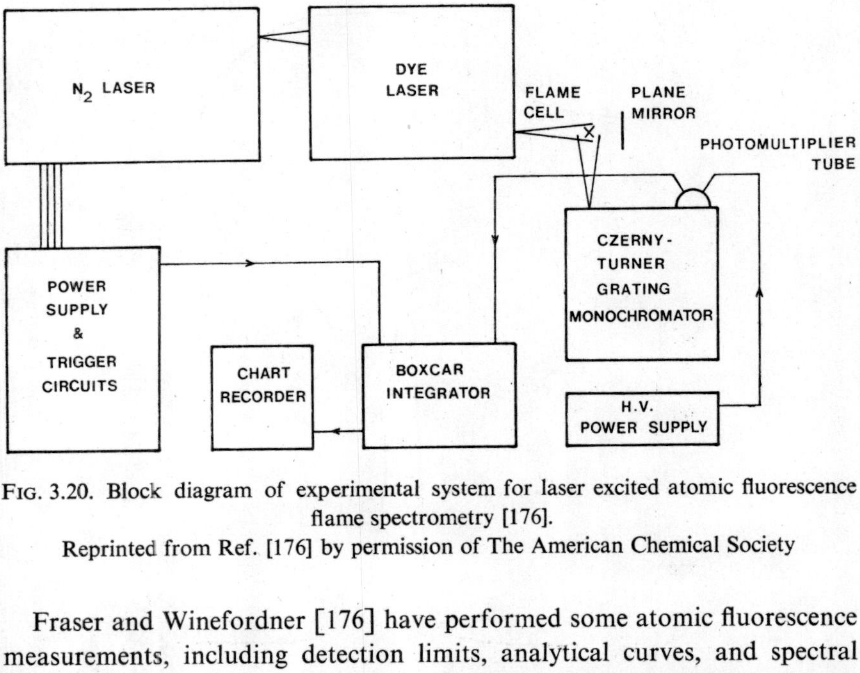

FIG. 3.20. Block diagram of experimental system for laser excited atomic fluorescence flame spectrometry [176].
Reprinted from Ref. [176] by permission of The American Chemical Society

Fraser and Winefordner [176] have performed some atomic fluorescence measurements, including detection limits, analytical curves, and spectral resolution for Al, Ca, Cr, Fe, Ga, In, Mn, Sr, and Ti in either air-hydrogen or nitrous oxide-acetylene flames with a pulsed, tunable dye laser pumped with a nitrogen laser (AVCO Everett Research Laboratory, Everett, Mass.) Their experimental configuration for laser-excited atomic flame fluorescence spectrometry is shown in Fig. 3.20. The dye laser system has a peak of greater than 10 kW at all wavelengths, a pulse repetition rate of about 1 – 25 Hz, a spectral half-width of about 1.0 – 10 Å, and a pulse half-width of about 2 – 8 nsec. By proper choice of dye and grating angle, any wavelength region (1.0 – 10 Å wide) between 3600 and 6500 Å can be selected to excite atomic fluorescence. A simple scan-drive system enables the grating angle in the dye laser to be varied and thus allows a continuous variation of the dye laser wavelength (i.e. continuous scanning) over a small wavelength range (limited by the spectral band-width of the monochromator employed (Fig. 3.21). Use of a pulsed-operation mode requires a fast response photomultiplier and a gated amplifier-integrator capable of operating at gate widths of the order of 10 nsec, but all noise except scatter of laser radiation can be essentially eliminated.

Analytical curves (log-log plots) for laser-excited atomic fluorescence are linear from the detection limits to concentrations at least 10^3 times the

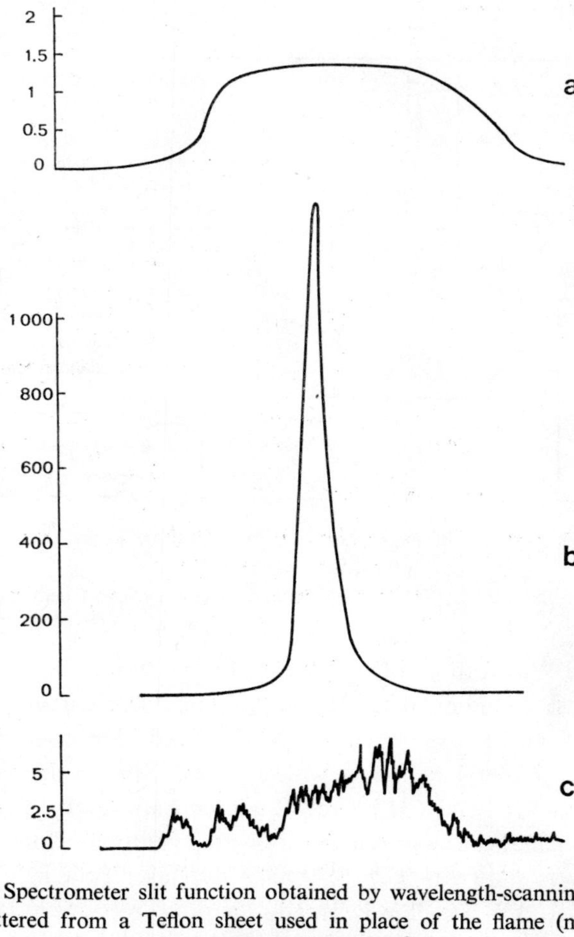

FIG. 3.21. (a) Spectrometer slit function obtained by wavelength-scanning the dye-laser radiation scattered from a Teflon sheet used in place of the flame (monochromator wavelength set at 4226.7 Å).
(b) Spectral distribution of fluorescence radiation of calcium (1000 μg/ml) obtained by varying the dye-laser wavelength over about 10 Å surrounding 4226.7 Å (monochromator wavelength set at 4226.7 Å).
(c) Spectral distribution of radiation scattered by the flame gases obtained by varying the dye-laser wavelength over about 10 Å surrounding 4226.7 Å (monochromator wavelength set at 4226.7 Å) [176].
Reprinted from Ref. [176] by permission of The American Chemical Society

detection limits (see Fig. 3.22). The relative standard deviations of fluorescence signals 10, 100, and 1000 times the detection limits were found to be about 8, 5, and 3%, respectively. Detection limits attained for all the elements examined are given in Table 3.7 (p. 110). Note that the laser-excited atomic fluorescence detection limits for aluminium, indium, and titanium are better than any previous atomic fluorescence flame spectrometric values obtained with other types of primary source.

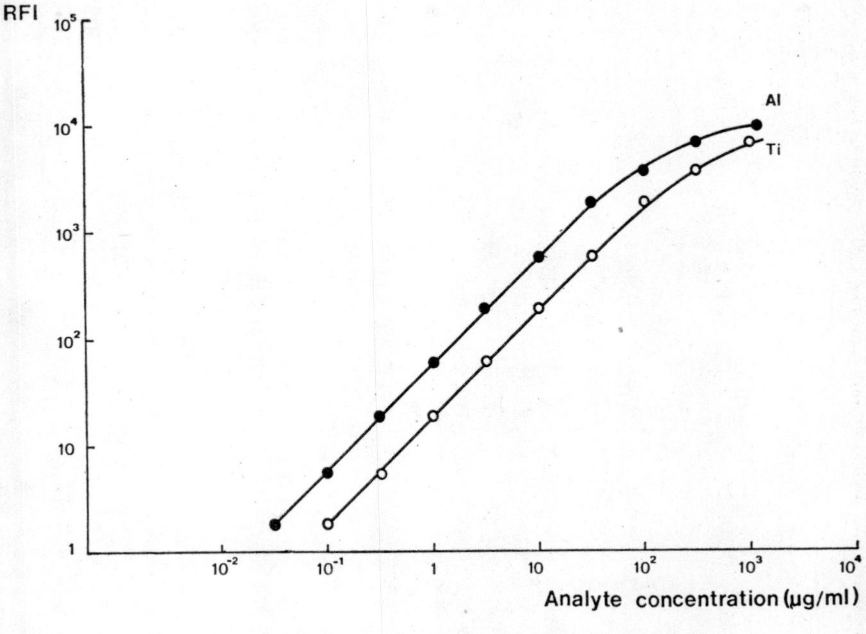

FIG. 3.22. Calibration curves for laser-excited atomic fluorescence of aluminium and titanium in a nitrous oxide-acetylene flame [176].
Reprinted from Ref. [176] by permission of The American Chemical Society

These initial experimental results demonstrate a possible use of tunable dyes lasers as excitation sources in atomic fluorescence flame or non-flame spectrometry. Their main advantages are very high total power output (kW), tuning capability over a wide wavelength region, and continuous scanning over a small wavelength range around the nominal wavelength chosen. To extend the wavelength range down to the ultraviolet region, future research in this field will probably be directed to the use of lasers with frequency-doubled lines [422]. However, the extremely high cost of these excitation devices will undoubtedly limit their general practical use.

TABLE 3.7. Atomic flame-fluorescence detection limits attained with different excitation sources

Element Wavelength (Å)	Source and detection limit (μg/ml)[a, b, c, d]									
	MVDL[e]	SHCL[f] [258, 265]	SHCL(p) [144, 145, 146]	BOHCL[g]	HHCL [152, 349]	EDL[h]	R.F. plasma [206]	Xe-arc, 150W	Xe-arc, 450 W	L [176]
Ag 3280.7	—	0.003	0.002	0.000 3 [245]	0.001	0.000 1 [483]	—	0.06 [190]	0.001 [164]	—
Al 3961.5	—	—	0.07	1.5[l] [259]	—	0.1[j] [118]	—	—	4.0 [190]	0.03
As 1972.6	—	6	—	1.0[k] [264]	—	0.07 [145, 146]	—	—	2.0 [190]	—
Au 2676.0	—	1.0	0.009[l]	0.005[l] [258, 278]	0.05	0.2 [58, 267]	—	3.5 [411]	—	—
Ba 5535.5	—	—	0.2	—	—	—	—	—	—	—
Be 2348.6	—	—	—	0.008[l] [259]	—	0.01[j] [191]	—	—	—	—
Bi 3067.7	—	5	—	0.25 [258]	0.1	0.004 7 [190]	—	0.3 [190]	0.17 [190]	—
Ca 4226.7	—	0.2	0.000 3	—	—	0.02 [483]	0.1	0.2 [190]	0.02[j] [142]	0.01
Cd 2288.0	0.0001 [253]	0.02	0.001	0.000 04 [245]	—	0.000 001[j] [483]	0.08	0.033 [190]	0.015 [190]	—
Co 2407.3	—	0.15	0.002	0.001 5 [258]	0.3	0.005 [173]	—	0.14 [190]	0.021 [190]	—
Cr 3578.7	—	20	0.000 3[m]	0.001 5. [258]	1.0	0.05 [398]	—	0.3 [264]	3.0[m] [96]	0.03

TABLE 3.7 (continued)

Element Wavelength (Å)	Source and detection limit (μg/ml)[a, b, c, d]									
	MVDL[e]	SHCL[f] [258, 265]	SHCL(p) [143, 145, 146]	BOHCL[g]	HHCL [152, 349]	EDL[h]	R.F. plasma [206]	Xe-arc, 150 W	Xe-arc, 450 W	L [176]
Cu 3247.5	—	0.01	0.000 3	0.000 3 [249, 264]	0.001	0.005[j] [483]	0.05	0.083 [190]	0.015 [190]	—
Fe 2483.3	—	0.4	0.000 6	0.003 [258]	2.0	0.005 [398]	—	0.67 [190]	0.035 [190]	0.3[n]
Ga 4172.1	0.02[o] [45]	0.5	—	—	1.0	0.007 2 [190]	—	0.16 [190]	0.14 [190]	0.3[o]
Ge 2651.6	—	—	—	0.6[i] [259]	—	0.1[j] [119]	—	—	1.8 [190]	—
Hg 2536.5	0.07 [258]	30	—	—	—	0.08[j] [73]	—	100 [65]	—	—
In 4101.8	0.015[p] [45]	0.2	—	—	0.3[p]	0.1 [483]	—	2.0 [265]	—	0.01
Ir 2544.0	—	4.0	—	—	—	—	—	100 [265]	—	—
Mg 2852.1	—	0.000 8	0.000 1	0.000 15 [258]	0.03	0.008 [483]	0.005	0.038 [190]	0.002 4 [190]	—
Mn 2794.8	—	0.03	0.000 5	—	0.05	0.001[r] [158]	0.1	0.03 [190]	0.002 5 [190]	0.3[s]
Mo 3132.6	—	—	0.2	2.0 [388]	—	0.46 [118]	—	—	—	—
Na 5890.0	—	—	—	—	—	—	—	—	0.007 5 [190]	— —

TABLE 3.7 (continued)

Element Wavelength (Å)	Source and detection limit (μg/ml)[a, b, c, d]									
	MVDL[e]	SHCL[f] [258, 265]	SHCL(p) [143, 145, 146]	BOHCL[g]	HHCL [152, 349]	EDL[h]	R.F. plasma [206]	Xe-arc, 150 W	Xe-arc, 450 W	L [176]
Ni 2320.0	—	0.1	0.001	0.003 [276]	0.5	0.006[j] [114]	—	3.0 [265]	1.0 [164]	—
Pb 4057.8	—	1.0	—	0.02 [390]	1.0	0.01[j] [74]	—	3.3[t] [190]	0.61[t] [190]	—
Pd 3404.6	—	2.0	—	0.04 [391]	—	—	—	50 [265]	—	—
Pt 2659.4	—	30	—	0.5 [388]	—	—	—	500 [265]	—	—
Rh 3434.9	—	0.3	—	—	—	—	—	10[u] [265]	—	—
Ru 3727.0	—	5.0	—	—	—	—	—	100 [265]	—	—
Sb 2311.5	—	20[v]	0.05	0.03[v] [248]	—	0.04 [399]	—	100 [65]	—	—
Se 1960.9	—	6.0	—	—	—	0.15[j] [114, 115]	—	134 [190]	3.9 [190]	—
Si 2516.1	—	—	—	—	—	0.55 [117]	—	—	—	—
Sn 3034.1	—	70	0.2	1.0 [245, 394]	—	0.12[j] [72]	—	5 [65]	—	—
Sr 4607.3	—	—	0.000 8	—	—	0.03 [483]	—	—	> 10[j] [142]	0.03

TABLE 3.7 (continued)

Element Wavelength (Å)	Source and detection limit (μg/ml)[a, b, c, d]									
	MVDL[e]	SHCL[f] [258, 265]	SHCL(p) [143, 145, 146]	BOHCL[g]	HHCL [152, 349]	EDL[h]	R.F. plasma [206]	Xe-arc, 150 W	Xe-arc, 450 W	L [176]
Te 2142.7	—	20	—	0.05 [245, 394]	—	0.05 [114, 115]	—	4.2 [190]	0.79 [190]	—
Ti 3199.9	—	—	—	4.0[i] [259]	—	—	—	—	—	0.1[w]
Tl 3775.7	0.005 [45]	0.1	—	—	0.1	0.04[j] [266]	—	0.13 [190]	0.073 [190]	—
V 3184.0[x]	—	—	0.25	—	—	0.07[j] [118]	—	—	—	—
Zn 2138.6	0.000 001 [253]	0.05	0.001	0.000 6 [245, 394]	—	0.000 04[j] [483]	0.05	0.080 [190]	0.015 [190]	—

[a] Abbreviations for sources: see Chapter 5.
[b] The detection limit refers to aqueous solutions.
[c] S : N ratio = 2 when not otherwise stated.
[d] References are given in parenthesis.
[e] Conventional Philips, Osram or Wotan metal vapour discharge lamps.
[f] Perkin-Elmer "Intensitron" or A.S.L. shielded hollow-cathode lamps.
[g] Boosted output (high-intensity) hollow-cathode lamps of the Sullivan-Walsh design when not otherwise stated.
[h] Single-element lamps.
[i] Löwe design
[j] S : N = 1.
[k] At the 1937.6 Å line.
[l] At the 2428.0 Å line.
[m] At the chromium triplet (3578.7, 3593.5, 3605.3 Å).
[n] At the 3719.9 Å line.
[o] At the 4033.0 Å line.
[p] At the 4511.3 Å line.
[r] At the manganese triplet (2794.8, 2798.3, 2801.1 Å).
[s] At the 4030.8 Å line.
[t] At the 2833.1 Å line.
[u] At the 3692.4 Å line.
[v] At the 2175.8 Å line.
[w] At the 3998.6 Å line.
[x] At the vanadium triplet (3183.4, 3183.9, 3185.4 Å).

3.2.3. CONTINUUM SOURCES

A stable and intense continuum source would be an ideal source for fluorescence spectrometry: it would not only be applicable to the simultaneous determination of many elements in a flexible multielement instrument, but, in comparison with line sources, would also make it possible to measure precisely light-scattering around the excited line. Sometimes discrimination between a spurious signal and a fluorescence signal is quite difficult, if only a line source is available. If right-angle illumination of the flame-cell is used, calibration curves obtained with line-source excitation exhibit a sharp maximum at a certain concentration, and after that a sharp decrease in fluorescence intensity occurs. For an ideal arrangement but with a continuum source, only a plateau occurs at high concentration. From the analytical standpoint, this is a distinct advantage. One principal disadvantage of the use of a continuum source is that it becomes essential to use a monochromator, since this is now the only means of attaining selectivity. Unfortunately, till now the practical comparison of line and continuum sources favours the former. The reason for this is the poor intensity of contemporary continuum sources. Although some experimental evidence exists that sources with much higher intensity in the ultraviolet region can be built [80], only high pressure xenon-arc discharge lamps have been used successfully [47, 96, 103 142, 162, 190, 264, 265, 332, 366, 411, 430, 445, 463, 488]. The Eimac short arc high-pressure xenon discharge lamp, which is similar to these lamps, has been used in several investigations [46, 65, 177]. The only other continuum sources investigated are a 90-W high-pressure mercury lamp [312] and the anode spot of a high-current graphite arc [382].

High-pressure xenon-arc lamps have been commercially available for over 30 years, and their technology is now well established. Probably no drastic improvements in performance can be expected in years to come.

The construction of these lamps is quite simple. Two tungsten electrodes are enclosed in a quartz envelope, the anode usually being the upper, more massive electrode. All common types use air-cooling; in an exceptional case the electrodes are water-cooled (1-kW lamp DKSSH 1000 [47]). If the lamp is not operated in a vertical position, dangerous overheating of the quartz envelope, and subsequent explosion may occur. The lamps must always be operated very carefully, according to the manufacturer's instructions. Their lifetime (1200 – 2000 hr) may be unfavourably influenced not only by the use of higher discharge currents, but also by the use of low currents or by excessive current ripple.

The spectral radiance of discharge lamps is determined by the axial temperature of the discharge, the pressure and composition of the discharge atmosphere and the transmission characteristics of the window or envelope.

The axial temperature may be increased by using commercial lamps in a pulsed mode. A 2200-W xenon lamp operated in a conventional mode (100 A, 22.5 W) had a spectral radiance at the cathode spot of around 3 W cm^{-2} sr^{-1} nm^{-1} at about 3000 Å; by an increase of the discharge current to 550 A (using 100-msec pulses), the intensity rose more than four times [239] at this wavelength. The typical diameter of the brightest area of a xenon lamp – the cathode spot – is about 1.5 mm. This is surrounded by the arc plasma, extending from cathode to anode. The plasma brightness is much lower than that of the cathode spot. Therefore, it is important to focus precisely this small part of the discharge into the atomic cell (flame) with high quality achromatic optics – e.g. a toroidal mirror. If only a simple spherical quartz lens is used, the spectral radiance of the blurred image of the cathode spot is substantially lower. Hollow-cathode tubes or high-frequency discharge lamps usually emit from a much larger area and therefore the quality of the optics between these primary sources and the flame is much less critical.

The quartz envelope of a xenon lamp is made from technical quartz as the lamps are predominantly used in the visible region – and only two or three types can be ordered with a "Spectrosil" quartz jacket. Therefore the useful radiant output drops sharply at wavelengths below 2500 Å. The envelope transparency also decreases owing to vaporization of the electrodes.

The deposited layer is probably similar to a grey filter – at least no indication could be found in the literature that its transparency depends on wavelength.

In the selection of a light-source, stability is an important criterion. Some types of lamps are more stable than others, for example the Hanovia 901 C is reported to be relatively stable with short-term fluctuations of less than 2%, but with the Hanovia 959 C, which has a particularly small electrode spacing, the arc sometimes jumps from one location on the cathode to another [80].

Arc-wandering can contribute markedly to observed intensity fluctuations (±10% in a few seconds, ±40% in 10 min); it has been found that considerable improvement is obtained after 200 – 300 hr of operation. However, the quantitative data on intensity fluctuations and their time frequency spectrum are very limited at the time of writing. It may be expected (as is the case with other arc sources) that the output noise decreases with increasing frequency and attains a stable level at frequencies of some hundred Hz or higher.

The Eimac arc sources are constructed so that the light emerges through a flat window, perpendicular to the lamp axis. The cathode spot of the arc burning in a xenon atmosphere is reflected by a metal mirror, mounted around the anode, so that either a quasi-collimated beam is formed or the light is focused about five inches from the lamp. In both cases the quality of the mirror seriously limits the attainable radiance. In all cases, when comparable experimental data are available for atomic fluorescence flame spectrometry with line and continuum sources [47, 162, 163, 265, 411], the limits of detection are substantially poorer with the continuum sources. This is somewhat surprising in view of the signal ratios and signal-to noise ratios calculated from theoretical formulae [466]. With a 450-W xenon lamp, limits of detection should be about the same for both line and continuum sources, as the spectral radiances are roughly the same [332] (see

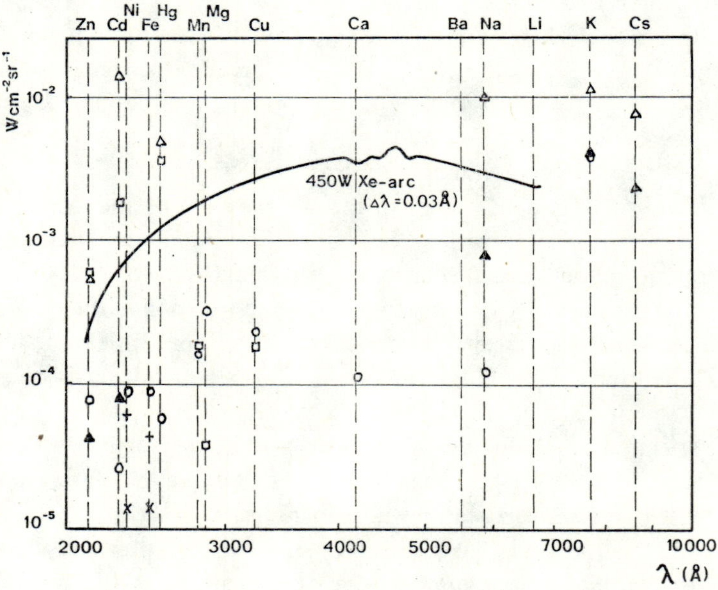

FIG. 3.23. Radiance of a 450-W xenon arc lamp continuum compared to radiances of some line sources [332]: (○) shielded hollow-cathode lamps; boosted-output hollow-cathode lamp (×) without secondary discharge and (+) with secondary discharge; (△) metal vapour-discharge lamps with specific operating current and (▲) diminished operating current; (□) electrodeless discharge lamps; (—) 450-W xenon arc lamp.

Fig. 3.23). The discrepancies between theoretical and experimental extimations of limits of detection are in excess of two orders of magnitude and the authors have no explanation for the relatively poor experimental detection limits with continuum sources [466].

3.2.4. PRACTICAL EVALUATION OF ATOMIC FLUORESCENCE EXCITATION SOURCES

Table 3.7 gives a list of the best detection limits which have been obtained with the individual excitation sources described in Section 3.2. The authors are well aware of the fact that it is not possible to compare the sources according to limits of detection obtained (in most cases) with different instrumentation. However, since satisfactory comparative absolute measurements on radiances of these sources are practically non-existent, this is the only way to give at least a rough comparison of the capabilities of the individual sources to excite atomic fluorescence.

3.3. ATOMIZERS

3.3.1. GENERAL CONSIDERATIONS

The provision of an atom cell, i.e. of a properly defined cloud of free atoms of the element of interest, is the major limitation in atomic fluorescence spectrometry, probably even more so than in emission or absorption spectrometry.

The main requirements of an atom cell in atomic fluorescence spectrometry are:

(1) high density of analyte atoms,
(2) long residence time of the analyte atoms in the optical path,
(3) high volatilization efficiency (to avoid light-scatter),
(4) low quenching properties,
(5) low radiance from analyte atoms or background,
(6) good homogeneity and/or structural stability,
(7) high reproducibility,
(8) simplicity and ease of handling.

It is evident almost at first glance that all these requirements cannot be fulfilled simultaneously. For simplicity and easy handling chemical flames are so far unrivalled. This explains their popularity, despite the fact that with flames it is difficult to obtain high density of free analyte atoms and long residence time because of the considerable dilution of the sample with flame gases. In flames, however, low analyte radiance is not compatible with volatilization efficiency for refractory elements which require rather high temperatures and reducing environments.

The conflicting nature of these requirements frequently makes the choice of sampling and atomizing techniques difficult and very often constitutes a major limitation.

In atomic fluorescence spectrometry chemical flames have been used most often. Their main advantage (besides simplicity and ease of operation) is low cost. Other atom cells, conveniently assembled under the heading "non-flame atomizers", form a rather heterogeneous group and include purely thermal atomizers, e.g. hot wire loops, graphite rods, as well as non-thermal atomizing devices exploiting cathodic sputtering, electron bombardment, etc.

Most of these devices are considerably more complicated, and require special power generators. Often transient signals must be measured, and the factors influencing atomization are even less understood than those in flames. It is hardly surprising, therefore, that the greater part of the work done on atomic fluorescence spectrometry has so far been performed with flames.

3.3.2. CHEMICAL FLAMES

3.3.2.1. Flame Structure

Chemical flames have a characteristic structure which is basically independent of the fuel and oxidant used. For convenience let us consider first a "premixed" type flame, which results when a laminar stream of intimately mixed fuel and oxidant from a burner is ignited (Fig. 3.24). Immediately above the burner port is a preheating zone: here the mixture is heated by thermal conductivity and radiation to the temperature of inflammability. At this point the combustion reaction starts.

To secure a steady flame and prevent flashback the velocity of the gas mixture through the burner ports (which can be calculated from the volume of the combustion mixture and the total area of the burner ports) must be greater than the burning velocity of the particular combustion mixture. For a laminar flow the lowest stream velocity is at the burner walls and flashback readily occurs at this point. The burner walls also

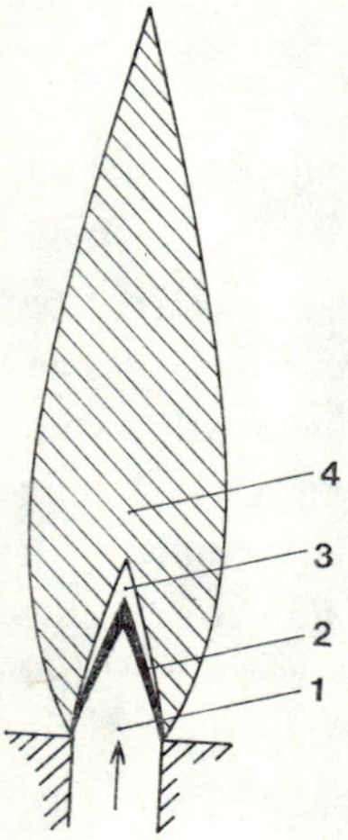

FIG. 3.24. Schematic representation of premixed flame structure: (1) preheating zone; (2) primary reaction zone; (3) interconal zone; (4) secondary reaction zone.

act as heat sinks, preventing the gas mixture from attaining the temperature necessary for the combustion reaction to start. The danger of flashback can therefore be decreased by cooling the burner head.

In the presence of a sufficient supply of oxygen in the combustion mixture, the reaction proceeds almost to completion. The flame volume where this energetic reaction takes place is called the primary reaction zone (or primary reaction cone if the burner port happens to be circular). Under the conditions used in flame spectrometry, the combustion time is of the order of μsec and the thickness of the primary reaction zone is around 0.1 mm. Gases in the primary reaction zone are far from both chemical and thermal equilibrium and emit strong molecular spectra of CN, C_2, NH, and CH species.

Adjacent to the primary reaction zone lies the so-called interzonal region or interconal zone. Here the flame gases have generally already reached a state of thermal and chemical equilibrium, although the concentration of some particles can still be at suprathermal values. This is the region of maximum temperature, and very often provides optimum conditions for atomization. From the analytical point of view it is often desirable to make measurements in this region. Unfortunately under stoicheiometric conditions, the interconal zone forms only a thin layer over the primary reaction zone. Its size can be somewhat extended however by increasing the fuel to oxidant ratio or even more effectively by separating the secondary reaction zone, as explained later.

Above the interzonal region comes the secondary reaction zone, where the composition of the flame gases is changing by diffusion and entrainment of the surrounding atmosphere. This results in a more complete combustion of the flame gases. In hydrocarbon flames, combustion proceeds to carbon monoxide in the primary reaction zone, and to carbon dioxide in the secondary combustion zone.

In consequence this secondary combustion zone emits a fairly strong continuous spectrum resulting from the reaction

$$CO + O \rightarrow CO_2 + h\nu$$

as well as strong bands due to the hydroxyl radical OH.

3.3.2.2. Separated Flames

By prevention of entrainment of the surrounding atmosphere and the oxygen it contains into the flame gases, the secondary reaction zone may be shifted away from the primary reaction zone, thus extending the region between the two zones. Such flames with separated zones are called separated flames.

The separation of zones, and measurement in the interzonal region, results for hydrocarbon flames in reduction of the intensity of the continuous spectrum by two or more orders of magnitude, and an almost complete suppression of the molecular OH band emission [192, 228]. This is very beneficial when fluorescence measurements are made in the air-acetylene flame [192] and becomes almost imperative in the nitrous oxide-acetylene flame [118, 191]. Extension of the interzonal region has the further favourable effect that the whole primary light beam can pass through the region where conditions for atomization are optimal (a highly reducing environment) so that the density of free atoms of refractory elements, as observed by atomic absorption spectrometry, is higher [117, 227]. The possibility of measuring fluorescence higher in the flame also gives more time for the volatilization of the aerosol particles.

The separation of zones may be achieved by means of a quartz tube placed above the burner head (Smithell's separator) or by a laminar stream of inert gas, generally nitrogen or argon, surrounding the flame (see Fig. 3.25).

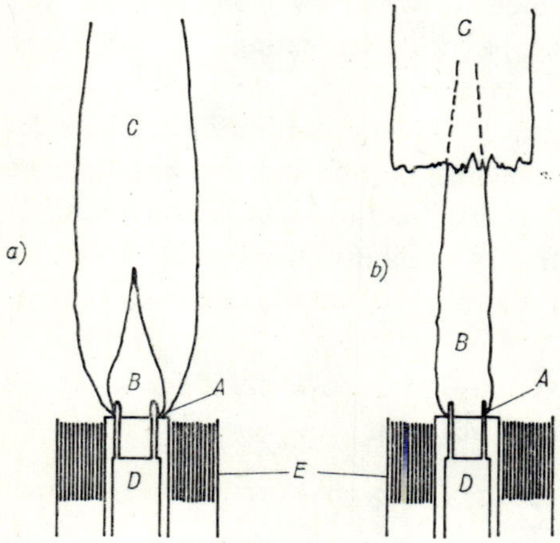

FIG. 3.25. Effect of nitrogen or argon shielding of the premixed nitrous oxide-acetylene flame [227]: *a* — conventional flame, *b* — separated flame. (A) primary reaction zone; (B) interconal red zone; (C) secondary reaction zone; (D) burner stem; (E) nitrogen or argon shield gas inlet.

"Mechanical" separation has the advantage that no cold gases dilute the interzonal flame gases. On the other hand, when fuel-rich hydrocarbon flames are used, carbon deposition on the separator walls is a nuisance. This may be avoided, at least within the optical path, by adding two side-arms with optical windows at right angles to the separator tube [191] (see Fig. 3.26).

The use of an inert-gas shield for prevention of atmospheric entrainment is simpler and more often employed. Typical burners with provision for flame shielding are shown in Figs. 3.27 and 3.28. Its disadvantage is the dilution of the flame gases by the cold shielding-gas and the corresponding temperature drop. Kirkbright and Vetter [234] have compared the tem-

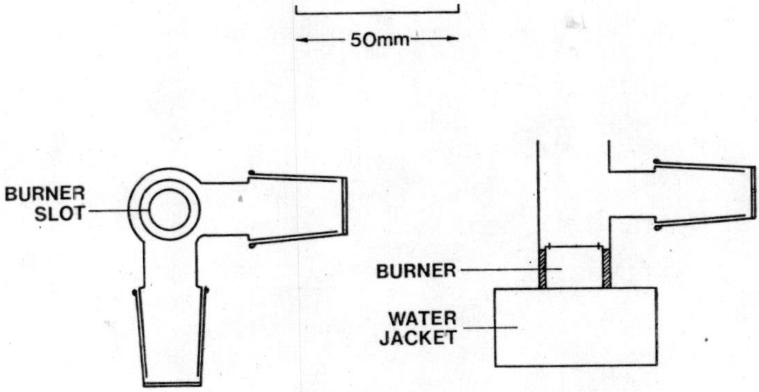

FIG. 3.26. Mechanical separator [191].
Reprinted from Ref. [191] by permission of The Society for Analytical Chemistry

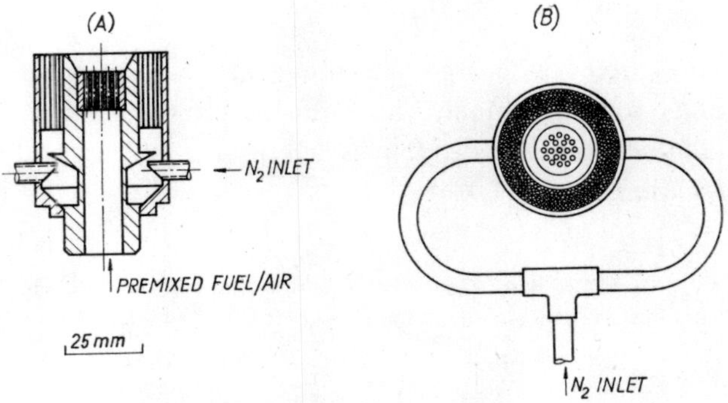

FIG. 3.27. Atomic fluorescence burner with shielding device: (A) Sectional drawing; (B) Top view showing the configuration of 19 holes (1.59 mm in dia.) in the titanium burner head [258].
Reprinted from Ref. [258] by permission of Pergamon Press Ltd.

perature profiles of nitrogen- and argon-shielded nitrous oxide-acetylene flames with those of unshielded flames at various fuel-oxidant ratios. As expected the temperature differences were smallest for a fuel-lean flame and highest for a fuel-rich flame, when the heat released by the secondary combustion reactions is eliminated. Nitrogen caused a greater temperature drop than argon, evidently due to its higher thermal capacity and conductivity. For fluorescence measurements argon has the further advantage of considerably lower quenching cross-section, so that higher fluorescence

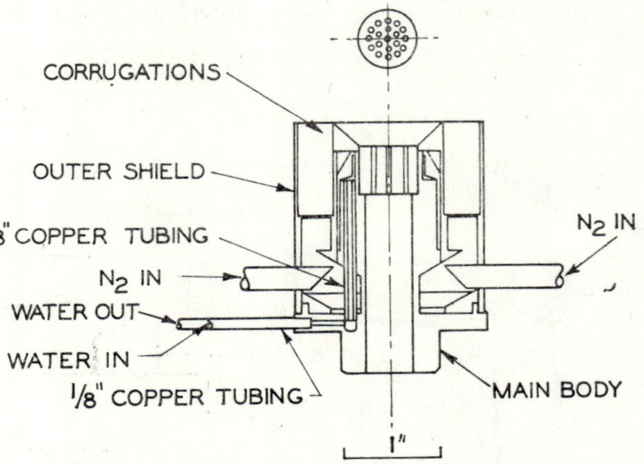

FIG. 3.28. Nitrogen-sheathed, water-cooled burner for atomic fluorescence measurements in nitrous oxide-supported flames [259]. (The titanium burner head drilled with 19 holes, 1.09 mm in dia.)
Reprinted from Ref. [259] by permission of Pergamon Press Ltd.

signals can be expected with argon-shielding than with nitrogen-shielding. Detection limits for aluminium, vanadium and molybdenum in a conventional and in nitrogen-shielded flames determined by Dagnall *et al.* [118] are given in Table 3.8.

TABLE 3.8. Improvements of detection limits (μg/ml) in a premixed nitrous oxide-acetylene flame with inert gas shielding [118]

Flame	Element and wavelength (Å)		
	Al 3961.5	V 3183.4, 3183.9, 3185.4 [a]	Mo 3132.6
Without shielding	0.20	1.200	10.20
N_2 - shield	0.12	0.088	0.75
Ar - shield	0.10	0.070	0.45

[a] Unresolved triplet.

3.3.2.3. Flames from Direct-Injection Burners

Combustion mixtures with very high burning velocities (e.g. acetylene-oxygen, hydrogen-oxygen) are difficult to handle with premix burners because of the danger of flashback. This danger can be avoided if the fuel and oxidant leave the burner head separately, usually from concentric nozzles. Mixing of the combustion gases proceeds rather quickly by turbulence and combustion starts at a small distance above the burner orifices. Such flames are turbulent and no distinct zones can be recognized in them. The combustion processes extend through a large volume of the flame.

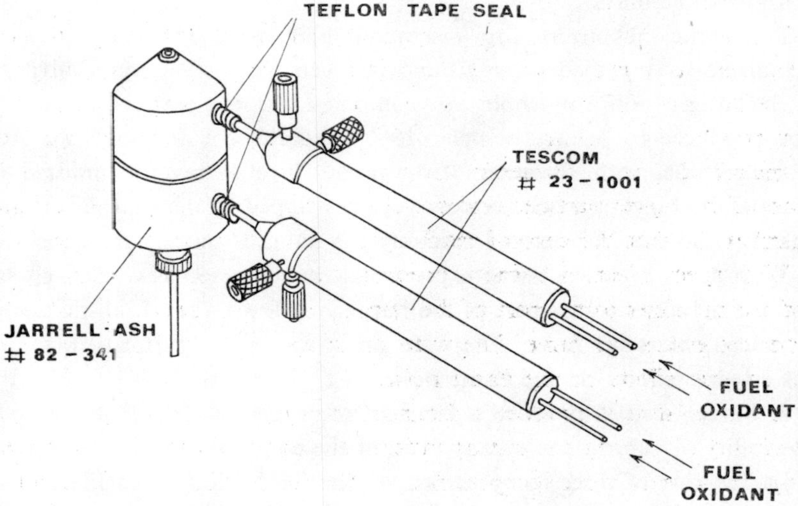

FIG. 3.29. Direct-injection burner for premixed flames [297].

In order to combine some advantages of premixed flames (greater uniformity of flame characteristics, lower flame noise) and of direct-injection burners (freedom from flashback, minimal carbon deposition with strongly reducing flames) premixed flames from direct-injection burners have been tried in emission flame spectrometry [297]. The gases were introduced into the burner body through small premixing assemblies as seen in Fig. 3.29.

Premixed flames from direct-injection burners have the advantage of simplicity and safety of operation even with nitrous oxide-acetylene and oxygen-hydrogen flames, and on the other hand secure good differentiation between the combustion zones. This is particularly important with the former flame.

Their use in atomic fluorescence spectrometry has been investigated by

Bratzel, Dagnall and Winefordner [62] for hydrogen as well as acetylene flames. Compared to unpremixed flames of the same combustion mixture they give better atomization efficiency and lower interferences, especially for the more refractory elements.

3.3.2.4. Burners and Associated Sampling Devices

To secure volatilization of the analyte in chemical flames the sample has to be introduced into the flame in the form of a very fine mist produced by the nebulizer. With premixed flames the nebulizer constitutes in general a separate unit connected to the burner through a spray chamber and possibly some ducts.

Pneumatic nebulizers are overwhelmingly used although ultrasonic nebulizers have gained some ground in recent years. The main advantage of the latter is uniform droplet size compared with the particle size distribution produced by pneumatic nebulizers. The latter covers about two orders of magnitude, with diameters from about five to several hundred μm. Most of the bigger particles coalesce or are trapped by the walls of the spray chamber, so that the aerosol reaching the flame is more homogeneous.

With direct-injection burners, pneumatic nebulization is exclusively used and the nebulizer forms part of the burner itself. As a result all the aerosol produced enters the flame. The wide size range of the particles introduced has adverse effects on the flame noise.

A burner should produce a laminar homogeneous flame providing the possibility of varying the mixing ratio of the combustion gases over a wide range. Mixing of three components, i.e. the fuel, oxidant and inert diluent is an advantage, although for simple analytical work mixing of two components (i.e. fuel and oxidant) should be sufficient. The burner should, however, be provided with some form of shielding from the atmosphere.

Méker burners are better suited than slot burners for provision of laminar flames. To secure an almost flat reaction zone perpendicular to the gas flow and homogeneous in the horizontal plane, the size of the openings should be small and their number high. This is most easily accomplished by constructing the burner head from a great number of syringe needles welded [218] or glued together with heat-resistant resin [7]. However, not all capillary burners are suitable for high-temperature flames with nitrous oxide when the combustion mixture is stoicheiometric, because they can be easily damaged. For such flames massive burners are preferable.

The flame itself should also be provided with shielding from the surrounding atmosphere. Most commercially produced burners for atomic absorption or flame emission spectrometry are not provided with gas shields.

For theoretical studies in atomic fluorescence spectrometry and quantitative determination of basic data, outer shielding by an inert gas is insufficient because its cooling effect has an adverse effect on the temperature homogeneity of the flame gases. For such studies the analytical flame, i.e. the flame into which the element measured is introduced, is surrounded by an outer flame mantle of identical flame-gas composition and supplied with the same combustion mixture but without introduction of the element studied. This whole composite flame is then shielded by an inert gas. The gas supply devices for such burners are more complicated.

In comparison direct-injection burners are of much simpler design. In general these are used as provided by the manufacturers, without change. The most frequently used direct-injection burners are the Beckman burner, the Zeiss burner and the HETCO burner, supplied by the Beckman, Zeiss and Jarrell-Ash Companies, respectively.

Bratzel and Winefordner [58] have compared limits of detection achieved under identical conditions with a Beckman burner (type 4020) and a Zeiss direct-injection burner, using hydrogen as fuel and oxygen, air or argon as support gases. The results are reproduced in Table 3.9.

TABLE 3.9. Limits of detection with different direct-injection burners under optimized conditions [58]

Wavelength (Å)	Beckman	Aspirant	Zeiss	Aspirant
Ag 3280.7	0.01	A[a]	0.02	A[a]
Au 2676.0	1.0	A	0.2	O
Bi 3067.7	0.7	A	1.0	Ar
Cd 2288.0	0.000 08	O	0.000 001	O
Co 2407.3	2.0	A or Ar	0.1	Ar
Cu 3247.5	0.04	A or Ar	0.005	Ar
Ga 4172.1	1.0	A	1.0	A
Hg 2536.5	0.1	A	0.1	A or Ar
In 4101.8	0.2	A	0.1	Ar
Mg 2852.1	0.2	A	0.008	Ar
Mn 2794.8	1.0	A or Ar	0.006	Ar
Pb 4057.8	0.5	A or Ar	10	Ar
Se 1960.9	0.8	A	0.4	Ar
Tl 3775.7	0.1	A	0.1	Ar
Zn 2138.6	0.000 1	A	0.000 04	A

[a] A — air, Ar — argon, O — oxygen.

The Zeiss burner gave at least ten times lower detection limits than the Beckman burner for Cd, Co, Cu, Mg and Mn. On the other hand, it gave a ten times higher detection limit for lead, whereas for the remaining elements the difference was less than a factor of ten. The lower detection limits achieved with the Zeiss burner are at the expense of almost twice the consumption of both hydrogen and the aspirant.

3.3.2.5. Comparison of Premixed Laminar Flames and Turbulent Flames

A comparison of the two different types of flames and their performance in atomic fluorescence spectrometry is rather difficult. Most of the atomic fluorescence work reported so far has been done with one type of burner only, and very few data are available on the performance of the two types of flames under otherwise identical conditions (i.e. primary light source, optical system, wavelength bandpass, detection and read-out systems).

Initially all the work done at the Florida Univ. used direct-injection burners. The advantage of attaining higher total analyte concentration in the flame gases was perhaps over-emphasized. The group at Imperial College exclusively used premix systems, emphasizing lower light-scattering in this flame, which results in improved selectivity of measurement. The lower flicker noise of the laminar flames helps to compensate for the lower analyte concentration in the flame gases, so that the detection limits in both flames are comparable. When the main disadvantage of the premixed flame, the high continuous background emitted from the secondary reaction zone, is eliminated by inert-gas sheathing, the premixed flame seems to be prefer-

TABLE 3.10. Limits of detection [μg/ml] in premixed shielded and unshielded flames and in turbulent flames from direct-injection burners under otherwise identical conditions

Element and wavelength (Å)	Flame					
	turbulent		premixed, unshielded		premixed, shielded	
	Air—C_2H_2	Ar—O—H	Air—C_2H_2	Ar—O—H	Air—C_2H_2	Ar—O—H
Zn 2138.6[a]	0.004 6[c]	—	0.003 2	—	0.000 79[d]	—
Fe 2483.3[b]	0.25	0.02	0.08	0.015	0.025	0.01
Co 2407.3[b]	0.15	0.01	0.05	0.01	N.M.[e]	0.005
Ni 2320.0[b]	0.10	0.006	0.03	0.003	0.007	0.001 2

[a] Reference [272].
[b] References [276, 277].
[c] S : N = 2 in all cases.
[d] Argon shield in all cases.
[e] N.M. = Not measured.

able to the turbulent flame if low limits of detection are sought (see Table 3.10).

Bratzel, Dagnall and Winefordner [61] compared premixed and turbulent hydrogen-air flames from a Perkin – Elmer nebulizing assembly and a Zeiss direct-injection burner, respectively. Measuring the elements Cd, Ga, Pb, Fe, Sn and Tl they found, as expected, that the fluorescence signals were about twice as high with the direct-injection burner. Unfortunately, quantitative data on noise levels are lacking so that no comparison of limits of detection can be made. When a 1 mg/ml solution of aluminium was used, light-scattering due to unevaporated aerosol particles was three times higher in the turbulent flame than in the premixed flame. Background emission increased twofold in both flames.

Martin, Hamm and Zeeman [272] investigated turbulent oxygen-acetylene and premixed unsheathed and sheathed air-acetylene flames with reference to the determination of zinc. They used a Zeiss direct-injection burner, and a Varian-Techtron nebulizing system with a circular flame-emission burner or a sheathed burner head constructed as described by West *et al.* [192]. Flame backgrounds measured without spraying of a solution were in the ratio 100 : 10 : 1 for the premixed, turbulent and sheathed flames respectively. The dependence of background emission at 2138.5 Å on the acetylene

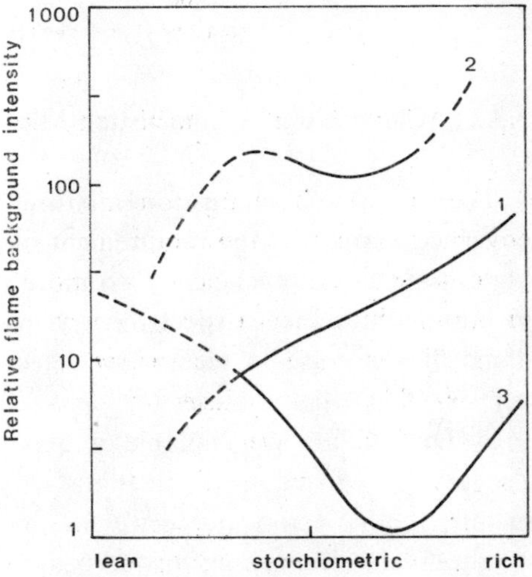

FIG. 3.30. Dependence of flame background at 2138.6 Å on acetylene flow at 5 cm height: (1) turbulent flame from a direct injection burner; (2) premixed flame; (3) premixed separated flame. (Broken line indicates region of instability) [272].

flow-rate and observation height is given in Figs. 3.30 and 3.31. Light-scattering measured with a zinc vapour-discharge lamp with spraying of only pure demineralized water was lowest in the separated flame at any observation height.

Detection limits reported by various authors for different flames but under otherwise identical conditions are listed in the Table (3.10).

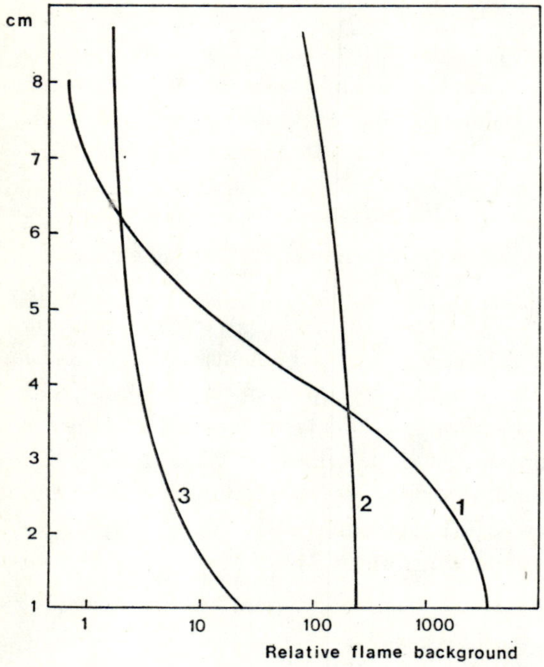

FIG. 3.31. Dependence of flame background at 2138.6 Å on observation height for stoichiometric mixtures: (1) turbulent flame; (2) premixed flame; (3) premixed separated flame [272].

3.3.2.6. Comparison of Combustion Mixtures

The choice of combustion mixture in flame spectrometry is generally governed mainly by the requirement of a high degree of volatilization and high atomization efficiency. Two more requirements have to be considered in atomic fluorescence spectrometry, namely, low flame emission and low flame flicker noise to secure low detection limits, and low quenching properties of the flame gases to secure high fluorescence yield.

So far few data are available on quenching cross-sections. Alkemade and Zeegers [12] listed available data which included figures for the alkali metals, thallium and mercury with the main components of flame gases. Qualitatively, quenching properties increase in the order $Ar < H_2 < H_2O < N_2 < CO < O_2 < CO_2$. Despite the scarceness of data, it is unlikely that by using argon instead of nitrogen as diluent of the combustible mixture the fluorescence yield could be improved by more than about one order of magnitude. This value is in agreement with the experimental data of Jenkins [218].

Lowest flame emission and flame flicker noise are associated with hydrogen flames, which have limited reducing properties and so refractory elements are not adequately atomized [440]. Strongly reducing, fuel-rich acetylene flames are efficient atomizers, but are associated with considerable background emission. The requirements are therefore conflicting, and the choice of combustion mixture to secure the best results is much less evident than in atomic absorption spectrometry.

For easily atomized elements, hydrogen flames are generally preferable. Replacing the nitrogen of an air-hydrogen flame by argon reduces quenching by the flame gases, and the fluorescence yield is improved. This leads to appreciable improvement in detection limits (see Table 3.11).

TABLE 3.11. Limits of detection [μg/ml] in premixed unshielded flames with different combustion mixtures [277][a]

Element line	Air—H_2	Ar—O_2—H_2	Air—C_2H_2
Fe 2483.3	0.025	0.015	0.08
Co 2407.3	0.02	0.01	0.05
Ni 2320.0	0.006	0.003	0.03
Au 2428.0	0.015	0.005	0.015
Pt 2659.4	0.7	0.5	1.0
Pb 4057.8	0.1	0.02	0.12

[a] S : N = 2 in all cases.

For some of the less readily atomized elements an increase of the fuel to oxidant ratio may also bring about improvement. Smyly *et al.* [371] measured relative atomization efficiencies in sheathed premixed argon-oxygen-hydrogen flames at different fuel to oxidant ratios. By changing from fuel-lean to fuel-rich flames the atomization efficiency could be improved as shown in Table 3.12.

For some of the more refractory elements, however, hydrogen flames may not give adequate atomization even when used with nitrous oxide as oxidant despite the higher temperature reached (around 2700 K) [440]. This may be seen from data reported by de Galan and Samaey [136] (see Table 2.5 p. 60).

For these elements fuel-rich acetylene-nitrous oxide flames must be used. Even other hydrocarbon fuels such as propane, butane or MAPP gas (a mixture of methylacetylene and propadiene) are not nearly as effective

TABLE 3.12. Atomization efficiencies in a premixed oxy-hydrogen flame diluted with argon [371]

Element	Fuel-lean flame $H_2/O_2 = 1.73$	Fuel-rich flame $H_2/O_2 = 9.77$	Improvement factor
Sr	0.000 32	0.078	200
Mg	0.003 9	0.052	45
Mn	0.005	0.17	35
Cu	0.78	0.6	0.77

as acetylene. All these fuels have a lower critical carbon to oxygen ratio, i.e. the C/O ratio of the combustion mixture at which soot begins to form. With acetylene as fuel the highest concentration of carbon atoms in the flame gases, i.e. strongest reducing conditions, may therefore be reached. Carbon in the form of soot is not effective in reducing metal oxides in the flame gases [136].

It would be pointless to replace nitrogen with argon in hydrocarbon flames as both CO and CO_2 molecules are even better quenchers than nitrogen.

3.3.2.7. Cool Flames

Flame emission and the associated flame flicker noise considerably increase the limits of detection in atomic fluorescence spectrometry. Flames with very low background emission are therefore desirable. From this point of view "cool flames" prove very useful. This term is applied generally to flames in which the oxidant necessary for combustion is fully supplied by diffusion and/or entrainment from the surrounding atmosphere. When a spray-chamber sampling system is used (which usually provides a premixed flame), diffusion prevails, and such flames are described as diffusion flames.

Argon or nitrogen is generally used to effect nebulization, and hydrogen is used as fuel. Hydrocarbon flames would evidently be luminous and not colourless, the main advantage sought.

The temperature of these flames ranges from 280 to 850 °C, depending on the exact region measured [109]. Despite this low temperature, cool flames are quite effective atomizers for several elements, but are highly susceptible to volatilization interferences.

The main advantages are low background emission, high transmittance in the ultraviolet region below 2000 Å and, when argon is used for nebulization, lower concentration in the flame gases of molecular species with high quenching cross-sections, resulting in higher fluorescence yields.

Thus Dagnall *et al.* [108] found background emission at 3025 Å 25 times higher in the premixed air-hydrogen flame and 200 times higher in the air-propane flame than in the argon-hydrogen diffusion flame. Typical figures for transmission (measured for a 7-cm long burner) are given in Table 3.13 [111].

Lower background emission and lower absorbance by flame gases results in improved detection limits for arsenic as may also be seen in Table 3.13. However, part of the improvement is also due to improved atomization as measurements of characteristic concentrations for 1% absorption suggest (last column).

TABLE 3.13. Comparison of some properties of air-acetylene and nitrogen-hydrogen diffusion flames for arsenic determination [111]

Wave-length (Å)	(*a*) Air—C_2H_2 flame		(*b*) N_2—H_2 diffusion		Ratio (*a*)/(*b*) of	
	Trans-mission (%)	Detection limit (μg/ml)	Trans-mission (%)	Detection limit (μg/ml)	detection limits (AFS)	characteristic con-centrations [μg/ml/1% absorption] (AAS)
1890.4	18.0	1.0	70	0.20	5	3.4
1937.6	33.5	1.0	87	0.25	4	3.6
1972.6	40.0	2.0	95	0.50	4	3.6
2349.8	—	2.0	—	0.15	13	—

The improvement due to less quenching when argon is used for nebulization may be best estimated by comparing fluorescence signals in the nitrogen-hydrogen and argon-hydrogen diffusion flames. Improvement factors are generally different for lines originating from different upper terms. Thus for arsenic [111] an increase by 33–80% of the fluorescence signals was observed, and for bismuth a 100% increase was found [108].

When direct-injection burners are used, entrainment caused by turbulence prevails in supplying the necessary oxidant. With these systems either an inert gas is used for nebulization [411, 482] or the fuel gas itself [140, 162]. Thus we speak about argon/hydrogen/entrained air flames or hydrogen/entrained air flames, respectively. Because of the more energetic mixing with the atmosphere and a more vigorous combustion, the temperature of the flame is not quite as low as that of laminar diffusion flames. Smith, Stafford and Winefordner [365] reported a temperature of 2122 K, for the argon-hydrogen-entrained air flame for example, whereas the temperature of the hydrogen-air flame is only slightly higher, 2287 K.

Demers [140, 142] and Ellis and Demers [162] found increased fluorescence signals and lower detection limits for a number of elements in hydrogen entrained air flames, despite the lower temperature. Their comparison of detection limits with those in the hydrogen-oxygen flame is reproduced in Table 3.14. According to these authors, for easily atomized elements such as silver and zinc, the improvement is probably due to the increased flame diameter of the hydrogen/entrained air flame. For the other elements higher atomization should also be involved. The third possible explanation,

TABLE 3.14. Comparison of limits of detection in a turbulent hydrogen/oxygen flame and a hydrogen/entrained air flame [142]

Element and wavelength (Å)	Limits of detection (μg/ml)	
	H_2—O_2	H_2 — entrained air
Ag 3280.7	0.003	0.001
Co 2407.3	3.5	0.2
Cu 3247.5	0.1	0.02
Fe 2483.3	25	2.0
Mg 2852.1	0.2	0.004
Mn 2794.8	35	0.04
Ni 2320.0	5.0	1.0
Tl 3775.7	0.3	0.07
Zn 2138.6	0.04	0.01

i.e. lower concentration of species with strong quenching properties in the flame gases is unlikely according to the work of Pearce, de Galan and Winefordner [323]. These authors found no appreciable difference in atomic fluorescence power efficiencies for Ag, Cu, Tl, Au and Pb in an argon-hydrogen-entrained air flame compared with the oxygen-hydrogen flame. Evidently the nitrogen entrained from the ambient atmosphere provides a sufficiently high concentration of effective quenchers to eliminate the difference. This observation is however valid only for flames from direct-injection burners. In premixed laminar flames lower quenching properties contribute significantly [212, 218].

The surprising properties of argon/hydrogen/entrained air flames or simple hydrogen/entrained air flames are probably due to free radicals (H and OH) which are formed in the primary reaction zone at concentrations exceeding equilibrium values. As their recombination rate is rather slow, free radicals survive for considerable periods. When oxygen is supplied only by air entrainment the above-equilibrium concentration of radicals generated by the combustion reaction is spatially extended compared with that in premixed flames.

3.3.3. OTHER FLAME SAMPLING DEVICES

One of the main disadvantages of aspirating sample solutions into flames is the considerable dilution of the sample by flame gases. This dilution may be partly avoided if the sample is introduced into the flame on a support, whence it volatilizes. If a speedy volatilization takes place a transient but relatively high analyte density is reached. Such a technique, known from atomic absorption spectrometry as the "sampling boat technique", has been applied by Sychra and Kolihová [393] in atomic fluorescence spectrometry.

Up to 100 μl of the sample solution is placed in a metal cup fixed in a holder (see Fig. 3.32). This is placed near to the flame until dry, and then inserted into the flame. The analyte volatilizes and atomizes, and its fluorescence signal is recorded. This technique is suitable only for the more volatile elements. The temperature of the flame used is limited by the cup

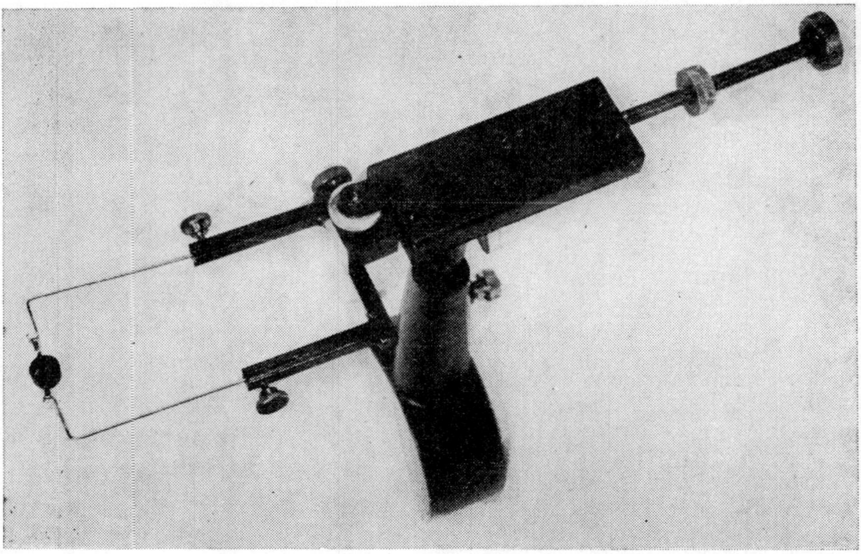

FIG. 3.32. "Sampling cup" assembly mounted in a rail saddle.

material, and high-temperature flames cannot be used. The cups were 9 mm o.d. and 6 mm deep, made of nickel or tantalum.

For the determination of lead, zinc, cadmium, and silver, best results were achieved with a fuel-lean air-acetylene flame. The cup reached about 900 °C within two seconds, which ensured a speedy volatilization. The fluorescence signals were almost independent of the height of observation within the range 0–25 mm above the cup rim. The absolute limits of detection for S/N = 2 were 10^{-12} g of Cd, 5×10^{-12} g of Zn, 1.3×10^{-11} g of Ag and 1.2×10^{-9} g of Pb.

Needless to say, the system needs careful adjustment to prevent any reflected radiation from the primary light source from reaching the detector.

3.3.4. NON-FLAME ATOMIZERS

3.3.4.1. General

As already mentioned, despite their popularity, chemical flames have serious limitations when applied for the production of free atomic vapour. The main disadvantages are as follows:

1. A high dilution effect of the sample by the flame gases which limits the atomic concentration attainable.

2. Precise control over the chemical environment of the analyte is not possible. In chemical flames the concentration of oxygen atoms cannot be arbitrarily lowered and diminishes the atomization of several elements forming thermally stable oxides.

3. Flames which are the most efficient atomizers and secure greatest freedom from interelement effects simultaneously contain effective quenchers that lower the fluorescence yield (CO, CO_2, N_2).

4. Flames generally have higher background emission than most non-flame atomizers. At least in some wavelength regions, high background emission causes deterioration of the signal to noise ratio.

5. Flames with continuous nebulization of the sample are not well suited to the analysis of small-volume samples.

These disadvantages should be avoided with non-flame atomizers. The composition of the gases present may be chosen freely to minimize quenching properties, oxygen can be completely excluded, and with discrete sampling a greater part of the sample is exploited for the actual measurement.

Because of these advantages considerable efforts have been devoted towards the development of effective non-flame atomizers which would be sufficiently simple to operate.

Three main types have been tried so far; thermally-heated graphite (carbon) furnaces, cups or rods, thermally-heated metal furnaces, wire loops or boats, and atomizing media produced by electric discharges. (At present the last category includes only the sputtering chamber, but other types of discharge already applied in atomic absorption spectrometry will undoubtedly also be tried in atomic fluorescence spectrometry. These could involve radiofrequency plasmas, and atomic vapour produced by ionic [48] or electron bombardment).

3.3.4.2. Graphite Furnaces

The first attempt to observe atomic fluorescence of vapours volatilized from a graphite crucible was made by Svoboda [382]: he was unsuccessful, mainly owing to the use of photographic detection. A workable model of a graphite furnace for use in atomic fluorescence spectrometry was described

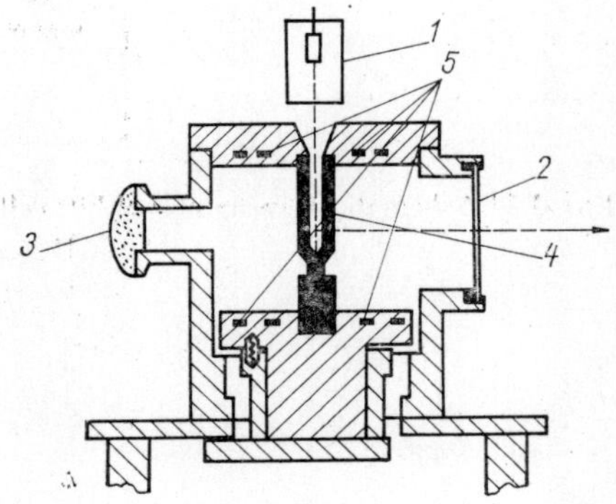

FIG. 3.33. Massmann graphite furnace for atomic fluorescence measurement [274]: (1) hollow-cathode lamp source; (2) quartz window; (3) argon inlet; (4) graphite cuvette; (5) heating.

by Massmann [274, 275]. His device consisted of a graphite cup of 6.5 mm internal diameter, and 40 mm high. Two vertical slots were cut into the wall on opposite sides to allow the fluorescence radiation to leave the furnace (Fig. 3.33). The primary radiation entered the cup from above (where the volatilized sample leaves the cup). The graphite cup was placed in a holder which also served to make electrical connections, and was surrounded by an argon atmosphere. The cup was heated by a current of up to 400 A. Sample volumes between 5 and 50 μl were used.

Massmann used as primary light sources hollow-cathode lamps operated continuously at normal currents but pulsed manually at several hundred mA for 10–15 sec during the measurement. He found the reproducibility of the light flux thus generated to be sufficient, i.e. the relative standard deviation was between 1 and 2%. Limits of detection achieved with the furnace are

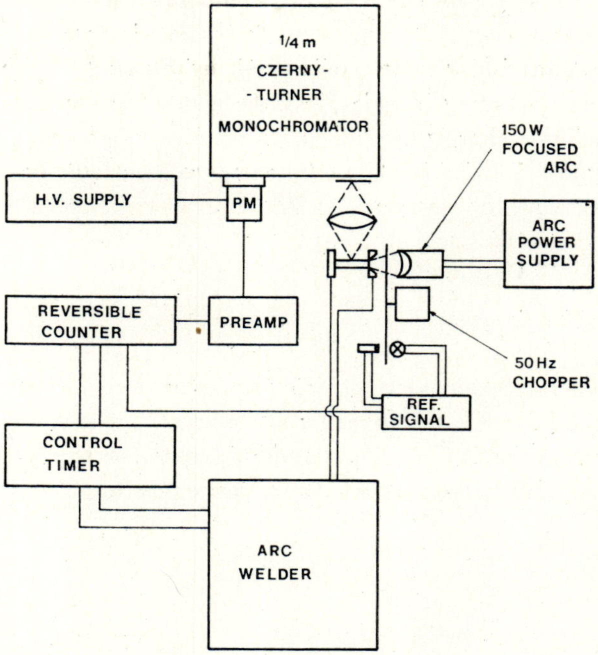

FIG. 3.34. Schematic drawing of graphite cell system for atomic fluorescence spectrometry [177, 459].

TABLE 3.15. Limit of detection (g) by atomic fluorescence spectrometry with the graphite furnace [274], the carbon filament [236] and the carbon rod [16]

Element	[274]	[236]	[16]	Element	[274]	[236]	[16]
Ag	1.5×10^{-12}	1×10^{-12}	5×10^{-13}	Fe	3×10^{-9}		2×10^{-12}
As			5×10^{-10}	Ga		5×10^{-11}	
Au		4×10^{-12}	1×10^{-11}	Mg	3.5×10^{-12}	1×10^{-12}	3×10^{-14}
Be			3×10^{-11}	Mn		5×10^{-12}	
Bi		1×10^{-11}		Ni		5×10^{-12}	
Ca			1×10^{-13}	Pb		1×10^{-11}	3×10^{-12}
Cd	2.5×10^{-13}	1.5×10^{-13}	3×10^{-14}	Sb	2×10^{-10}	1×10^{-9}	
Co		2×10^{-11}		Tl	2×10^{-9}	2×10^{-11}	
Cu	4.5×10^{-10}	1×10^{-12}	5×10^{-12}	Zn	4×10^{-14}	2×10^{-14}	1×10^{-13}

reproduced in Table 3.15 where they are compared with those of other non-flame systems.

A similar graphite furnace was constructed by Fraser [177, 459], who used a 150-W xenon-arc lamp as primary light-source and a photon-counting detection system with a low-level discriminator (see Fig. 3.34).

In both graphite furnaces the sample is introduced in the form of a solution which is dried before measurement. If organic matter is present, it is carbonized before atomization.

3.3.4.3. Thermal Impulse Atomizers

Russian workers have introduced a device called the impulse atomizer [40–43]. It consists of a supporting graphite electrode (as used in emission spectrography) loaded with the powdered solid sample and heated by a current of short duration (see Fig. 3.35). The current and the duration

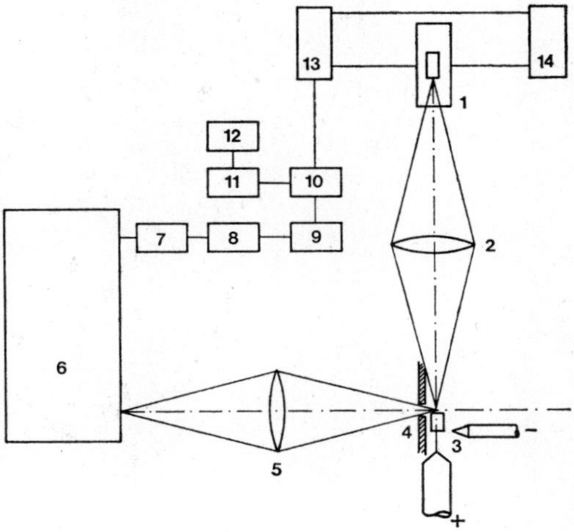

FIG. 3.35. Schematic diagram of atomic fluorescence installation for impulse-arc atomization of powdered samples: (1) excitation source; (2.5) lenses; (3) wineglass electrode; (4) screen; (6) monochromator; (7) photomultiplier; (8) cathode follower; (9) narrow-band amplifier; (10) synchronous phase-sensitive detector; (11) d.c. amplifier; (12) pen recorder; (13) audio-oscillator; (14) power supply.

of the pulse depends on the volatility of the element to be determined, on the material analysed and on the form and size of the electrode. In general the duration is somewhere between 1.5 and 2.5 sec. Pulsed hollow-cathode lamps were used as excitation sources. This device was used for the determination of cadmium, manganese, silver, thallium, bismuth and mercury in graphite powder [40, 324, 325]. The guaranteee of purity reached was $3\times10^{-8}\%$ for Cd and $10^{-6}\%$ for Mn. However, the reproducibility at concentrations 10 times higher than these was only 20–30%. The limit of detection may be improved if the sample is evaporated into an argon atmosphere [324, 325].

So far this technique has been used only for the analysis of graphite powder and silicon carbide. The elements determined evaporate completely from these matrices whereas the matrix itself stays in the electrode. This considerably reduces the possibility of applying this technique to other materials.

3.3.4.4. Carbon Filament or Carbon-Rod Atomizers

These devices, first described by West and Williams [430], were later simplified [1] and modified to become simple enough to warrant commercial production under the name carbon rod atomizer [16] (see Fig. 3.36).

FIG. 3.36. Carbon rod atomizer model 61 (Varian-Techtron Pty. Ltd.) Reprinted by permission of Varian-Techtron Pty. Ltd.

The original carbon filament was a graphite rod 2 mm in diameter about 20 mm long supported by water-cooled stainless-steel supports, which also serve as electrical connections (Fig. 3.37). The graphite rod was heated by passing a current of about 100 A. To prevent burning, the first devices were placed in an argon atmosphere under a glass cover provided with optical windows. Later it was found sufficient to surround the rod with a steady laminar flow of an inert gas. This considerably simplified their use for analysis. A small volume of the sample solution (about 1 μl) was placed on a depression in the centre of the rod, the solvent was evaporated, organic matter (if present) was ashed and then the rod was heated to 2000 – 2500 °C within about 5 sec. As with the other devices where pulse atomization is effected, it is important that the response time of the measuring electronic circuit be fast enough to permit accurate recording of the transient signals.

Although the carbon rod is simpler than the furnace assemblies, it has the disadvantage that the atomic vapour leaving the hot rod surface enters

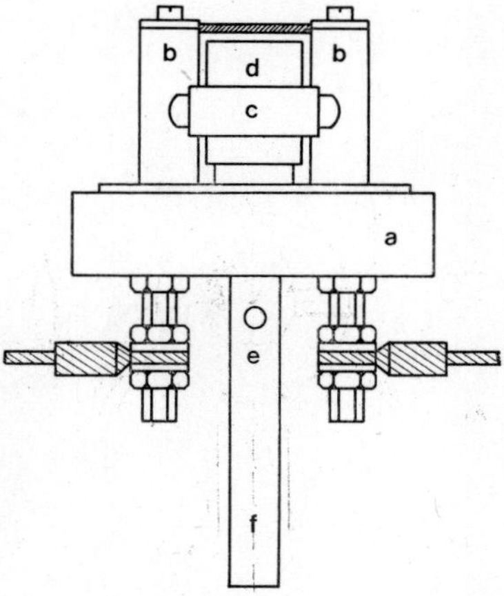

FIG. 3.37. Carbon-filament atomizer modified for use in open atmosphere: (a) base; (b) water-cooled electrodes; (c) water link between electrodes; (d) laminar flow box; (e) inlet for shield gas; (f) support stem for reservoir [3].

a relatively cool surrounding atmosphere where condensation immediately starts. The decay of the free atom population is considerably accelerated in comparison with that in the furnaces. The condensation rate of the analyte atoms depends moreover on the presence of other concomitants and this causes rather strong interelement interferences. Hence it is necessary to view the atomic vapour as close to the carbon rod as possible [1].

In order to raise the temperature of the surrounding atmosphere the carbon rod may be placed within a hydrogen diffusion flame by adding hydrogen to the shielding gas or by replacing the inert gas by hydrogen. This partly eliminates interferences caused by vapour condensation and light-scattering [16].

With repeated heating the porosity of the graphite material increases. The extent to which the sample solution is absorbed may then influence the volatilization rate and the reproducibility of measurements. To avoid this source of error the graphite rod may be impregnated with xylene, thus preventing absorption of sample solutions. With this precaution the reproducibility is claimed to be better than 2% [16]. The limits of detection are quoted in Table 3.15 (p. 136).

3.3.4.5. Metal Loops and Furnaces

Metal may be used instead of graphite as the supporting material and heating element in non-flame atomizers. An example of such devices is the hot wire loop first tried in atomic fluorescence spectrometry by Bratzel, Dagnall and Winefordner [63, 64]. This consists of a thin platinum or tungsten wire, of 0.25 mm diameter, in the form of a loop positioned in

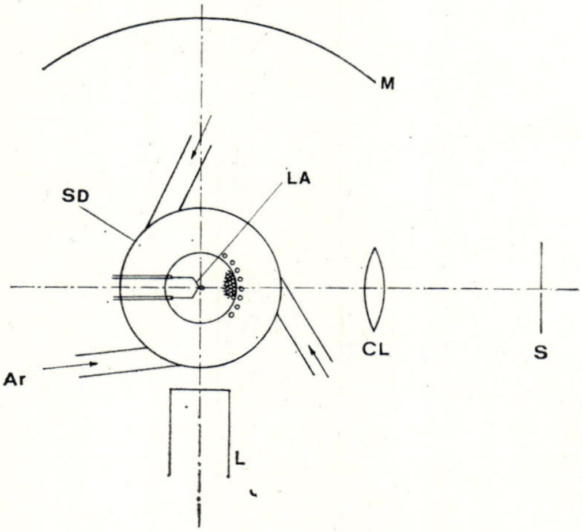

FIG. 3.38. Metal loop atomizer assembly [393]: (LA) loop atomizer; (SD) shielding device; (L) lamp; (M) reflecting mirror; (CL) condensing lens; (S) monochromator slit.

a laminar flow of an inert shielding gas (Fig. 3.38). The sample solution is placed on the loop, and dried and volatilized by passage of an electric current. The power necessary is only about 25 W, i.e. considerably less than with the carbon rod. The power supply circuit recommended by Sychra *et al.* [393] allows controlled and very reproducible heating of the loop.

The hot wire is a very simple device, but unfortunately its use is subject to serious interelement effects. As with the carbon rod the atomic vapour enters a cool atmosphere where condensation takes place. The presence of a number of concomitants may considerably influence this process. However, for pure solutions the reproducibility is fairly good (being limited by the rather large range of sample volumes taken up by the loop) and the limits of detection remarkable. Some applications could be expected in connection with electrolytic separation or amalgamation (see the work of Ulfvarson [405], Brandenberger [56, 57] on the determination of mercury, cadmium, zinc, lead, tellurium, copper, silver, gold and platinum by atomic absorption spectrometry).

Recently Winefordner *et al.* [49] have described the use of a platinum furnace for atomic fluorescence spectrometry. Contrary to all non-flame atomizers so far described the platinum furnace works with continuous nebulization of the sample solution. The aerosol produced is conducted to the resistance-heated furnace. Because of the low melting point of platinum the furnace could be operated only at 1600 °C. Tantalum metal sheets which would allow temperatures around 2500 °C could not be used because they oxidized readily and crumbled after heating even though argon was continually flowing through the system. Therefore, only relatively volatile elements were studied, namely Cd, Zn, Hg, Cu and Fe. With the exception of mercury the limits of detection with the furnace were inferior to those reached with flames.

3.3.4.6. Cathodic Sputtering Chambers

The cathodic sputtering effect has been exploited for the production of atomic vapour in atomic absorption spectrometry by several workers. In

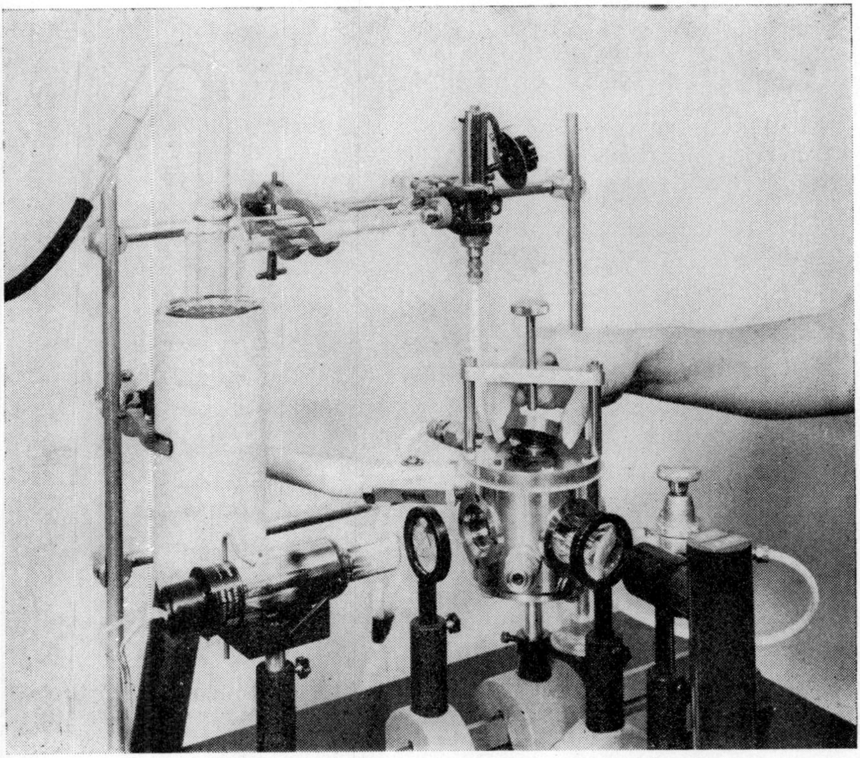

FIG. 3.39. Sputtering chamber modified for atomic fluorescence. Reprinted from Ref. [99] by permission of C.S.I.R.O.

recent years the design of the devices has been modified to suit fluorescence measurements (see Fig. 3.39) and to allow rapid interchange of metallic samples in the form of flat specimens. The sample is connected as the cathode, the chamber is filled with a rare gas at pressures between 1 and 10 torr and a d.c. glow discharge is started. Positive ions striking the cathode sputter metal atoms and comparatively large steady-state concentrations of the metal vapour can accumulate in the region near the cathode. Measurements are then made just beyond the negative glow where relatively noise-free fluorescence signals may be obtained [99, 419].

The sputtering effect depends on the pressure and the rare gas used. Generally a xenon discharge is used, as this favours high sputtering yield, high concentration of neutral ground-state atoms and weak background emission.

Use of a cathodic sputtering chamber in conjunction with non-dispersive spectrometry is a promising tool for the analysis of alloys (see Chapter 6.2).

3.3.4.7. Non-Flame Systems for the Determination of Mercury

Mercury is an exceptional element in that it has an appreciable atomic vapour-pressure at room temperature. This gives rise to the possibility of measuring mercury without the additional thermal energy supplied by a flame or resistance heating. The only problem is the isolation of mercury from other components of the sample which could give spurious readings.

In principle there are two possible ways of doing this. If the sample is present in solution, mercury may be liberated by reduction with tin(II) chloride or sulphate and expelled by blowing gas through the solution [300]. From solid samples mercury may be driven off simply by heating. However, in this case it must be isolated before measurement [300]. This is best done by absorbing it on gold, silver or iodized charcoal. The absorbing substance is then heated and the mercury liberated is carried to the measuring cell. This isolation method may be combined with the first method, but it is generally unnecessary in that instance.

These methods of isolating mercury are also used for atomic absorption measurements. However, atomic fluorescence has several advantages: because no closed measuring cell is used the danger of fogging of the windows and the necessity of including drying columns in the system are absent: no signal from molecular band absorption by organic compounds may be expected, and the concentration range which can be measured by atomic fluorescence is considerably wider. Also the limits of detection in atomic fluorescence spectrometry are better than in atomic absorption spectrometry [401].

3.4. OPTICAL AND DETECTION SYSTEMS

3.4.1 THE INSTRUMENTAL SYSTEM AND THE FLUORESCENCE SIGNAL

The instrumental system, i.e. the optical and electrical systems, converts a portion of the fluorescence flux emanating from the atomizer into an instrumental signal. The arrangement of a typical optical-electrical measurement system is shown schematically in Fig. 3.40. Even if more than one

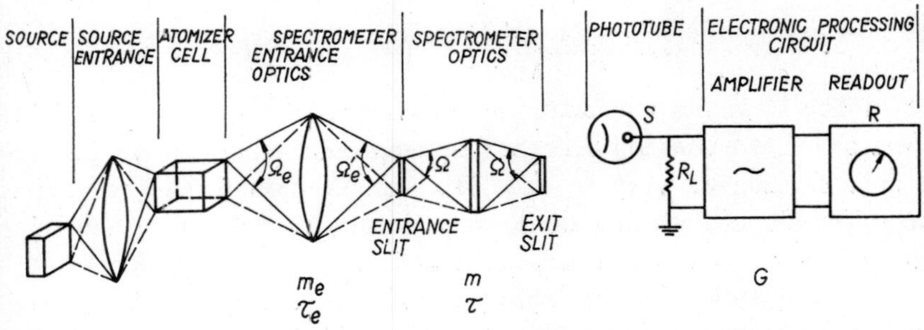

FIG. 3.40. Principal of the arrangement of an optical-electrical measurement system for an atomic fluorescence spectrometer.
Reprinted from Ref. [539] by permission of The Society for Applied Spectroscopy

lens or mirror is used in the spectrometer, the following general treatment will still apply, except that certain factors must be changed to account for the additional optical components (e.g. the transmission of the additional components). The assumptions and conditions necessary for the fluorescence signal expressions derived to be valid are as follows.

(*i*) The solid angle of the entrance beam from the fluorescence cell Ω_e'' must exceed, or at least be equal to, the solid angle collected by the spectrometer, Ω'

$$\Omega_e'' \geqq \Omega'$$

In other words, the spectrometer optics limit the linear aperture.

(*ii*) The image of the fluorescent cell on the entrance slit of the spectrometer (or aperture of the filter) must exceed the area of the entrance slit (or aperture), i.e., the image height must exceed the spectrometer slit height h and the image width must exceed the spectrometer slit width w.

(*iii*) Several trivial assumptions must also be made.

(*a*) The transmission factors (τ_e of the entrance optics, and τ of the spectrometer) are constant over the measured spectral frequency range between ν_u, the upper frequency cut-off, and ν_l, the lower frequency cut-off. The frequency interval $(\nu_u - \nu_l)/2$ is the spectrometer spectral bandpass, s_ν in frequency units. Generally, analytical spectroscopists are concerned with wavelength rather than frequency, primarily because spectrometers have wavelength read-outs. In this general discussion, frequency (or relative frequency) will be utilized as the variable for consistency with the general expressions derived in Chapter 2 for fluorescence flux. A limiting but analytically useful expression will be given later, however, in terms of the more analytically useful wavelength units. To convert from frequency ν into wavelength λ, the equation $\nu = c/\lambda$ is applied, where c is the speed of light. To convert from a frequency interval $d\nu$ into a wavelength interval $d\lambda$, it can be shown that $d\nu = d\lambda\left(\frac{c}{\lambda^2}\right)$.

(*b*) The magnification factors m_e of the entrance optics and m of the spectrometer are constant over the frequency range from ν_u to ν_l. Generally a magnification of unity is used. The magnification factors account for magnification in both directions, perpendicular to the optical axis of the spectrometer (parallel and perpendicular to the entrance slit length).

(*c*) The photoanodic sensitivity γ of the photodetector is constant over the frequency range $\nu_u - \nu_l$. The photoanodic sensitivity has units of AW^{-1} (1 watt of power $= 10^{-7}$ erg s^{-1}).

(*d*) The spectrometer has identical entrance and exit slit widths. If the slit widths are not the same, loss in resolution results with greater chance of spectral interferences and generally loss in signal-to-noise ratio (with poorer limits of detection and precision of measurement). Usually the focal lengths of the collimator and decollimator are identical, so that $m = 1$.

(*e*) The fluorescence cell is optically aligned with the source of excitation and the spectrometer.

If the spectrometer is set at a fixed frequency, ν_s, the measured photodetector signal, I_s (amperes), is given by

$$I_s = \frac{\gamma\tau\tau_e h w}{m m_e A_F} \int_{\nu_l}^{\nu_u} \Phi_{F_\nu} F_\nu \, d\nu \tag{3.4}$$

where A_F is the surface area of the fluorescence cell, in cm^2, and F (no units) is the spectrometer or apparatus distribution function, which is triangular except when very narrow slits (diffraction-limited case) are used, when the distribution becomes more complicated [286]. If the source of excitation s pulsed or modulated, then I_s is the signal level during the source on-time. A triangular distribution results as long as assumption *iii*(*d*) is valid. If the slit widths are identical and greater than the diffraction-limited slit widths, then:

$$w = \frac{2cf_1}{\nu_s h}$$

where f_1 is the collimator focal length, in cm. If this condition is not valid (see the article by Mielenz [286]), Eq. (3.4) is expressed in terms of frequency ν, whereas in Chapter 2 on theoretical growth curves, the relative frequency v was used; to convert from v to ν, use Eq. (2.5) and to convert from dv to dν use the expression $\mathrm{d}v = 2\sqrt{\ln 2}\,\mathrm{d}\nu/\Delta\nu_D$. The photodetector signal is generally processed by passing the photoanodic current through a load resistor R (ohms) and then through an electronic circuit with a gain G_A (no signal units). Therefore, the instrumental read-out R_A (volts) is given by

$$R_A = RG_A I_S \tag{3.5}$$

The product of Φ_F and F_ν is said to be convoluted. If γ, the τ's, and the m's are also frequency-dependent, then these terms must also be convoluted and the expression becomes more difficult to solve. Fortunately, γ, the τ's, and the m's are not usually frequency-dependent, and since F_ν goes to zero for frequencies outside the range $\nu_u - \nu_l$, the integration limits can be set from zero to infinity. Therefore, Eq. (3.4) is more readily solved although it is still necessary to use numerical methods for an exact solution.

Fortunately, it is possible to solve Eq. (3.4) for a limiting but analytically useful case. Assuming the fluorescence line of concern is much narrower than the spectral bandpass of the spectrometer (spectral line halfwidth, $\Delta\nu \leq s_\nu/5$), then the integral in Eq. 3.4 can be written as

$$\int_{\nu_l}^{\nu_u} \Phi_{F_\nu} F_\nu \,\mathrm{d}\nu = F_{\nu s} \int_0^\infty \Phi_{F_\nu} \,\mathrm{d}\nu \tag{3.6}$$

where $F_{\nu s}$ is the peak value of the apparatus function at the frequency ν_s, and is equal to unity. For a triangular spectrometer distribution

$$F_\nu = F_{\nu_s}\left[1 - \frac{(\Delta\nu)}{s_\nu}\right] = 1 - \frac{(\Delta\nu)}{s_\nu} \tag{3.7}$$

where $(\Delta\nu) = (\nu - \nu_s)$ and s_ν is the spectral bandpass in frequency units. The integral in Eq. (3.7) is simply the fluorescence flux (in erg s^{-1}) produced by the fluorescence line of the analyte within the atomizer cell and so

$$\int_0^\infty \Phi_{F_\nu} \,\mathrm{d}\nu = \Phi_F \tag{3.8}$$

Therefore, the complete expressions for I_S and R_A are

$$I_s = \frac{\gamma\tau\tau_e hw\Phi_F}{mm_e A_F} \tag{3.9}$$

$$R_A = \frac{RG_A\gamma\tau\tau_e hw\Phi_F}{mm_e A_F} \tag{3.10}$$

where Φ_F can be evaluated for any specific case by reference to Chapter 2.

Exactly the same equations hold for non-monochromated instruments using solar-blind photomultipliers or a combination of filters. Instead of hw, the area of smallest exit or input field stop has to be inserted. All observed lines must then be convoluted with F_ν, which in this case is represented by the transmittance of the filter at various frequencies.

Because Φ_F depends upon the analyte concentration n_i in the atomizer (n_0 if the ground state is the lower state involved in the absorption process), the *shapes* of *experimental growth curves* (linear-linear or log-log plots of I_s or R_A *vs.* C (or q) *will be identical to theoretical growth curves* (linear-linear or log-log plots of Φ_F or some function of Φ_F *vs* n_i) *except for the non-linearities between* n_i *and* C *(or* q*)* mentioned in the section on the atomizers.

If instead of measurement of a continuous atomic-fluorescence signal at a constant frequency ν_s, the total radiant energy emerging from the spectrometer exit slit and reaching the photodetector during frequency scanning is measured, this is given by

$$Q_D = \int_{t_1}^{t_2} \Phi_{Dt}\, dt = \left(\frac{\tau\tau_e hw}{mm_e A_F}\right) \int_{t_1}^{t_2} \left[\int_{\nu_l}^{\nu_u} \Phi_{F\nu} F_\nu \, d\nu \right] dt \tag{3.11}$$

where Φ_{Dt} is the radiant flux from the spectrometer. All terms before the integral signs are assumed to be independent of frequency and of time between the start and stop times, t_1 and t_2, of the wavelength scan. If it is now assumed that the scanning speed is constant, then the spectrometer setting, in Hz, ν_s, is given by

$$\nu_s = y + zt \tag{3.12}$$

where y and z are constants of the scanning system. Therefore

$$\frac{d\nu_s}{dt} = z$$

Now, if a triangular slit (spectrometer or apparatus) function is again assumed, and if z is assumed constant, then

$$Q_D = \frac{\tau\tau_e hw}{mm_e A_F z s_\nu} \int_{\nu_l}^{\nu_u} \left[\int_{-\infty}^{\infty} \Phi_{F\nu} \, d\Delta\nu \right] d\nu_s \tag{3.13}$$

where ν_l and ν_u are the lower and upper cut-off frequencies of the frequency scan (these values correspond to the upper and lower cut-off wavelengths λ_u and λ_l, respectively, of the wavelength scan), and s_ν is the frequency spectral bandpass of the spectrometer. To convert Q_D into a charge, q_D (coulombs), stored in a capacitor connected to the photoanode, we use the equation:

$$q_D = Q_D \gamma \tag{3.14}$$

3.4.2. ENTRANCE OPTICS

The role of the entrance optics is twofold, to project an image of the primary source into the atomic cell so that the highest radiant flux is achieved in the area observed by the monochromator, and, further, to project the excited radiation onto the entrance slit of the monochromator so that the image of the fluorescing atomic vapour covers the whole slit height, and at the same time the whole solid angle of the collimator is filled.

Several typical arrangements are illustrated in Fig. 3.41. All, with the exception of *e* and *g*, use right angle illumination.

The disadvantages of the lens system (Fig. 3.41*a*) are chromatic and spherical aberration and the limited transmittance of the lenses below 2100 Å. This system was used in most of the early studies (e.g. [266]). The system shown in Fig. 3.41*b* is a variation of *a*, employing mirrors. It is quite popular [23, 74, 345, 390] (a lens-mirror variation is used in the Techtron AA-5 flame-fluorescence accessory). Both mirrors should be elliptical, since spherical mirrors produce an image with serious aberrations. A similar system is illustrated in Fig. 3.41*c*. This can be used to advantage when the primary source height is bigger than the slit height, since with this system a diminished image of the source is formed in the flame (contrary to the previous case). Another variation is a combination of systems *b* and *c*, with an additional mirror placed at the monochromator slit. This spherical mirror reflects back into the flame the light not used by the monochromator. Another spherical mirror on the primary source – flame axis has the same function, reflecting the light transmitted by the flame back into the flame. The two remaining mirrors are elliptical, and arranged as in *b*. A similar

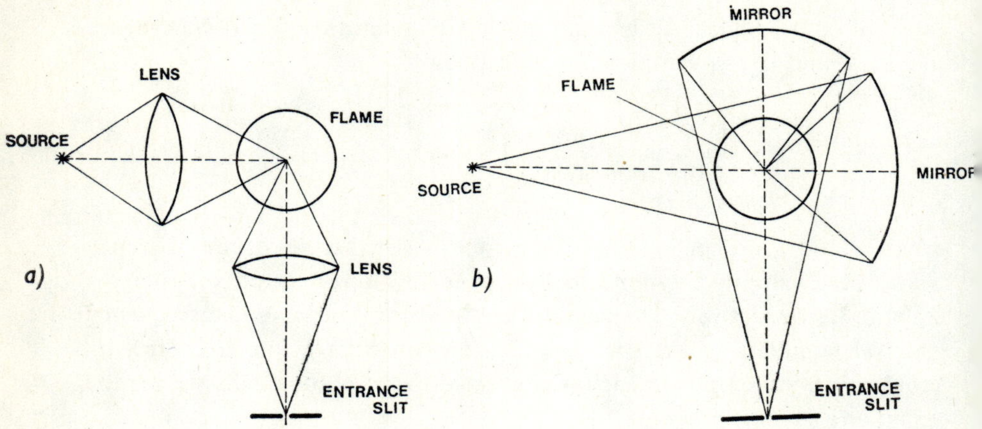

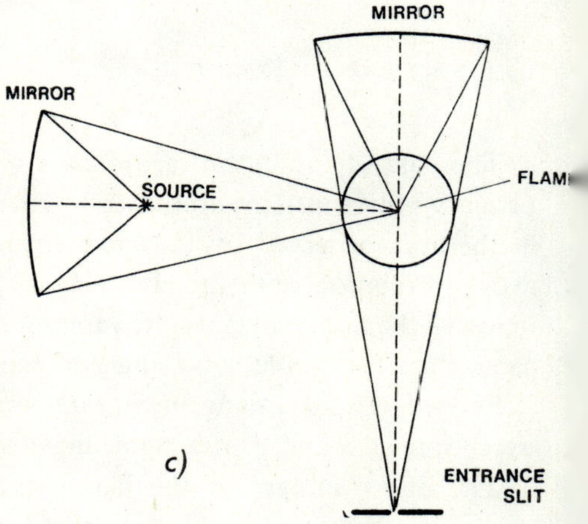

FIG. 3.41. Schematic diagram of entrance optics used in atomic fluorescence spectrometry: (a) all-lens system; right-angle illumination; (b, c, d) all-mirror (or mirror-lens combination); right-angle illumination; (e) focusing ellipse; (f) Cassegrain mirror system; right-angle illumination; (g) all-mirror system; front-surface illumination.

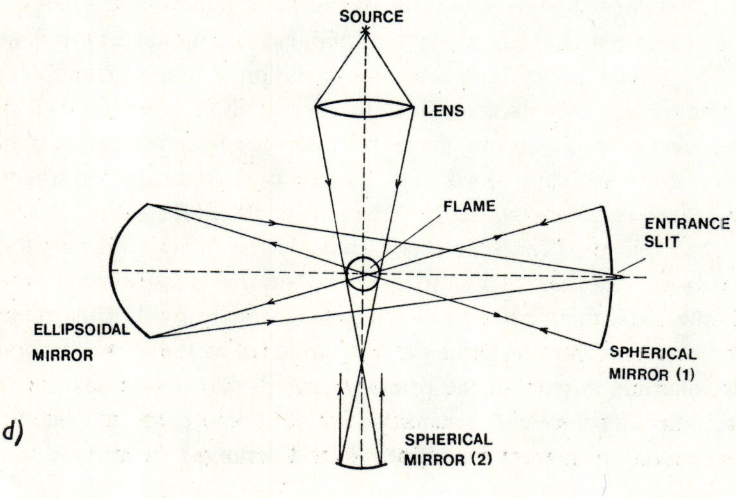

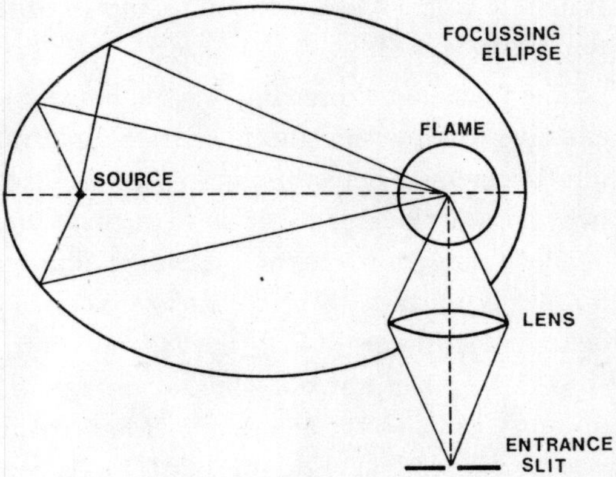
FOCUSSING
ELLIPSE
FLAME
SOURCE
LENS
ENTRANCE
SLIT

e)

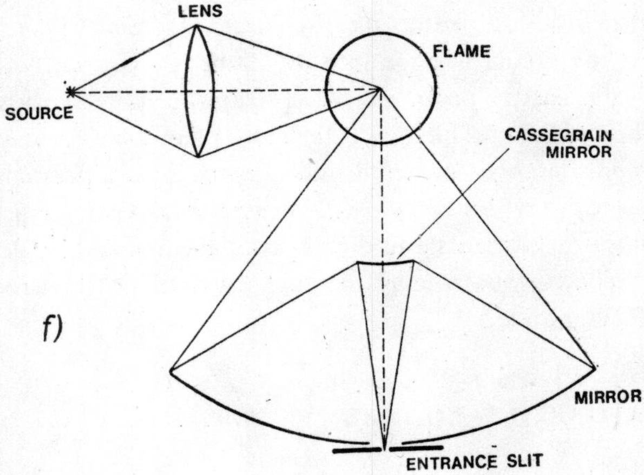
LENS
FLAME
SOURCE
CASSEGRAIN
MIRROR
MIRROR
ENTRANCE SLIT

f)

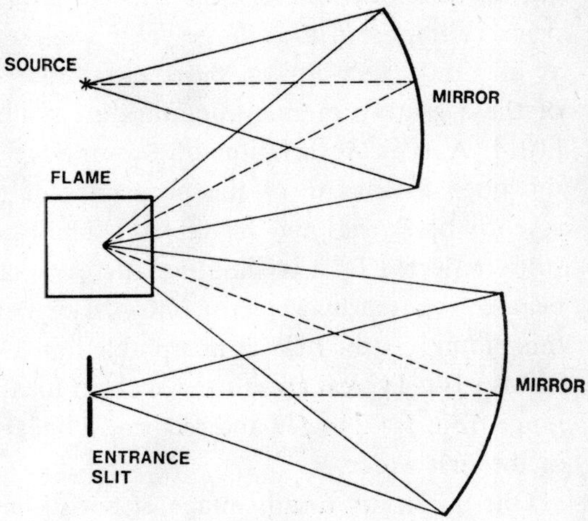
SOURCE
MIRROR
FLAME
MIRROR
ENTRANCE
SLIT

g)

system to Fig. 3.41*d* is claimed to improve the limits of detection for thallium tenfold [95].

The focusing ellipse (Fig. 3.41*e*), made from aluminium foil [356] is very effective for focusing the primary source radiation into the flame. Tenfold improvement in detection limits was reported. However, scattered light may become a serious problem with such an arrangement.

The Cassegrain mirror system of illumination (Fig. 3.41*f*) is compact, but has practically the same performance as a simple elliptical mirror placed behind the flame [143]. The all-mirror front-surface illumination [165] (Fig. 3.41*g*) has not only the advantage of great light-gathering power at low cost, but also the advantage of improving the shape of calibration curves (see Section 2.2.1). For the best results, both mirrors should be toroidal. To improve the performance of the system, three spherical mirrors could be added: one behind the source, and two behind the flame. One plane mirror on the slit would reflect the radiation back into the flame. Of course, with such a complicated system scatter problems could again be quite serious.

When light-scatter becomes a problem, strategically placed baffles may improve the situation considerably. Scattered light may become a problem in all systems where multiple reflection is used. Minute scratches, surface imperfections, dust on optical elements and even optical mounts and baffles covered with partially reflecting paint may increase the level of scattered radiation by orders of magnitude.

3.4.3. MONOCHROMATORS AND FILTERS

In most analytical studies of atomic fluorescence spectrometry, small grating monochromators have been used, because resolution is not of primary importance. A large solid angle of acceptance and low light loss at all (and especially at short ultraviolet) wavelengths is required. Most of these grating monochromators are built with Czerny–Turner mounts [101]. A typical variation of this mount is shown in Fig. 3.42. A very thorough discussion of the properties of this mount has been published recently by Barnes and Jarrell [31]. The beam passing through the entrance slit is reflected by a collimating mirror onto the grating and the diffraction beam is focused by a mirror onto the exit slit. Coma and astigmatism with this mount can be held at acceptable limits even in fast, commercial instruments [31]. Typical apertures for short focal-length instruments (below 1 m) range from $f/5.2$ to $f/9$ and reciprocal linear dispersions from 4 to 20 Å/mm in the first order.

One important disadvantage of both Ebert and Czerny–Turner mounts

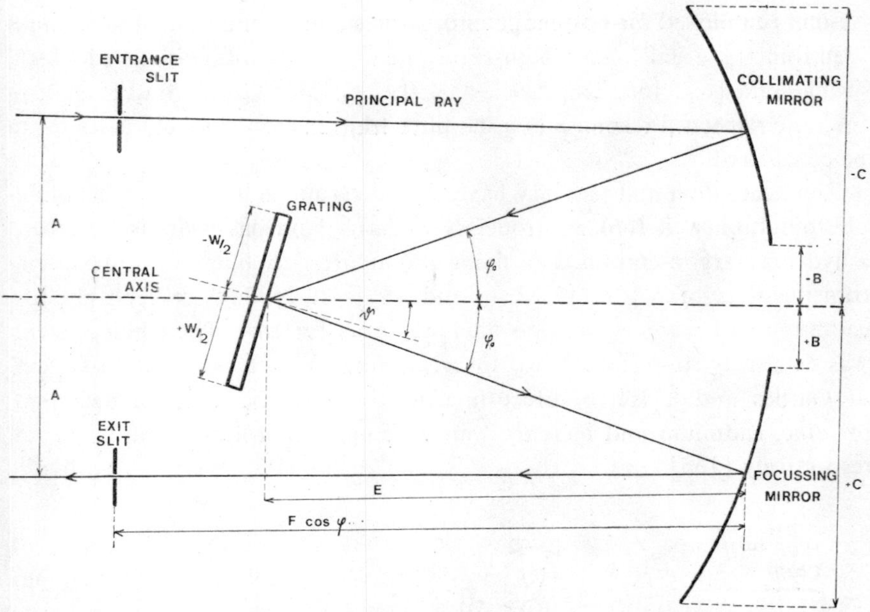

FIG. 3.42. Basic layout of a Czerny–Turner monochromator.

is their ability to present at the exit slit diffracted stray light which has been returned by the reflecting optics for a second or higher multiple diffraction [295, 351]. The problem of stray light arises when wavelengths other than the desired wavelength can follow a path to the exit slit. Discrete ranges of grating angle for this mount occur at which these reflections take place. Pribram and Penchina [351] and Mitteldorf and Landon [295] calculated the wavelength ranges of multiple diffracted stray light as a function of grating setting for Spex model 1400 or 1700 monochromators. Their equations are quite general and only simple calculations are necessary to identify the stray light. Usually only double diffractions in the first and zero orders need to be determined, and only basic information about the spectrometer (see Fig. 3.42), is necessary.

The spectrometer examined by Mitteldorf and Landon passed light of wavelengths from 4410 to 5130 Å and 3700 to 3950 Å when set at 1800 Å, and from 5420 to 5710 Å and 2950 to 3850 Å when set at 2300 Å.

Once the doubly diffracted light has been identified, it can be eliminated either by using a more finely ruled grating, or filters or by masking the part of the grating to which a singly-diffracted spectrum is returned in focus by either concave mirror [351]. This method has been found consistently effective in eliminating doubly and triply diffracted light in all orders, while reducing the efficiency of the instrument by only about 30%.

Filters employed for non-monochromator methods of atomic fluorescence spectrometry usually have high transmission in the ultraviolet and block visible light [e.g. 165, 258, 259, 290, 291, 292, 293, 420]. Both the photomultiplier spectral response and the filter transmission characteristics must be considered.

Zinc, cadmium and mercury have been determined by use of solar-blind photomultiplier (R 166), electrodeless discharge lamps as excitation sources, a hydrogen-argon-entrained air flame and interference filters with maximum transmissions of 6% (at 2288 Å) and 18% (at 2537 Å) respectively for cadmium and mercury, and no filter was used for zinc. The whole system was extremely simple; only two spherical mirrors were used, with a system of Gaffles and a Hetco direct injection burner. The limits of detection for zinc, cadmium and mercury were 0.45 ng/ml, 0.1 ng/ml and 2.5 ng/ml respectively [165].

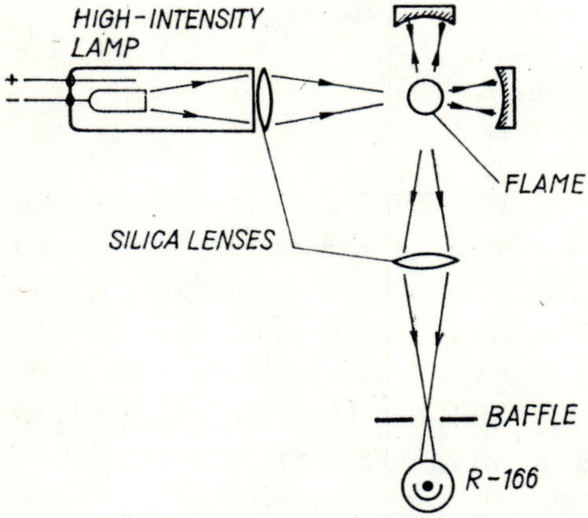

FIG. 3.43. Schematic layout of non-dispersive atomic fluorescence instrument. Reprinted from Ref. [258] by permission of Pergamon Press Ltd.

Similar results were achieved without any filter, using only an R 166 photomultiplier [416]. A schematic diagram of such a non-dispersive system is shown in Fig. 3.43. The same system, supplemented only by a chlorine gas filter, which blocks the radiation of OH bands around 3000 Å, has been used for determination of arsenic and mercury. It is typical that the signal itself increased on changing from an instrument with a monochromator to one without by a factor of 10^3-10^4, but the signal to noise ratio improved by factors of only 20 and 6 for arsenic and mercury respectively. The advantage of higher throughput (larger solid angle, exit and input pupils and higher transmission) and measurement of several lines simultaneously is widely offset by the increased flame background noise. Combination of an RCA 1P28 photomultiplier with an ultraviolet broadband-filter (Baird-

Atomic Prominent Ultraviolet Line Filter A 1) gave results very similar to that of a single R 166 solar-blind photomultiplier. In comparison with determinations made with a monochromator instrument, limits of detection for zinc and mercury were lowered by one or two orders of magnitude [420].

3.4.4. DETECTORS

Although in theory various detectors could be used for detection of radiant signals in atomic fluorescence spectrometry, only photomultiplier tubes have become widely established.

The basic principle of operation of a photomultiplier is quite simple: a photon striking the photocathode may be either reflected, transmitted or absorbed. In the latter case there is a certain probability that a photoelectron will be emitted. The ratio of the number of photoelectrons to the number of photons impinging on the photocathode is called the quantum efficiency Q_λ; it is a function of wavelength. A related quantity, the ratio of the electrical current from the photocathode to the incident radiation power is called radiant sensitivity, R_λ. It is simply related to Q_λ by the equation

$$R_\lambda = \frac{Q_\lambda e_0}{hc} \tag{3.15}$$

An electron emitted by the photocathode may be collected (by the application of a favourable electrical field) by the first dynode. The average collection efficiency is usually between 50 and 90%; it is usually higher for electrons emitted from the central parts of the cathode than for those from the edges. For each primary electron collected by the first dynode, several secondary electrons are subsequently emitted, and are accelerated towards the second dynode, and so on.

The current gain of a photomultiplier would ideally be given by

$$G = (g\eta)^n \tag{3.16}$$

where η is the average secondary emission coefficient, g is the transfer efficiency between dynodes and n is the number of dynodes. The value of g under normal conditions lies between 2 and 6; it varies with the interdynode voltage applied.

The chain of dynodes is followed by an anode which, of course, is kept at the most positive (relative) potential of all the photomultiplier electrodes (Fig. 3.44) and made from material with low electronic sensitivity. The anode is coupled to the amplifier by a capacitor (a.c. mode) or by a resistor (d.c. mode).

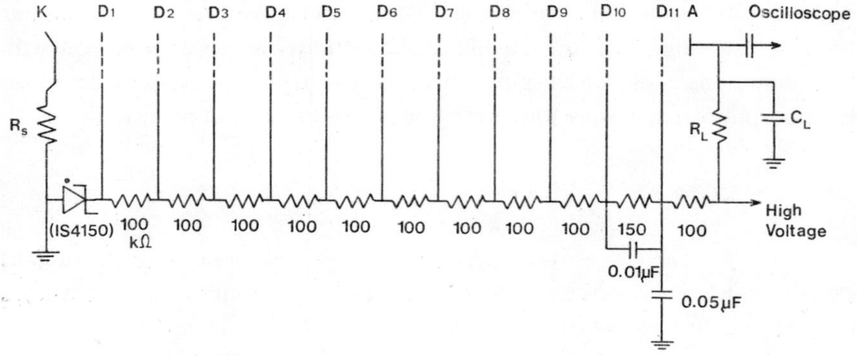

FIG. 3.44. The wiring scheme of a photomultiplier: (R_L) anode load; (R_S) protective resistor (~100 kΩ); (C_L) capacitance of anode circuit to earth (including parasitic anode capacitance).
Reprinted by permission of EMI Electronics Ltd.

The most important characteristics of a photomultiplier in fluorescence spectrometry are the NEP (noise equivalent power) and the dynamic range. NEP is defined as the radiant power giving the same response as the average amplitude of dark-current noise or, in terms of pulses, the power producing the number of pulses per unit of time equal to the number produced by the photomultiplier without illumination.

Combining Eqs. (3.15) and (3.16), we obtain for the anode current I_A of a photomultiplier

$$I_A = \Phi R_\lambda \Psi G + I_D = I_S + I_D \tag{3.17}$$

where Φ is the radiant power reaching the photomultiplier, I_S is the signal current, Ψ is transmittance of tube window and I_D the dark-current. Because the value of the dark-current depends not only on the properties of the photocathode (and hence on R_λ), but also on the gain G, it is obvious that NEP is a complex function. For any particular photomultiplier tube the optimum NEP (i.e. the value measured at the optimum gain setting at the wavelength of maximum R_λ) is usually tabulated (see Table 3.16). A useful characteristic of photomultipliers, which is sensitive to noise-current value, is photoanodic sensitivity, defined as I_S/Φ.

The cathodes of photomultipliers are made from a special type of material (e.g. antimony or bismuth) impregnated with caesium, and occasionally with other elements in addition. This material is deposited either on a metallic support (in side-window photomultipliers) or on the inner surface of the tube envelope (in end-window photomultipliers). In the latter case the cathode must be semi-transparent, i.e. very thin. This may cause complica-

TABLE 3.16. Characteristics of photomultipliers

Photomultiplier type	Spectral response	Window material	Projected cathode size (mm)	Cathode construction	Nominal cathode material	No of stages	Dynode structure	Supply potential (V)	Anode radiant sensitivity (A/mW)	Wavelength of A.R. sensitivity measurement (Å)	Current amplification	Anode dark current (nA, 25 °C)	Dark current equivalent (W at max. sensitive wavelength)
RCA 8571	S-4	Lime glass	9.5×1.5	S	CcSb	9	C	1000	73	4000	2.1×10^{6}	3.8	5.2×10^{-14}
RCA 1P21	S-4	Lime glass	24×8.5	S	CsSb	9	C	1000	78	4000	2.0×10^{6}	10	1.3×10^{-13}
RCA 1P28	S-5	UV-trans. glass	24×8.5	S	CsSb	9	C	1000	120	3400	5.0×10^{6}	24(5)*	2.0×10^{-14} (4.2×10^{-15})*
RCA 7200	S-19	Fused silica	24×8.5	S	CsSb	9	C	1000	65	3300	1.0×10^{6}	18	2.7×10^{-13}
RCA 7102	S-1	Lime glass	DIA 31.4	H	Ag0Cs	10	C	1000	0.061	8000	3×10^{5}	260	2.3×10^{-9}
HTV R 106	S-19	Fused silica	24×8	S	SbCs	9	C	1000	150	4300	3×10^{6}	50	3.3×10^{-13}
HTV R 166	Solar blind	Fused silica	24×8	S	CsTe	9	C	1000	5.6	2200	2×10^{5}	0.5	8.9×10^{-14}
HTV R 306	S-19	Fused silica	23×4	S	CsSb	9	C	1000	72	3300	2×10^{6}	7 (max)	1.0×10^{-13}
HTV R 300	S-5	UV-trans. glass	23×4	S	CsSb	9	C	1000	66	3400	2×10^{6}	7 (max)	1.1×10^{-13}
EMI 6255	S 13*	Fused silica	DIA 44	H	CsSb	13	V	1900	5000**		1×10^{8}	25	1.7×10^{-16}
EMI 6256 B	S-13	Fused silica	DIA 10	H	CsSb	13	V	1700	5000**		1×10^{8}	8	2.4×10^{-16}
EMI 6256 S	S-13*	Fused silica	DIA 10	H	CsSb	13	V	1900	5000**		1×10^{8}	1.3	4.1×10^{-17}
EMI 9592 B	S-10	UV-trans. glass	DIA 23	H	BiAg0Cs	11	B	1500	2000**		—	100	1.4×10^{-14}
EMI 9601 B	S-11	UV-trans. glass	DIA 23	H	CsSb0	11	B	1350	2000**		1.5×10^{7}	20	2.8×10^{-16}

* Special.
** Luminous sensitivity A/lm.
V Venetian blind.
B Box and grid.

C Circular cage.
S Side-window.
H End-window.

tions when high-radiance levels have to be measured, because of the inherently higher cathode resistance.

The spectral response of different photocathode materials varies over a wide range for wavelengths below about 8000 Å. As the upper wavelength limit for sensitivity increases, the energy necessary for electron emission decreases. Cathodes with extended red sensitivity, therefore, exhibit high thermionic emission, which constitutes a substantial component of the dark-current [166]. This may be virtually eliminated by cooling to about −40 °C (with the exception of a photocathode with S 1 response, which has to be cooled to about −160 °C). Decrease in temperature also changes the spectral response of photocathodes (at least of S 4 and S 11 cathodes studied in the visible region): the short-wavelength response increases, and long-wavelength sensitivity decreases [54]. To achieve 1% stability of response, temperature regulation to 1 °C or better is generally required.

Although the technology of cathode preparation has been improved substantially in recent years, severe fluctuations of cathode sensitivities over

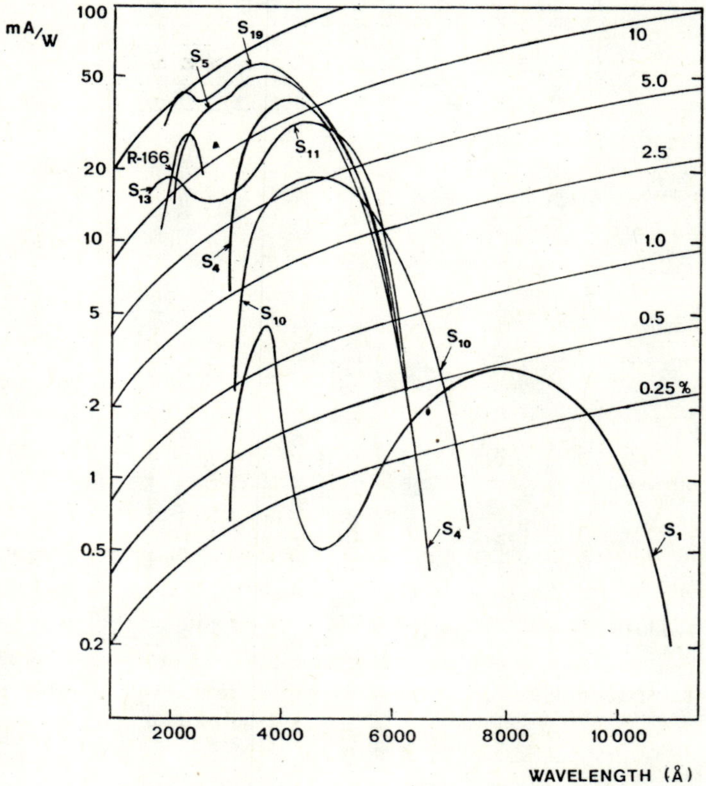

FIG. 3.45. Sensitivity of various cathodes used in photomultipliers.

the total area of the cathode have still been reported [481]. The inhomogeneous response of semi-transparent cathodes is strongly influenced by the focus electrode voltage.

One of the most recent improvements in commercially available photocathodes is the CsTe photocathode, used in the "solar blind" photomultiplier (see Table 3.16 and Fig. 3.45). Maximum spectral response for this cathode occurs at around 2200 Å, and for wavelengths above 3500 Å it is practically insensitive. Although its maximum quantum efficiency seems to be lower than that of classical CsSb photocathodes, its thermionic emission current is also about two orders of magnitude lower. Because of its exceptional characteristics it has been successfully used for multielement non-dispersive fluorescence analysis [165, 258, 259, 415, 416, 420].

The thermionic emission current is directly proportional to the cathode area. For applications where limits of detection should be as low as possible, tubes with a cathode area equal to that of the fluorescence radiation beam cross-section [347] are preferable.

With conventional transmissive photocathodes, an appreciable amount of the incident light passes through the photocathode layer without contributing to photoelectron emission. These light losses can be reduced by use of a reflective photocathode instead of a transmissive one [307].

The operation of a photomultiplier under irradiation with relatively high radiation fluxes can cause saturation of the cathode or fatigue [224, 481]. These effects are of course very unusual in fluorescence spectrometry, but if the potential drop between the dynodes is not lower than that recommended by the manufacturer and if the anode current is not higher than about 0.1% of the maximum allowable, they may occur.

The construction of the cathode is determined by the construction of the dynode chain. If the cathode is non-transparent, only the circular-cage system of dynodes is used. This arrangement is very compact, but the number of dynodes is limited and only lower interdynode voltages may be used. This again limits the gain attainable. If the cathode is semi-transparent, the tube may be made long enough for other dynode constructions to be used, and the total attainable gain is higher. The lowest values of NEP were therefore reported for end-on tubes (see Table 3.16).

Windows for photomultipliers are made either from lime glass, ultraviolet transparent glass or quartz (see Fig. 3.46). For most applications, ultraviolet-transparent glass is satisfactory, and ultraviolet-glass tubes are a great deal less expensive than those with quartz windows. Only for work below 2000 Å does the use of tubes with fused quartz windows become essential.

The use of quartz brings one other advantage. Glass always contains some potassium, and one of its naturally occurring isotopes (^{40}K) is radioactive. Its disintegration causes fluorescent flashes in the glass and so increases the photomultiplier dark-current.

The sensitivity of photocathodes is usually characterized by their radiant sensitivity or quantum efficiency, or the whole photomultiplier tube by its

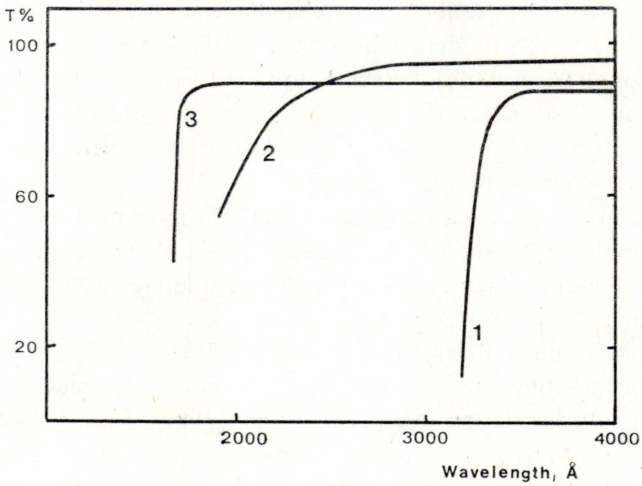

FIG. 3.46. Ultraviolet transmission of typical photomultiplier windows: (1) lime-soda glass, 2 mm; (2) ultraviolet transmitting glass; (3) Spectrosil-quartz glass. Reprinted by permission of EMI Electronics Ltd.

spectral response. Unfortunately, some manufacturers still use the luminous sensitivity (electric current for unit light flux). It has been shown [33] that conversion of luminous sensitivity into radiative response is quite complicated and large ambiguities arise which are difficult to avoid. This is especially true when measurement of ultraviolet radiation is of prime interest.

The construction of the dynodes [circular cage, venetian blinds, box and grid, focused type or others (Fig. 3.47)] may have some influence on the stability of photomultiplier or electron collection efficiency.

The secondary emitting surface of the dynodes is usually made from BeOCs, CsSb, MgOCs or some such material. Although these materials may show some thermionic emission, only the first dynode should appreciably influence the dark-current. Because the thermionic emission of the photocathode is probably somewhat higher than that of the dynode, probably less than 10% of the dark-current originates from the dynodes [308].

Another possible cause of spurious dark-current may be loss of electrons from the dynode system (causing glass fluorescence) if the photomultiplier wall is at a positive potential relative to the cathode. Operation of the tube with the cathode at earth potential gets over this problem; alternatively the difficulty may be eliminated by arranging that any material which comes into contact with the tube wall is at the same potential as the cathode of the tube [355].

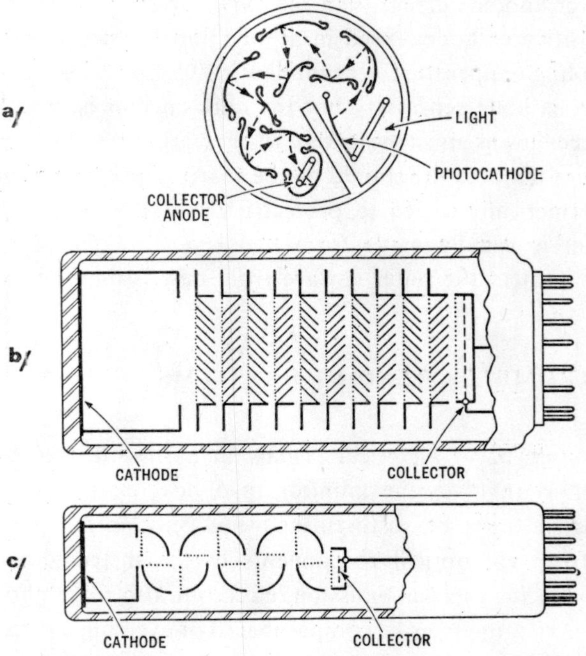

FIG. 3.47. Electrostatic dynode systems: (a) compact focused structure (circular-"squirrel"-cage); (b) Venetian-blind structure; (c) box-and grid structure. Reprinted by permission of EMI Electronics Ltd.

The NEP of a photomultiplier depends *inter alia* upon the total voltage applied to the dynode chain. It is approximately constant over a wide range of gain, until a limiting value is approached which causes the equivalent dark-current to rise very rapidly. This occurs primarily because the leakage-current noise increases very rapidly above a certain limit. If only one thermionic current is considered, theoretically constant NEP is obtained for all gains [83].

The leakage current occurring between electrodes in the tube socket, base, stem and internal parts is a particularly troublesome source of spurious dark-current noise in low-noise multiplier tubes. If this leakage current were purely ohmic in nature, the noise contribution would of course be negligible. This is usually not the case, and the leakage often has a high noise current characterized by sudden discharges [160]. Special care must always be taken, therefore, when the socket of the photomultiplier is handled or when cooling is used.

The cooling of photomultipliers in atomic fluorescence spectrometry may be advantageous if (under optimal conditions) the measured radiant flux evokes a lower anodic current than the dark-current at room temperature. For blue-sensitive cathodes, used in atomic fluorescence spectrometry, the optimum cooling temperature is around -30 °C. Lowering the temperature decreases the cathode sensitivity but the dark-current is constant.

Special precautions must be taken when working at low temperatures. The decrease of temperature should not be sharp. The whole photomultiplier has to be hermetically sealed to protect it from dew. The window of the cooling jacket is usually made from two layers of quartz, insulated by a vacuum to protect the outer surface from dew formation.

3.4.5. ELECTROMETER READ-OUT SYSTEMS

The measurement of electrical signals in atomic fluorescence spectrometry invariably involves the monitoring of low electrical currents. The analytical signal must be distinguished not only from spurious signals originating from the optical components of the instrument (scatter, reflection, non-specific thermal emission, etc.), but also from photomultiplier noise, which may sometimes be comparable to or even higher than the useful analytical signal.

The three methods most widely used are: direct-current measurement, alternating-current measurement with a lock-in amplifier and pulse (photon)-counting techniques.

3.4.5.1. Direct-Current Measurement

The mean amplitude of a d.c. amplifier input circuit using a shunt circuit is obviously

$$E = RI \qquad (3.18)$$

where I is the total current flowing from the photomultiplier and R is the anode (shunt) resistance. Johnson noise (thermal agitation) originates in the shunt resistance and sets the ultimate limit of voltage resolution. Its value is given by the formula

$$E_j = \sqrt{4kT\Delta fR} \qquad (3.19)$$

where Δf is the bandwidth of measurements in Hz, k is the Boltzmann constant (1.38×10^{-23} J/K) and T is the absolute temperature of the resistance.

A third limiting condition is set by the relationship between the bandwidth limit, Δf, and values of the shunt resistance, R, and the input capacitance, C:

$$\Delta f = \frac{1}{2\pi RC} \tag{3.20}$$

These three equations determine the lowest detectable signal in terms of current.

If only Johnson noise is present, and the shunt resistance and input capacitance determine the bandwidth of the whole system, combination of these equations gives for the signal to noise ratio:

$$\frac{E}{E_j} = I\sqrt{\frac{R}{4kT\Delta f}} = IR\sqrt{\frac{2\pi C}{4kT}} \tag{3.21}$$

At constant bandwidth the theoretical signal to noise ratio increases with the square root of input impedance. If the bandwidth is decreased by increasing the resistance, the signal to noise ratio would increase linearly with the increasing shunt resistance [second variant of Eq. (3.21)]. It should be stressed, however, that under almost all practical conditions the bandwidth – or time of measurement – is one of the most important characteristics of an analytical method, which must be held constant for any valid comparison between various instrumental arrangements.

Values must be allocated to the input resistance and bandwidth before an approximate estimate of the lowest theoretically detectable current by shunt-type measurement under typical conditions met in fluorescence spectrometry can be made. Assuming that the input capacitance is somewhere around 50 pF we get from Eq. (3.20) for a 1-Hz bandwidth, a shunt resistance R_3 equal to $3.2 \times 10^9\,\Omega$. For measurement at room temperature (293 K) and a signal/noise ratio (E/E_j) of 6, we get from Eq. (3.21) $I = 1.3 \times 10^{-14}$ A. This value lies well below the anode dark-current of typical photomultipliers (Table 3.16). Taking as an example an RCA 1 P 28 photomultiplier with an anode dark-current of 5×10^{-9} A and typical current amplification $G = 5 \times 10^6$, and assuming that the dark-current I_D is caused only by thermionic emission from the photocathode, we get for the number of electrons emitted from the cathode per second

$$n_e = \frac{I_D}{G \,.\, e_0} \tag{3.22}$$

a value of about 600 per sec. Because of the statistical nature of thermionic emission, this value will fluctuate with a standard deviation equal to the square root of the number of emitted photoelectrons, i.e. in our case $\sim$ 25 per sec. In turn, this will cause a dark-current noise of about 2×10^{-10} A. Again this is well above the theoretical limit of d.c. measurement noise. It is obvious that under these conditions the only source of noise obscuring the signal is the shot-noise and no further deterioration of the signal should be expected from the measuring chain.

This is of course strictly true only for the typical example mentioned above. When either a high-speed measurement is required [82] or a cooled photomultiplier with only a few electrons emitted per second is used, then a critical examination of amplifier noise may become necessary. The signal to noise ratio of an ideal d.c. photomultiplier system may be computed from the simple equation:

$$\frac{I_S}{I_N} = \frac{I_S}{\sqrt{2e_0G(I_S + I_D)\,\Delta f}} \tag{3.23}$$

This equation takes into account only the shot-noise of the photomultiplier tube. If Johnson noise has also to be considered, the equation takes the form

$$\frac{E_S}{E_N} = \frac{I_S\sqrt{R}}{\sqrt{\Delta f(2kT + Re_0G(I_S + I_D))}} \tag{3.24}$$

When the performance of commercially available instruments is tested, usually noise levels not far from theoretical levels may be found, except for the most sensitive ranges. Sometimes a "noise figure" is used for comparison of the theoretical and actual performance of an amplifier; it is defined as

$$NF = 20\log_{10}\left(\frac{\text{Total actual noise}}{\text{Theoretical Johnson noise}}\right) \tag{3.25}$$

and is expressed in decibels (dB). So if $NF = 20$ dB, the amplifier has a noise amplitude ten times that of an ideal instrument.

Other very important characteristics of d.c. amplifiers are the dynamic range and the zero stability. The dynamic range of a d.c. amplifier is not theoretically limited and it is not unusual to find instruments covering 10–15 decades, which is more than would ever be necessary in fluorescence spectrometry. Even a home-made d.c. amplifier used for fluorescence measurement and constructed simply from three solid-state operational amplifiers exhibits a range of $10^{-11} - 10^{-5}$ A. Zero drift is a more difficult problem. It produces an error by adding algebraically to the input signal. It may be a function of time and temperature and may not be linear. It may not be predicted theoretically and is determined solely by the components and construction used. It becomes troublesome particularly when measurements are made over a long period of time. The same is also true for gain instability.

Other problems quite often encountered in fluorescence are the measurement of transient signals and compensation for primary light-source intensity fluctuations. Both can be solved in principle by an identical approach: the signal current is integrated and the resultant electrical charge is either directly displayed, recorded or stored in a memory, or divided by another electrical charge collected from a monitoring beam – usually directly reflected from the primary light-source to the detector [309]. This integration technique has been successfully used for many years in commercial emission spectrometers.

3.4.5.2. Lock-In Amplifiers

The d.c. systems have one principal disadvantage – to distinguish between thermal radiation and optically-excited radiation, the primary light-source has to be turned on and off during the aspiration of analyte into the flame cell, and the emission subtracted. This may be unnecessary if there is virtually no background radiation (e.g. at short wavelengths in the hydrogen-argon entrained-air flame). A more general solution is to use a phase-sensitive amplifier. This functions by demodulating the a.c.-coupled input signal via a switch that is operated in precise synchronization with the phase and frequency of the periodic fluorescence signal. The radiation from the primary light-source is modulated either electronically or by a mechanical chopper. The result is a d.c. input of a magnitude proportional to the input signal. This is superimposed on the noise. The d.c. signal may be "cleaned" of higher frequency components with a low-pass filter of time-constant τ_c. The overall filter characteristic given by this system is that of a bandpass filter centred on the reference frequency, the equivalent noise bandwidth (for white noise) being

$$\Delta f = \frac{1}{4\tau_c} \tag{3.26}$$

So, for a time-constant variable from 0.1 to 100 s the equivalent noise bandwidth can be varied between 2.5 and 0.0025 Hz. This approach has several advantages. By choosing higher frequencies for the signal, the signal/noise ratio may be improved, if the noise is not white, but "pink", i.e. increases towards lower frequencies. (Although no precise information exists about the frequency spectrum of flame flicker noise, this may well be the case.)

Because only the a.c. component of the signal is amplified, background subtraction is automatically achieved. Zero drift, which may be very serious for d.c. amplifiers, is almost completely eliminated. One small drawback is the slight theoretical decrease of resultant signal/noise ratio caused by the chopping of the exciting light [21]. This is of course strictly true only if mechanical chopping is used. If the primary light-source is electronically modulated, so that the average light output equals the light output of a d.c.-operated discharge lamp, the value of the signal/noise ratio remains constant. It should be stressed that the practical performance of a lock-in system approaches much closer to the theoretical limit than that of most d.c. systems; this is the reason why practically all commercial atomic absorption spectrometers use some modification of phase-sensitive detection.

3.4.5.3. Photon Counting

The advantages of photon counting in comparison to conventional d.c. current measurement are believed to be the possibility of direct digital processing of observed spectral intensity, precise integration of transient signals, accurate measurement of low levels of radiation, elimination of reading error and ability to reduce effective signal/noise ratio via pulse-height discrimination [90].

The basic principle is very straightforward: every electron impinging upon the first dynode produces at the anode a current pulse. These pulses are comparatively narrow—their shape is determined by the transit-time spread limits of a particular photomultiplier construction. The pulses are then amplified, discriminated and counted.

The first attempts to use this principle—which is well established in nuclear physics—for measurement of low light levels, were made in the fifties [e.g. 363]. Since then, a number of papers have been published dealing with this theme [e.g. 90, 125, 130, 159, 160, 161, 175, 253, 346, 404, 480]. Although originally a drastic improvement in sensitivity for digital systems compared with d.c. measurement was expected [159, 363], it is now evident that both systems are about equally efficient [132, 404, 480]; this has been proved both theoretically and experimentally. No effective way exists for discriminating between the dark-current pulses and the photoelectron pulses. Their height is the same, and only very high pulses, probably originating from radioactive decay of the glass envelope or Cherenkov radiation can be eliminated by this technique. These pulses may introduce a noticeable error into current measurement only if refrigerated photomultipliers are used and no background radiation is present [404].

To compare quantitatively the inherent noise of pulse-counting and d.c. techniques, we can, to a first approximation, use Eq. (3.23) where

$$\Delta f = \frac{1}{2t} \tag{3.27}$$

t being the counting time.

One serious disadvantage of pulse-counting techniques is the pile-up of pulses at higher counting rates. As the pulse-rate increases, the pulses overlap and the counter is not able to count all of them. At present, the highest measurable count-rates are around 10^6 counts/sec [132, 175]; this is about one or two orders of magnitude lower than the light levels measurable without difficulty with analogue systems.

However, the digital system is less sensitive to changes in photomultiplier gain, because discrete events are counted [132]. If modern high-stability high-voltage power supplies and amplifiers are used, the practical stability of both systems is approximately the same.

One simple way of combining the advantages of both approaches is to use a simple analogue to digital converter after current amplification. This may help to overcome difficulties arising with analogue outputs or analogue methods of integration. At the same time such a combined system would preserve the high dynamic range of a current-measuring technique. Modern analogue to digital converters are very reliable, and relatively inexpensive.

3.5. INSTRUMENTS USED FOR ATOMIC FLUORESCENCE MEASUREMENTS

Atomic fluorescence measurements published to date have been carried out almost exclusively either with modified commercial atomic absorption/flame emission spectrometers or with specialized research instrumentation built in the laboratory (cf. Chapter 5). The similarity of atomic-absorption, atomic emission, and atomic fluorescence spectrometry allows practically all commercial spectrometers to be relatively easily converted to make atomic fluorescence measurements [97, 107, 111, 190, 203, 248, 264, 312, 399, and cf. Chap. 5]. This modification usually involves the addition of an excitation source and its power supply, changes in entrance optics, and the use of a sensitive multi-range recorder. Versatile instruments such as the Varian-Techtron AA-4 and Jarrell-Ash instruments equipped with an optical rail are more convenient for atomic fluorescence work, because they offer free access for use of a variety of additional optical components and sampling devices. When using boosted-output hollow cathode lamp sources, for example, the transition from the atomic absorption to the atomic fluorescence mode with the Techtron AA-4 spectrometer takes only a few minutes. Compact instruments (e.g. Perkin-Elmer, Southern Analytical) usually require a more complex modification. The modification necessary to convert the Model 403 Perkin-Elmer spectrometer for fluorescence studies is shown in Fig. 3.48.

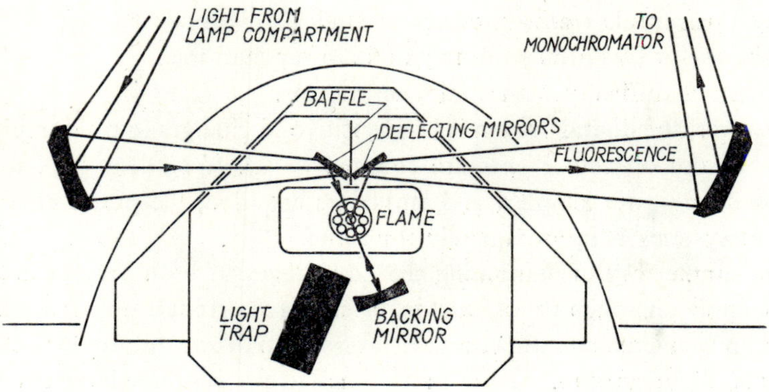

FIG. 3.48. Schematic representation of the accessory optical system designed to convert the Perkin–Elmer 403 for fluorescence studies [32].
Reprinted from Ref. [32] by permission of The American Chemical Society

The main advantage of instruments built in the laboratory specially for atomic fluorescence spectrometry is the possibility of choosing individual optical and electrical units with optimum parameters for the fluorescence measurements of interest. Simple non-dispersive instruments (see Fig. 3.49) offer a definite cost advantage.

Atomic fluorescence spectrometry should be the simplest and best method to use for multielement analysis with multiplex spectrometers [291–293, 419]. Recently Technicon Instruments Corporation (Tarrytown, N.Y. 10591, U.S.A.) introduced a prototype six-channel atomic fluorescence spectrometer (Model AFS-6) [143–147]. This instrument involves the use of pulsed hollow-cathode lamps, a rotating interference filter wheel, a Cassegrain mirror system, a flame cell, and logic circuitry to measure the fluorescence of each metal when the proper interference filter is in place. A schematic diagram of the instrument is shown in Fig. 3.50. Each hollow-cathode lamp emits a 32-msec pulse of radiation, square-wave modulated at 500 Hz, at a repetition rate of 1 Hz.

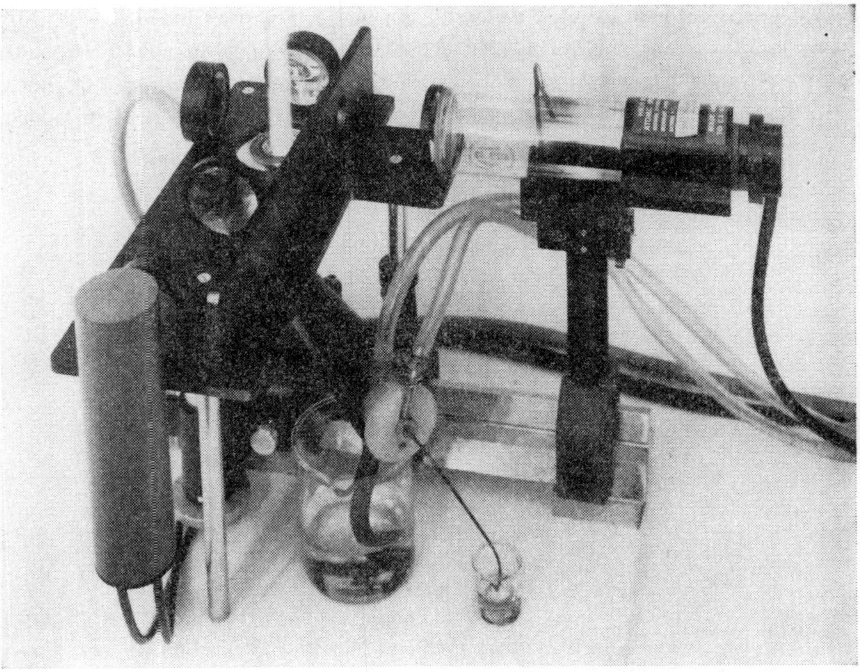

FIG. 3.49. Simple monochromatorless atomic fluorescence instrument [258]. Reprinted from Ref. [258] by permission of Pergamon Press Ltd.

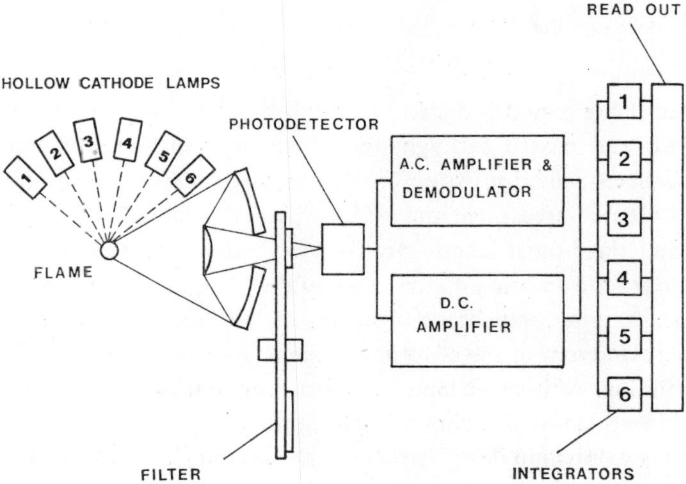

FIG. 3.50. Schematic diagram of the Technicon AFS-6 multichannel atomic fluorescence flame spectrometer [143].
Reprinted from Ref. [143] by permission of Technicon Instruments Corporation

A cross-sectional view of the optical system is shown in Fig. 3.51. The excitation sources are mounted at 60° to each other in a vertical plane with radiation from each source collected by a toroid mirror and focused in the flame where a magnified image (×1.25) is formed. Radiation from the

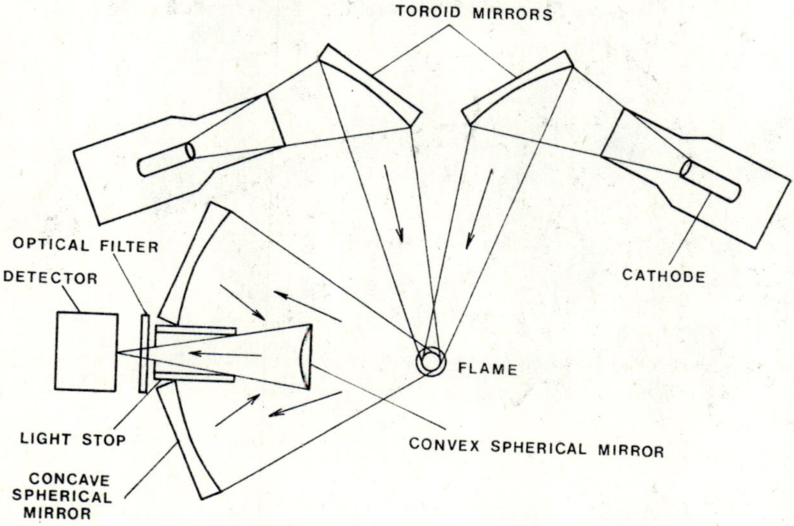

FIG. 3.51. A cross-sectional view of the optical system of the Technicon AFS-6 instrument [143].

Reprinted from Ref. [143] by permission of Technicon Instruments Corporation

irradiated flame region is collected by an inverse Cassegrain mirror system, magnified, and passed through optical filters to the photodetector. This system collects radiation within a solid angle of 0.82 ster. An all-reflective optical system is used since mirrors give higher radiation transmission than lenses, and their focal lengths are independent of wavelength. The toroid mirrors have the additional advantage of deviating the light beams through 90°, permitting a compact arrangement of the excitation sources.

A block diagram of the electronic system, showing signal processing and system timing, calibration and read-out functions is shown in Fig. 3.52. Signals generated at the photodetector pass to:

(*a*) an a.c. preamplifier, where signals arising from flame background radiation are rejected, and fluorescence signals, modulated at 500 Hz, are amplified;

(*b*) a d.c. preamplifier, in parallel with the a.c. preamplifier, to amplify d.c. signals generated by atomic or molecular emission in the flame;

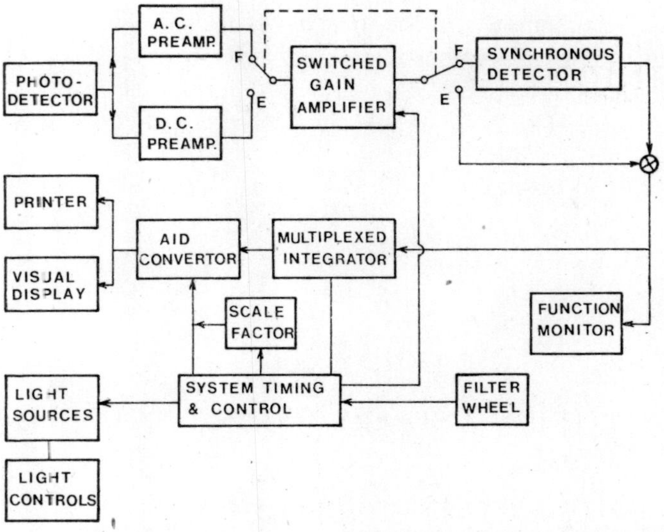

FIG. 3.52. A block diagram of the electronics system used in the Technicon AFS-6 instrument [143].

Reprinted from Ref. [143] by permission of Technicon Instruments Corporation

(*c*) selector switches to select the a.c. channel for fluorescence measurements or the d.c. channel for emission measurements, as required;

(*d*) a synchronous detector, to demodulate the amplified fluorescence signal;

(*e*) a multiplexed integrator, to separate signals arising from each atomic species in the flame and integrate a preselected number of signals;

(*f*) a read-out system, a nixie tube display and printer, to read out these signals in digital form.

Excellent results were obtained for a number of elements with this unique new instrument. They are included in Chapter 5. Some applications are discussed in Chapter 6.

Chapter 4

TECHNIQUE

4.1. SENSITIVITY AND PRECISION

4.1.1. DEFINITION OF DETECTION LIMIT

When only subjective tests were used in analytical chemistry for the detection of trace amounts of an element, obviously no misunderstanding of the term "detection limit" could arise. As early as 1923 Feigl presented a very concise and clear definition of this term [169]; "Es währe wohl zweckmässig... als Erfassungsgrenze die kleinste absolute Menge Substanz zu definieren, die durch eine Reaktion oder Methode noch nachweisbar und bestimmbar ist". ("It would be reasonable... to define the detection limit as the smallest absolute amount of a substance which is detectable and recognizable by a reaction or a method.") This half-century old definition could, after slight modification, also be used for modern instrumental analysis. When the measurement of small amounts or concentrations* of an element is performed in atomic fluorescence spectrometry, both background and analytical signals are measured. Measurement of the analytical signal involves read-out of a deflection on a measuring instrument (recorder, digital read-out, meter). We then have two values, X_A for the total signal and X_B for the background. Let us now suppose that a linear calibration curve has been established:

$$\bar{X} = gC \tag{4.1}$$

where $\bar{X}$ is the net deflection on the measuring instrument, g is a constant and C is the concentration. A straightforward computation shows that the

* For simplicity, in this paragraph the term "concentration" will be used for both concentration and amount.

best estimate of the concentration of the element determined in single measurement is

$$C = \frac{X_A - X_B}{g} \tag{4.2}$$

Of course, both X_A and X_B are to some extent arbitrary: if a series of measurements is made under exactly the same experimental conditions, the results fluctuate around a mean value. These fluctuations (noise) are caused by flame and nebulization instabilities, detector noise, etc. Suppose that the statistical distribution of both X_A and X_B is normal (Gaussian), with mean values $\bar{X}_A$ and $\bar{X}_B$ and standard deviations σ_A and σ_B. Suppose also, for simplicity, that both standard deviations are equal ($\sigma_A = \sigma_B = \sigma$).

If the difference of two statistical variables is computed, then the variance (σ^2) of the difference is the sum of the individual variances.

If

$$S = \bar{X}_A - \bar{X}_B \tag{4.3}$$

then

$$\sigma_S = \sqrt{\sigma_B^2 + \sigma_A^2} = \sqrt{2}\sigma \tag{4.4}$$

We can now estimate the probability of false statements connected with the estimation of some very low ("just detectable") concentrations. When only pure solvent is aspirated (instead of sample), then the probability that the value of X_S will be higher than $\sqrt{2}\sigma$ is 15.8%, higher than $2\sqrt{2}\sigma$ is 2.27% and higher than $3\sqrt{2}\sigma$ only 0.14%. Generally, the higher the factor k in the $k\sqrt{2}\sigma$ criterion, the lower is the probability of a false statement, and the higher the confidence level. Usually the criterion $2\sqrt{2}\sigma$ is used. *If the net signal measured* ($X_A - X_B$) *is higher than* $2\sqrt{2}\sigma$ *the statement "element is detected" is made. If it is lower, the statement "not detected" is made.*

Using this rule, what is the lowest concentration that can be distinguished from zero? Let us suppose that we have a set of samples, with equal numbers of samples with zero concentration and with the lowest discernible concentration. A simple computation shows that if this concentration is $2\sqrt{2}\sigma/g$, then the frequency of false and true statements would be:

	Frequency of Statements	
	true	false
Concentration zero	48.86%	1.14%
Concentration $2\sqrt{2}\sigma/g$	25%	25%
Total	73.86%	26.14%

If we choose as the lowest discernible concentration $4\sqrt{2}\sigma/g$ then the results would be improved markedly:

	Frequency of Statements	
	true	false
Concentration zero	48.86%	1.14%
Concentration $4\sqrt{2}\sigma/g$	48.86%	1.14%
Total	97.72%	2.28%

The frequencies of false statements for both zero and the lowest detectable concentration are now equal. Generally, the value quoted for the lowest detectable concentration should be $4\sqrt{2}\sigma/g$, since the risk of false statements for both concentrations should then be equal and small enough for most practical purposes [100, 383].

Unfortunately, in the analytical literature much confusion exists about these terms. "Detection limit" is used to describe both the criterion of detection and the lowest detectable concentration [221]. In German literature the term "limit of guarantee" (Garantiegrenze) [222] has been used for the lowest detectable concentration.

It cannot be overemphasized that no "general" experimental limit of detection can exist for atomic fluorescence spectrometry. It may be defined only for a particular, completely specified method. Also the estimation of standard deviation has to be done very carefully. Under normal conditions, 80–120 readings is a practical compromise. When the number of readings is decreased, the probable error of estimation of the standard deviation increases rapidly.

4.1.2. PRECISION AND ACCURACY

The precision of an analytical procedure is most conveniently expressed in terms of the relative standard deviation. It is defined as the ratio of the standard deviation to the mean, both in the same units of concentration or amount. It should be stressed that when the calibration curve is non-linear, the relative standard deviation of the signal is different from that of the analytical result, (it is usually lower). This is very pronounced if the calibration curve forms a maximum.

One special difficulty exists when using atomic fluorescence with a line source placed at an angle of 90° to the atom-cell – detection axis. The calibration curves have two branches, one with positive slope and the other with negative slope. To every measured signal over a certain range two concentrations may be assigned. This problem may be overcome by diluting the sample and taking a second measurement. Precision is usually expressed in per cent, although the presentation as a decimal fraction is preferable. The concentration range for which the precision figure is valid should be specified.

Accuracy is related to the difference between the analytical result obtained by a given procedure and the true analyte concentration in the sample. It is therefore much more difficult to assess than precision.

If no standard material with certified analyte concentration is available, and the results cannot be compared with results of other reliable methods of analysis, then recovery and dilution tests may be used to check for some systematic errors in the procedure. Although the correct outcome of these tests is essential, it does not prove categorically the absence of systematic errors in a given analytical procedure. There have been several publications in atomic fluorescence spectrometry where light scatter has been mistaken for analytical signal.

Sensitivity is the slope of the calibration curve (dx/dC). In atomic fluorescence spectrometry, it may be positive, negative or zero. Again, it is defined only for some particular method and concentration (or concentration range).

4.1.3. FACTORS AFFECTING THE LIMIT OF DETECTION

By use of the results obtained from the theory of analytical curves, expressions can be derived for calculation of the lowest detectable atomic concentration in atomic fluorescence spectrometry for a given arrangement of optical components, detection system, atomic cell and excitation source [132, 133]. Similar attempts have been made in the past [132, 133, 298, 321, 322, 451, 456, etc.]. Unfortunately, there are many poorly defined factors, which drastically influence the results of such computations. While the noise of the detector may be estimated quite precisely, the flame-flicker noise, flame background, source flicker and radiance are usually not known precisely. This is particularly true for noise levels in flames and sources. Often theoretical estimates and practical results differ by several orders of magnitude. Normally only one variable, i.e. slit width, remains to be experimentally optimized for each determination. All other components can be chosen by using the simple general rules given below. Readers interested in the derivation of the theoretical expressions for the calculation of detection limits are referred particularly to the papers by Winefordner, Parsons, Mansfield and McCarthy [321, 322, 451].

When the maximum detectability has to be attained, these simple rules should be followed.

(*a*) Maximum radiance. The radiance from the primary source in the atomic cell should be as high as possible. The primary source flicker should be low, but is not usually critical.

(*b*) Minimal scatter. Scattered light from the source reaching the detector should be minimal. This is usually one of the factors which limits detectability in atomic fluorescence spectrometry and great care should be taken to avoid it. Not only droplets, solid particles and inhomogeneities in the flame, but also dust and imperfections on the optical surfaces placed around the flame may increase scatter.

(*c*) Selection of the optimum line. Not only the absolute fluorescence yield, but also the source intensity and the detector response must be taken into account.

(*d*) Flame gas composition and temperature. These may influence both the fluorescence yield and the degree of atomization. Premixed flames with a high content of rare gas and a temperature just high enough for effective atomization are probably the best compromise.

(*e*) Nebulizer yield and solution transport rate. Both of these factors may be optimized independently of other variables. Non-flame cell methods (hot wire, graphite cell, cup, etc.) are in a special category where rules (*d*) and (*e*) are concerned. These methods require careful optimization, which is independent of other parameters.

(*f*) Entrance optics and spectrometer. The highest possible solid angle of acceptance and a reasonable resolution are required. Again, poor quality of optical surfaces can decrease the detectability by some orders of magnitude when working in the ultraviolet region, by scatter of visible light. The optimum slit width may be found by a simple set of measurements and evaluation of the signal/noise ratios. The radiation flux of continuum background radiation reaching the detector increases proportionally with the square of the slit width, but the detected flux of line radiation is directly proportional to the slit width. Generally speaking, the optimum slit width is in the range over which the photocurrent from the continuum background radiation equals the dark-current of the photomultiplier, or, if the continuum background radiation is extremely weak, in the range over which the fluctuations of the continuum background radiation are equal to the fluctuations of the scattered line radiation.

(*g*) Detector optimization is dealt with in Section 3.4.4. It is again mostly independent of other variables.

4.2. INTERFERENCES

4.2.1. GENERAL

An important property of all analytical methods is undoubtedly their susceptibility to interference. In atomic fluorescence spectrometry, as in all atomic spectroscopic methods of analysis, interferences occur as a result of either inadequate isolation of the radiation carrying the analytical information (analytical radiation) or of the preparation of a cloud of free atoms which is not representative of the sample composition. Interferences of the first type are generally considered under the heading "spectral interferences". For the second type of interferences there is no general term because the different mechanisms causing unwanted deviations in the composition of atomic vapour depend very much upon the technique by which the atomic vapour is produced. As flames are most often used as atomizing devices in atomic fluorescence spectrometry we shall deal first with these and mention interferences occurring with non-flame devices separately later. In atomic fluorescence flame spectrometry interferences are identical to those encountered in the other flame spectrometric methods of analysis. Only their relative magnitude and frequency of occurrence differ.

4.2.2. SPECTRAL INTERFERENCES

Spectral interferences, as mentioned previously, are caused by incomplete isolation of the analytical fluorescence signal from the rest of the radiation falling on the detector. Isolation is generally achieved by spectral isolation, i.e. with the help of a monochromator, filter or simply a detector which responds over a limited spectral region (e.g. solar blind photomultiplier) or possibly a combination of these. It may be further improved by applying synchronous detection in conjunction with modulated primary light-sources emitting either continuous or, even better, line spectra.

The interfering radiation may originate from various processes. It may be scattered radiation from the primary light source, or it may be overlapping fluorescence radiation from molecular bands or atomic lines or indeed bands or lines thermally emitted by the flame itself. These different types play different roles according to the experimental conditions.

4.2.2.1. Light-Scattering

The most serious problem which cannot be eliminated by improving spectral resolution, is light-scattering by non-volatilized aerosol particles. The contribution of light-scattering to the fluorescence signal depends on many factors. So far very few generally valid data are available and workers using atomic fluorescence spectrometry mostly have to estimate the contribution of light-scattering to the fluorescence signal measured for their particular experimental conditions.

Light-scattering depends on the size and number of the unevaporated particles per unit volume, and to a much lesser degree on their optical properties. Therefore, in general, greater scattering is encountered with direct-injection burners than with premixed flames [61]. Light-scattering also decreases with increasing height of observation or increasing flame temperature, or when solvents with higher volatility are used [62]. All these factors improve volatilization of the aerosol particles. Light-scattering may therefore be particularly troublesome when working with "cool flames" which, as already mentioned, are very suitable for atomic fluorescence spectrometry from other points of view.

The light-scatter signal is proportional to the intensity of the light-source and so is the atomic fluorescence signal. The most suitable way of reporting the magnitude of the scatter signal is therefore to express it in terms of concentration of the analyte element producing a signal of equal magnitude. This method of presentation was used by Smith, Elser and Winefordner

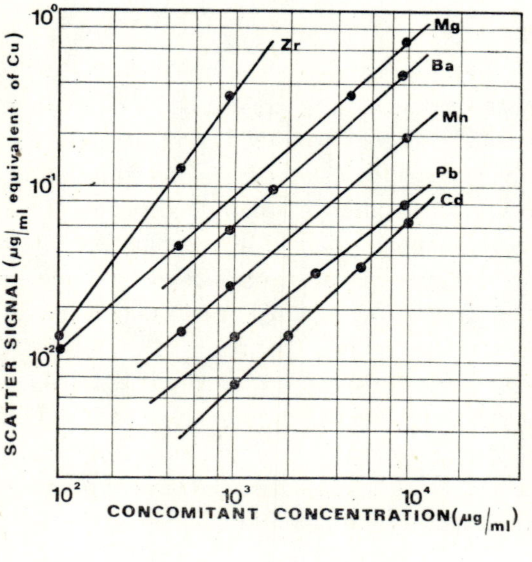

FIG. 4.1. Scatter caused by high concentrations of dissolved solids expressed as μg/ml of copper required to give the same fluorescence signal [368].

[368] when determining copper in an argon-hydrogen-entrained air flame on a direct-injection burner. From their results, reproduced in Fig. 4.1, it follows that light-scatter caused by the different concomitants studied increased in the order Cd < Pb < Mn < Ba < Mg < Zr. This is the same as the order of volatility of the particular concomitants.

The most elegant way to avoid interferences from scattered light is to measure direct-line or stepwise fluorescence, simultaneously filtering off the measured line from the exciting radiation. In this way atomic fluorescence from quite "turbid" media may be measured. The method has been successfully applied to lead [16, 390, 393], bismuth [108] and antimony [107].

Unfortunately not many elements emit non-resonance fluorescence lines with sufficient intensity, and the deterioration of detection limits can become prohibitive [278]. Also it is not always easy to choose suitable filters which transmit at the wavelength of the exciting radiation and absorb at the wavelength of the line measured. This difficulty increases as the wavelength difference of the two particular lines decreases [316]. For some elements excitation by an intense overlapping line of another element may be exploited. Thus Dagnall *et al.* [108] used an iodine electrodeless discharge lamp to excite bismuth at 2061.7 Å with the iodine line at 2061.6 Å and observed direct line fluorescence of bismuth at 3024.6 Å. Some other possible line pairs [312] could be

Tl 3775.72 Å excited by	Hg 3776.26 Å
Mg 2852.13	Hg 2852.42
Cr 3593.49	Hg 3593.48
Fe 2483.27	Hg 2482.72
As 2288.12	Cd 2288.02

Another possible way of dealing with scattered light is to estimate its contribution and subtract it from the signal measured. If a continuous primary source is used for excitation, the correction is simply done by scanning the wavelength region on both sides of the analytical line, and the background thus measured is subtracted. However, since large monochromator entrance and exit slits are generally used in atomic fluorescence work, care must be taken not to include adjacent fluorescence lines of other elements present in the sample, in the interval measured [64].

If a line source of excitation is used, the contribution of light-scattering is generally lower but correction is more difficult. A control sample similar in concentration and in matrix composition may be prepared and the corresponding scattered light measured. The control sample must be free from the element determined, which for analyses at trace levels may be rather difficult or even impossible.

In such cases the amount of scattered light may be estimated by using a different line for correction. The line used for measuring the scattered radiation must be either a non-fluorescence line or the sample must not contain the particular element. However, the procedure is somewhat more complicated than in atomic absorption spectrometry. It is first necessary to establish conditions of equal detector response with the "analytical" and "correcting" line, keeping the optics geometry constant.

Such a procedure has been used by Sychra and Matoušek [389] when determining nickel in oils. Scattered light measured with the Au 2428 Å line was subtracted from the fluorescence signal measured at the Ni 2320 Å line. The current of the gold hollow-cathode lamp was set to give the same detector signal as the nickel lamp when pure xylene was sprayed with the flame turned off. The same procedure was also applied by Larkins [258].

4.2.2.2. Overlapping Line and Band Interferences

The danger of overlapping line and/or band interference depends markedly on the width of the wavelength interval measured with the particular equipment. It should be borne in mind that this interval may be reduced either by reducing the width of the bandpass of the dispersive system (e.g. by reducing the entrance and exit slits) or by reducing the wavelength interval of the exciting radiation (e.g. by using line primary sources).

With a continuous primary source spectral interferences are very common. Possible interfering lines for Cu, Ag, Ga, In, Tl, Zn, Cd, Mg, Ca, Cr, Mn, Co and Ni were reported by Cresser and West [96].

Similarly, when non-dispersive systems are used, selectivity may only be achieved by using line primary sources which are free from other elements having fluorescence lines in the spectral response interval of the detector [258]. Even so, spectral interferences can occur. Larkins [249] reported iron fluorescence excited by a platinum lamp for non-dispersive atomic fluorescence spectrometry and no doubt other spectral interferences will be found in future applications.

When line sources and dispersive systems are combined the danger of spectral line interference diminishes, mainly with the width of the excitation lines, and should therefore be smaller with hollow-cathode lamps than with vapour-discharge lamps. Hollow-cathode lamps boosted with an auxiliary discharge are preferable to normal hollow-cathode lamps.

Very little is known about molecular band fluorescence interferences. From the results reported by Cresser and West [96] it follows that molecular band interferences are greatly reduced compared with flame emission spectrometry. These authors found no interference by SnO bands in indium determination and by BaO bands in thallium determination although in flame emission these interferences were very strong. Such a result could indeed be expected because the energy absorbed by the particular molecule will be spread throughout the various vibrational and rotational levels before re-emission, leaving little in the particular bandpass measured.

Where molecular band absorption interference occurs in atomic absorption spectrometry, some degree of interference is to be expected in atomic fluorescence spectrometry, because of the removal of some of the exciting radiation by absorption.

Elimination of fluorescence-line interference is not always feasible. The fluorescence signals are generally weak and a wide bandpass has to be used in most instances. Reduction in bandpass to exclude the interfering line might reduce the signal to an impractically low level. For some elements a choice of different fluorescence lines is available so that interference may be avoided by choosing another line.

In some cases it should be possible to reduce atomization of the interfering species. This may be achieved by using a flame of lower temperature which would just suffice to atomize the analyte but not the interferent. Some particular species may be bound in non-volatile compounds by the addition of suppressors. This technique is well established in flame emission spectrometry, but has not so far been applied in flame fluorescence work. Non-volatilized particles could evidently cause light-scattering and thus replace line interference by (probably) less serious light-scattering.

4.2.2.3. Thermal Emission Interferences

When non-modulated measuring systems are used, interference by thermal radiation from the flame may become relevant, particularly when working in the visible wavelength region. The obvious way to avoid this sort of interference is to modulate the exciting radiation, and use an a.c. or phase-sensitive amplifier/read-out system.

Direct-current measuring systems are mainly used in simple inexpensive equipment which is suitable for elements easily atomized in low-temperature flames and having lines in the near ultraviolet region. Often d.c. systems are applied in combination with a sampling loop where transient signals of short duration are generated.

If such equipment is used the interfering flame emission may be corrected by repeating the measurement without the exciting radiation and subtracting the signal observed from the signal with the exciting radiation on.

The elements most likely to cause this sort of interference are those which most readily emit atomic lines or molecular bands.

4.2.3. TRANSPORT INTERFERENCES

Transport interferences are identical in all flame spectrometric methods, and as such are discussed in all monographs on flame spectrometry. It is therefore sufficient to deal only briefly with them here.

They are associated with the production, transport and desolvation of aerosol particles in the nebulizer-flame sampling system. The most important factors are the rate of liquid uptake, the efficiency of nebulization and the rate of desolvation. All these affect the amount of desolvated sample passing per unit time through the horizontal cross-section of the flame at the observation height. Thus transport interferences occur even when direct-injection burners are used, as some of the aerosol droplets leaving the burner orifice are ejected sideways and do not pass through the flame at the particular height of observation. Some of the bigger droplets may even not be completely desolvated.

The main factors involved comprise viscosity of the solution aspirated, its surface tension and the vapour pressure of the solvent. Because all these are physical properties these interferences are sometimes described as "physical interferences".

4.2.4. SOLUTE VOLATILIZATION INTERFERENCES

This group of interferences is due to changes in the rate or degree of volatilization of desolvated particles and is the same in all flame spectrometric methods. Solute volatilization interferences are in general caused by formation of a new solid phase with different thermal stability in the flame in the presence of the interferent. Thus they depend mostly upon the chemical properties of the analyte and interferent and are, therefore, sometimes called "chemical interferences".

All of these interferences are well known and documented in flame emission or atomic absorption work. So far where they have been checked in atomic fluorescence spectrometry, general agreement with previously reported results has been found. Thus, for example, the effect of aluminium on magnesium was found to decrease with increasing flame temperature and in the general order: turbulent flames from direct-injection burners, premixed flames from direct-injection burners, premixed flames from spray-chamber laminar burners [62]. Matoušek [277, 278] compared the interferences of 20 elements in the determination of gold, lead, platinum and molybdenum by atomic fluorescence spectrometry and atomic absorption spectrometry in premixed air-hydrogen, argon-oxygen-hydrogen and air-acetylene flames under otherwise identical experimental conditions. He found general agreement of the resulting interferences as may be seen in Table 4.1. All the interferences observed could be eliminated by the addition of lanthanum to give a final concentration of 0.1%.

TABLE 4.1. Interferences in the determination gold and lead [5 μg/ml] by atomic fluorescence spectrometry and atomic absorption spectrometry in different premixed flames. (Interferences expressed in %[a]. Fluorescence corrected for light-scattering [277])

Interferent	Gold					Lead				
	Air – H_2		Ar – O_2 – H_2	Air – C_2H_2		Air–C_2H_2		Air – O_2 – H_2	Air – C_2H_2	
	AFS	AAS	AFS	AFS	AAS	AFS	AAS	AFS	AFS	AAS
Ca	0	0	0	+2	0	+10	+5	+5	0	0
Zn	–16	–25	0	0	0	0	0	–4	0	0
Al	–7	0	0	0	0	–3	0	–12	0	0
Cr	0	0	+8	0	0	0	0	0	0	0
Mo	–36	–30	–10	0	–5	–57	–65	–97	–91	–95
W	–8	–5	0	–2	0	–11	–15	–3	0	0
Pd	+3	+20	0	–43	–50	N.M.	N.M.	N.M.	N.M.	N.M.
Bi	–9	–10	0	–2	0	–2	0	0	0	0

[a] Differences greater than ±3% are considered as due to interference.
[b] All interferents as chlorides, except for Mo [molybdate], W [tungstate], Bi [nitrate].
N.M. Not measured.

One point worthy of discussion is the occurrence of these interferences in the low-temperature flames which are favoured in atomic fluorescence spectrometry. Smith *et al.* [364] studied interferences in the determination of calcium in an argon-hydrogen-entrained air flame on a Beckman medium-bore burner. Although they measured thermal emission, the results should not be different for fluorescence measurements. Bratzel *et al.* [62] studied atomic fluorescence flames on a Zeiss direct-injection burner and compared interferences in the argon-hydrogen-entrained air, the oxygen-hydrogen and the nitrous oxide-hydrogen flames, operated in the premixed and unpremixed modes. Interferences in the "cool flame" were prohibitively strong.

Ellis and Demers [140] studied interferences in the atomic fluorescence determination of calcium and magnesium in a hydrogen/entrained air flame on a HETCO burner, and again found considerably greater volatilization interferences than occur in hotter flames. Their findings are summarized in Table 4.2. Some of the interferences (Al^{3+}, Cr^{3+}, PO_4^{3-}, SiO_3^{2-}, NO_3^-,

TABLE 4.2. Interferences in the determination of magnesium and calcium observed in hydrogen-entrained-air flames [140].
(Magnesium 2 μg/ml, calcium 4 μg/ml, all interferents 200 μg/ml.)

Interferent	Suppresion of signal %		Interferent	Suppresion of signal %	
	Magnesium	Calcium		Magnesium	Calcium
Al	—38	—65	Pb	—88	—70
Ba	—9	0	Zn	—40	0
Ca	—18	—	HCO_3^-	—64	—75
Co	—22	—	citrate	—51	—70
Cr	—53	—75	oxalate	—34	—100
Cu	0	0	PO_4^{3-}	—77	—100
Fe	—21	—25	SiO_3^{2-}	—100	—100
K	—13	0	NO_3^-	—33	—65
La	0	0	SO_4^{2-}	—53	—75
Mg	—	—10	I^-	—53	—80
Mn	—27	0	F^-	—68	—85
Na	—10	0	S^{2-}	—70	—80

SO_4^{2-}) are well known, and take place in hotter flames. According to the authors all these interferences except that of S^{2-} with magnesium and of S^{2-} and SiO_3^{2-} with calcium could be removed by the presence of 0.1% strontium. However, in atomic fluorescence spectrometry addition of relatively large amounts of releasing agent is objectionable because of the danger of light-scatter.

4.2.5. VAPOUR PHASE INTERFERENCES

This group includes interferences due to processes in the gaseous phase, i.e. after the solute has been volatilized. The possible processes controlling the final density of free unexcited atoms are dissociation, ionization and the excitation/de-excitation equilibrium. Because in atomic fluorescence spectrometry we are concerned only with atoms excited by absorption of radiation, only the de-excitation process is relevant. It is therefore common in atomic fluorescence spectrometry to talk about quenching interferences.

Both dissociation and ionization interferences are generally exactly the same in atomic fluorescence spectrometry as in the other flame methods. The only difference is that in atomic fluorescence spectrometry quite often cool flames are used, where ionization interferences are less serious but dissociation interferences are more marked. Thus for example, Dagnall *et al.* [111] found interferences in the determination of arsenic at relatively low interferent concentrations ($10^{-2}M$) for Al, Cu, Mg, Zn, Na and NH_4^+ in the nitrogen-hydrogen-entrained air flame, and suggested that these are due to incomplete dissociation of the metal arsenite. However, the observed depression may equally well be due to incomplete volatilization of the solute.

As mentioned in Section 2.2.2., the fluorescence yield depends markedly upon the composition of the flame gases [218]. The quenching of excited atoms by effective quenchers could thus decrease the fluorescence signal. Smith [369], using experimental results reported by Jenkins [218], calculated (after making some plausible simplifying assumptions) the necessary amount of an interferent with a quenching cross-section equal to that of nitrogen, which would be necessary to decrease the fluorescence yield by one unit in the last significant figure (from 0.039 to 0.038). He found that the interferent concentration would have to be in excess of about 10^4 μg/ml, to cause a measurable change. However, this calculation was made for an air-propane flame. Under special conditions when working with flames of high fluorescence yield such as the hydrogen-oxygen-argon flame, the necessary interferent excess would be less.

4.2.6. FLAME GEOMETRY INTERFERENCES

Some concomitants may change the flame shape or volume (i.e. the flame geometry). If a limited region is observed this may cause changes in the observed signal. For atomic fluorescence spectrometry the possible changes depend upon the illumination and entrance optics used. If, as is preferable for atomic fluorescence spectrometry, complete illumination of the flame is used with relatively wide-angle entrance optics, these changes should not influence the fluorescence signal observed. However, one should be aware of the possibility that these factors may eventually cause interferences. They can be reduced by changing the optical path in the flame.

4.2.7. INTERFERENCES IN NON-FLAME ATOMIZERS

Non-flame atomizing devices have only comparatively recently been used in analytical applications, and therefore much fewer data are available for these devices than for flames. However, it is evident that spectral interferences are in general similar to those in flames, as are the possible approaches for their elimination. On the other hand interferences connected with the production of free atoms are quite different.

Let us first consider interferences in resistance-heated atomizers. The possible interferences should be qualitatively the same for all types of atomizer. In order of increasing temperature these devices include the hot wire loop [63, 64], the platinum furnace [467], the carbon rod [430] and graphite furnaces of various designs [16, 274]. With the exception of the platinum furnace shorter or longer transient fluorescence signals are produced, and the interferences observed may depend on whether the peak values or integrated fluorescence signals are measured. Changes in the volatilization rate of the analyte should influence only the peak values but not the integrated fluorescence signals.

In furnaces of limited internal volume, the peak absorption will also decrease if the amount of sample inserted is greater than the furnace volume when completely volatilized. Again this should affect the integrated signal much less than the peak signals. In Fig. 4.2 the effects of zinc and lead on the peak and integrated fluorescence signals of silver in a graphite furnace, as reported by Massmann, are reproduced [274].

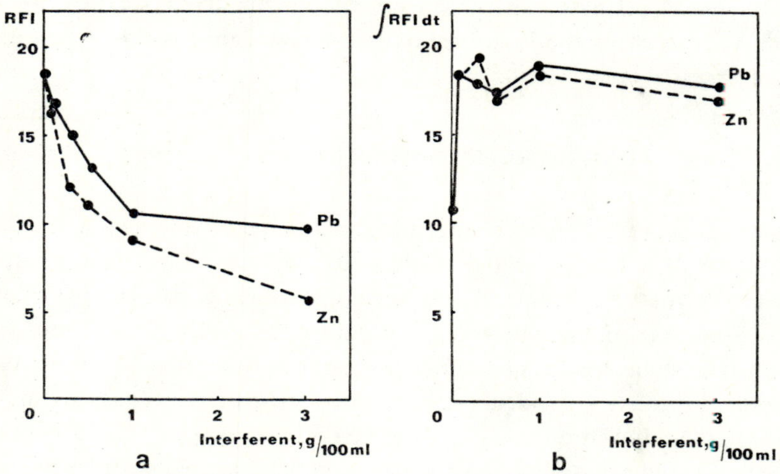

FIG. 4.2. Effect of lead and zinc on peak (a) and integrated (b) relative fluorescence intensity of silver [274].
Reprinted from Ref. [274] by permission of Pergamon Press Ltd.

With the "open" type atomizers, e.g. the hot wire loop or the carbon rod atomizer, the analyte atoms leaving the heated support surface enter the relatively cool surrounding inert gas atmosphere. Here condensation takes place and, as Aggett and West have shown [1], this may be considerably accelerated by the presence of foreign salts. Using two carbon rods they were able to show that the interference also takes place when the analyte and interferent are volatilized independently. At least for gold as analyte it seems that the gold atoms are simply physically occluded in the condensing particles and no intermetallic compounds are formed in the vapour phase. The relative depression was found to be higher with greater amounts of gold atomized and with increasing height of observation (Fig. 4.3).

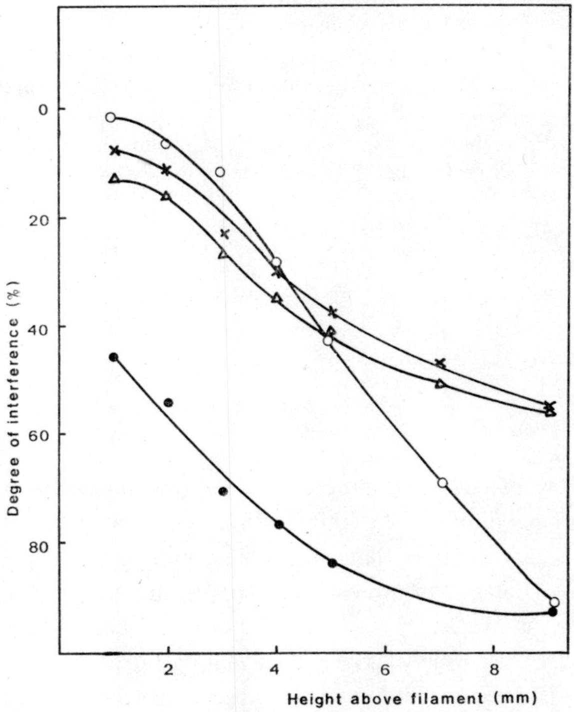

FIG. 4.3. Variation of degree of interference with height above the filament: (○) 0.1 ppm Au + 100 ppm W; (×) 0.1 ppm Au + 100 ppm Cu; (△) 0.1 ppm Au + 100 ppm Co; (●) 0.1 ppm Au + 100 ppm Cr.
Reprinted from Ref. [1] by permission of Elsevier Publishing Company

The condensation is particularly serious with the hot wire loop atomizer where the temperature difference between the support and the atmosphere is enormous. Thus Sychra and Kolihová [393] found strong depression of zinc and cadmium fluorescence in the presence of sodium, potassium, magnesium, calcium and phosphates. With the carbon rod atomizer, which has a greater heat capacity, the temperature difference is smaller and the interferences are correspondingly less. They can be further reduced by surrounding the carbon rod by an argon-hydrogen diffusion flame [16]. This is illustrated by a study of interferences in the determination of lead with a carbon rod atomizer with and without the hydrogen flame mantle (see Table 4.3).

Condensation of salts also causes light-scattering, which again can be greatly reduced by surrounding the rod by a hydrogen diffusion flame, as the results in Table 4.3 indicate.

TABLE 4.3. Interferences in the determination of lead with the carbon rod atomizer (Pb 5 ng, interferent 5 μg)

Interferent	Without hydrogen flame		With hydrogen flame	
	fluorescence peak signal	light scatter	fluorescence peak signal	light scatter
None	100	—	100	—
H_3PO_4	32	19	103	0
NaCl	110	51	92	1
KCl	87	37	87	1.5
$CaCl_2$	0	46	92	1.5
$MgCl_2$	6	11	20	0

The possibility of choice of different inert atmospheres with non-flame atomizers may be exploited for making corrections for light-scattering. This is based on the assumption that the fluorescence signal depends on the quenching properties of the atmosphere whereas light-scattering does not. First the ratio of fluorescence signals of the analyte, free from light-scattering, in two gases with widely different quenching properties is determined, e.g. $S_{Ar}/S_{N_2} = P$. If the signals including light-scattering measured in the two gases are S'_{Ar} and S'_{N_2}, it follows from the equality of light-scattering in both gases that

$$S'_{Ar} - S_{Ar} = S'_{N_2} - S_{N_2}$$

and substituting from the ratio for S_{Ar} we get after simple rearrangement

$$S_{N_2} = \frac{S'_{Ar} - S'_{N_2}}{P - 1}$$

Among non-thermal atomizers, one device which has found some application is the sputtering chamber where atomization is achieved by cathodic sputtering in the glow discharge. Some work with these devices has been done in atomic absorption [e.g. 250] but the results may be equally well applied to atomic fluorescence. As cathodic sputtering is dependent upon many factors, these devices have found only limited application to the analysis of metals and alloys, identical alloys of known composition being used as reference samples.

Chapter 5

ANALYTICAL POTENTIAL OF ATOMIC FLUORESCENCE SPECTROMETRY FOR INDIVIDUAL ELEMENTS

In this chapter the atomic fluorescence characteristics of 39 elements are described. For each element (if the necessary data are available) the atomic fluorescence spectrum, calibration curves, detection limits in various flames and interferences are discussed. Results obtained with non-flame atomizers are summarized at the end of the chapter in tabular form. Practical analytical applications are not included, because they are discussed separately in Chapter 6. For each element, all papers contributing to its determination by atomic fluorescence are also briefly reviewed. Since most of the information is presented in tabular form, a complete list of abbreviations used in the tables is given at the beginning of the chapter.

At the time of writing, no atomic fluorescence has been reported for boron, scandium, yttrium, zirconium, hafnium, niobium, tantalum, tungsten, uranium, rhenium, osmium, and the lanthanides. This is primarily due to the lack of sufficiently intense excitation sources and the unfavourable atomization characteristics of these elements in the atomizers used so far. No analytical atomic fluorescence data have been reported for the alkali metals (except for sodium).

List of Abbreviations Used in Tables in Chapter 5

Abbreviation	Full expression
Instruments:	
S.A. A 3000	Southern-Analytical Model A 3000
S.A. A 1740	Southern-Analytical Model A 1740
L.I.	Laboratory instrument (i.e. instrument assembled from commercial components)
N.S.	Non-dispersive atomic fluorescence system
M.S., filter	Non-monochromator atomic fluorescence system with an interference filter

Abbreviation	Full expression
Excitation sources:	
EDL	Electrodeless discharge lamp
EDL (mod.)	Electrodeless discharge lamp (electronically modulated)
MVDL	Metal vapour discharge lamp
HCL	Hollow-cathode lamp (conventional or shielded)
HCL(p)	Hollow-cathode lamp operated in a pulsed-mode
BOHCL (S-W)	Boosted-output (high-intensity) hollow-cathode lamp (Sullivan-Walsh design)
BOHCL(L)	Boosted-output (high-intensity) hollow-cathode lamp (Löwe design)
HHCL	Hot hollow-cathode lamp
L	Laser
150 W Xe-arc	High-pressure 150-W xenon arc lamp
450 W Xe-arc	High-pressure 450-W xenon arc lamp
R.F. plasma	Radiofrequency plasma source
Hg-pen	Low-pressure mercury pen lamp
Hg-g. lamp	Mercury germicidal lamp (Phillips type OZ 4W)
Non-flame atomizers:	
GFA	Graphite furnace atomizer
MFLA	Metal filament loop atomizer
CFA	Carbon-filament atomizer
GEA	Graphite electrode atomizer
CRA	Carbon rod atomizer
Pt-F	Platinum furnace
Flames and burners:	
A-H	Air-hydrogen flame
A-Ac	Air-acetylene flame
A-P	Air-propane flame
A-T.G.	Air-town gas flame
No-H	Nitrous oxide-hydrogen flame
No-Ac	Nitrous oxide-acetylene flame
Ar-H(d)	Argon-hydrogen diffusion flame
N-H(d)	Nitrogen-hydrogen diffusion flame
O-Ar-H	Oxygen-argon-hydrogen flame
O-H	Oxygen-hydrogen flame
O-Ac	Oxygen-acetylene flame
H-e.a.	Hydrogen/entrained air flame
Ar-H-e.a.	Argon/hydrogen/entrained air flame
(p)	Premixed flame
(t)	Turbulent flame
(s)	Separated flame (shielded flame)
(p.d.i.)	Premixed flame burning on a direct-injection burner
(c)	Capillary burner

Abbreviation	Full expression
Other abbreviations :	
S : N	Signal-to-noise ratio
R.S.D.	Relative standard deviation
Σ	Sum of lines detected with a non-dispersive system
N.M.	Not measured
N.D.	Not detected
N.I.	No interference
λ_{max}	Wavelength of peak transmission of an interference filter
$\Delta\lambda_{(1/2)}$	Half-width of an interference filter

5.1. ALUMINIUM

5.1.1. GENERAL CHARACTERISTICS AND FLUORESCENCE SPECTRUM

Early reports [152] of aluminium fluorescence in the air-hydrogen flame were later shown to be attributable to scatter. Dagnall *et al.* [114] reported aluminium fluorescence in a nitrous oxide-hydrogen flame which might also be due to scatter rather than fluorescence, because of the low population of aluminium atoms in this flame. Manning and Heneage [265] were able to detect weak fluorescence at 3961.5 Å in a nitrous oxide-acetylene flame, using a 150-W xenon arc excitation source. The intense thermal emission, and the subsequent photomultiplier shot noise, limit the analytical applicability under these conditions. Recently several authors have reported analytically useful fluorescence for aluminium in the separated nitrous oxide-acetylene flame, using an electrodeless discharge lamp [118, 190, 433] or a high-intensity hollow-cathode lamp [259] as excitation source. A recent study employing laser excitation yielded a detection limit of 0.03 µg/ml [176]. Details of the fluorescence spectrum excited by a high-intensity hollow-cathode lamp of the Löwe design are given in Table 5.1.

Atomic absorption measurements indicated that the $^2P_{3/2}$ level (0.014 eV above the ground-state) is appreciably thermally populated, and therefore fluorescence lines corresponding to transitions both to the ground-state and to the $^2P_{3/2}$ state are at least partially due to resonance fluorescence. Direct-line fluorescence probably contributes to the resonance fluorescence in the case of fluorescence transitions having the $^2P_{3/2}$ level as a lower term. For example, the strongest aluminium fluorescence line at 3961.5 Å is probably a mixture of resonance fluorescence and direct-line fluorescence.

TABLE 5.1. Atomic fluorescence spectrum of aluminium in the nitrogen-shielded (separated) nitrous oxide-acetylene flame [259]

Wavelength (Å)	Energy levels (eV)	Transition	Relative source[a, b] emission intensity	Relative fluorescence[a, b] intensity
2367.1	0—5.235	$^2P^0_{1/2} - 5d\,^2D_{3/2}$	2	0.03
2373.1[c] 2373.4[c]	0.014—5.235	$^2P^0_{3/2} - 5d\,^2D_{5/2}$ $^2P^0_{3/2} - 5d\,^2D_{3/2}$	4	0.05
2568.0	0—4.827	$^2P^0_{1/2} - 4d\,^2D_{3/2}$	3	0.05
2575.1[c] 2575.4[c]	0.014—4.827	$^2P^0_{3/2} - 4d\,^2D_{5/2}$ $^2P^0_{3/2} - 4d\,^2D_{3/2}$	6	0.05
3082.2	0—4.020	$^2P^0_{1/2} - 3d\,^2D_{3/2}$	56	12
3092.7[c] 3092.8[c]	0.014—4.020	$^2P^0_{3/2} - 3d\,^2D_{5/2}$ $^2P^0_{3/2} - 3d\,^2D_{3/2}$	100	24
3944.0	0—3.143	$^2P^0_{1/2} - {}^2S_{1/2}$	52	50
3961.5	0.014—3.143	$^2P^0_{3/2} - {}^2S_{1/2}$	78	100

[a] Spectral band-width 5 Å.
[b] Values uncorrected for detector spectral response.
[c] Lines unresolved both in emission and fluorescence spectrum.

5.1.2. EXPERIMENTAL CONDITIONS AND DETECTION LIMITS

Papers concerned with the atomic fluorescence flame spectrometric determination of aluminium are summarized in Table 5.2. The 3961.5 Å line gives the lowest detection limit: with this line, the upper linearity limit of the calibration curve is 250 μg/ml. The most intense fluorescence signal is observed in the red interconal zone of the nitrous oxide-acetylene flame, immediately above the primary reaction zone [118]. It is not surprising that the limit of detection with the non-dispersive instrument is as high as 35 μg/ml, since the R-166 photomultiplier has relatively poor response in the 3090 Å region.

5.1.3. INTERFERENCES

Dagnall *et al.* [118] found no effect from 100-fold amounts of As, Cd, Co, Cr, Cu, Fe(III), K, Mg, Mn, Mo, Na, Ni, Pb, Si, Ta, Th, V(V), W, Zn, Cl^-, NO_3^-, and PO_4^{3-} on the fluorescence signal from aluminium at the 20 μg/ml level.

TABLE 5.2. Papers contributing to the atomic fluorescence flame spectrometry of aluminium

Ref.	Spectrometer Detector Band-width (Å)	Excitation source	Flame (burner)	Wavelength (Å)	Detection limit (μg/ml) S : N = 2
[190]	Beckman 440 RCA 1P-28	450 W Xe-arc	No-Ac(p, s) Autolam	3092.7, 3092.8 [a]	4.0
		EDL	No-Ac(p, s) Autolam	3092.7, 3092.8 [a]	3.0
[118, 433]	Techtron AA4 HTV R-106 10	EDL	No-Ac(p, s) No-Ac(p)	3961.5 3961.5	0.1[b,c] 0.2[b]
[143, 145, 146]	Technicon AFS 6 filter, $\lambda_{max} = 4030$, $\Delta\lambda_{(1/2)} = 180$	HCL(p)	No-Ac(p)	3944.0, 3961.5 [a]	0.3 0.07[d]
[259]	Techtron AA4 HTV R 106 10	BOHCL(L)	No-Ac(p, s)	3961.5 3944.0 3092.7, 3092.8 [a] 3082.2	1.5 2.5 3.5 5.5
	HTV R 106 N.S.	BOHCL(L)	No-Ac(p, s)	3082.2, 3092.7, 3092.8 [a]	35
	RCA 931A filter, $\lambda_{max} = 3130$ $\Delta\lambda_{(1/2)} = 220$	BOHCL(L)	No-Ac(p, s)	3082.2, 3092.7, 3092.8 [a]	45
[176]	L.I. RCA 1P-28 14.4	L	No-Ac(p, c)	3961.5	0.03

[a] Unresolved lines.
[b] S : N = 1.
[c] R.S.D. 1.15% at the 100 μg/ml level.
[d] Reference [145, 146]; linear range 0 – 50 μg/ml.
[e] Without monochromator.

Abbreviations: See p. 187.

5.2. ANTIMONY

5.2.1. GENERAL CHARACTERISTICS AND FLUORESCENCE SPECTRUM

The basic atomic fluorescence characteristics and the determination of antimony were described by Dagnall, Thompson, and West [107], who excited atomic fluorescence in an air-propane flame with an antimony microwave-excited electrodeless discharge lamp source. Antimony hollow-cathode lamps give a relatively low radiation output at the resonance wavelengths, but some recent results have shown that good fluorescence detection limits can be expected even with this type of excitation source [143, 245, 258, 265]. Results from our laboratory indicate [248] that antimony fluorescence may be easily generated in all types of low-temperature flames with a boosted-output hollow-cathode lamp source. The use of continuum sources to excite antimony fluorescence is practically without analytical utility [65, 265] at present.

TABLE 5.3. Atomic fluorescence spectrum of antimony [248]

Wavelength (Å)	Energy levels (eV)	Transition	Relative source[a] emission intensity	Relative[a, b] fluorescence intensity
2068.3	0—5.992	$^4S^0_{3/2} - ^4P_{5/2}$	52	62
2175.8	0—5.696	$^4S^0_{3/2} - ^4P_{3/2}$	63	100
2311.5	0—5.362	$^4S^0_{3/2} - ^4P_{1/2}$	87	60
2528.5	1.222—6.123	$^2D^0_{5/2} - ^2P_{3/2}$	90	2
2598.05[c]	1.055—5.826	$^2D^0_{3/2} - ^2P_{1/2}$	100	19
2598.09[c]	1.222—5.992	$^2D^0_{5/2} - ^4P_{5/2}$		
2670.6	1.055—5.696	$^2D^0_{3/2} - ^4P_{3/2}$	8	4
2770.0	1.222—5.696	$^2D^0_{5/2} - ^4P_{3/2}$	18	3
2877.9	1.055—5.362	$^2D^0_{3/2} - ^4P_{1/2}$	24	8

[a] Excited with BOHCL (S-W); values corrected for detector spectral response.
[b] In an A-H flame.
[c] Lines unresolved both in emission and fluorescence spectrum.

The atomic fluorescence spectrum of antimony consists of nine lines (Table 5.3). As expected from atomic absorption measurements, the resonance lines at 2068.3, 2175.8 and 2311.5 Å give the strongest fluorescence signals due to resonance fluorescence. The fluorescence signals due to direct-line fluorescence at 2670.6, 2770.0 and 2877.9 are very weak. The $^2P_{1/2}$ state lies only 0.13 eV above the $^4P_{3/2}$ state (see Table 5.3) and as the $^4S_{3/2} - {}^4P_{3/2}$ transition line gives the greatest absorption sensitivity, the $^2P_{1/2}$ state should be appreciably thermally populated resulting in fluorescence emission at 2598.05 Å. This is not resolved from the direct-line fluorescence at 2598.09 Å, however. The atomic fluorescence at 2528.5 Å, on the other hand, involves a purely thermally-assisted fluorescence process. Radiative excitation at 2068.3 Å is followed by thermal excitation to the $^2P_{3/2}$ level (which lies only 0.13 eV above the $^4P_{3/2}$ level). These mechanisms were not confirmed experimentally, but seem to be very probable because the relative fluorescence emission intensity at 2528.5 and 2598.05 Å increases with increasing flame temperature. The disagreement in the relative fluorescence intensities of antimony lines listed in Table 5.3 with those reported by Dagnall *et al.* is caused by the different relative emission intensities of the high-intensity hollow-cathode lamp and the electrodeless discharge lamp sources used.

5.2.2. EXPERIMENTAL CONDITIONS AND DETECTION LIMITS

All resonance lines yield comparable detection limits (Table 5.4). The 2175.8 and 2311.5 Å lines are preferred because of the better spectral response of photomultipliers at these wavelengths. Slightly better detection limits are obtained in argon-hydrogen diffusion and argon-oxygen-hydrogen flames, as expected, owing to lower quenching and a more favourable signal-to-noise ratio compared with that in other conventional premixed flames [248]. To avoid possible interferences and light-scattering, use of a separated air-acetylene flame is recommended. Linear calibration curves are obtained for aqueous solutions of antimony over the range *ca.* 0.1 – 100 μg/ml, at the resonance lines. The linearity of the calibration curves for some lines corresponding to direct-line fluorescence extends up to *ca.* 160 μg/ml. Very good detection limits have been reported with a non-dispersive fluorescence system [258].

5.2.3. INTERFERENCES

No effect was found from 100-fold molar ratios of Cd, Co(II), Cu(II), Fe(II), Hg(II), K, Mg, Mn(II), Mo(VI), Na, NH_4^+, Pb, Zn, AsO_4^{3-}, Cl^-, NO_3^-, PO_4^{3-}, and SO_4^{2-} on the fluorescence signal of antimony at 2775.6 Å in an air-propane flame [107]. Thompson and Wildy [399] reported that, in an air-acetylene flame, a 100-fold amount of aluminium raised the fluorescence signal of antimony by about 12%, while the same amounts of sodium and calcium reduced the signal by about 10%.

TABLE 5.4. Basic papers contributing to the atomic fluorescence flame spectrometry of antimony

Ref.	Spectrometer Detector Band-width (Å)	Excitation source	Flame (burner)	Wavelength (Å)	Detection limit (μg/ml) S:N = 2
[107]	Unicam SP 900A EMI 9601 B 9	EDL	A-P(p)	2175.8	0.05[a]
[483]	L.I. RCA IP-28 32	EDL	A-H(t) Zeiss	2311.5	0.4
[114]	L.I. EMI 6256 B 17	EDL	A-H(p)	2175.8	0.08[a]
[179]	Techtron AA-4 HTV R 106 10	EDL	A-Ac(p)	2175.8	0.9
[399]	S.A. A 3000 HTV R-213 30	EDL	A-Ac(p)	2311.5	0.04
[143] [145] [146]	Technicon AFS6 filter, $\lambda_{max} = 2175$ $\Delta\lambda_{(1/2)} = 225$	HCL(p)	A-Ac(p)	2175.8 2311.5	0.1[c] 0.05
[248]	Techtron AA4 HTV R 106 10	BOHCL(S-W)	Ar-H(p)	2068.3 2175.8 2311.5 2598.05 } d 2598.09 }	0.1 0.03 0.1 0.5
			A-Ac(p, s)	2068.3 2175.8 2311.5 2598.05 } d 2598.09 }	0.2 0.15 0.5 2.0
[268]	Techtron AA-4 HTV R 106 10	BOHCL(L)	A-Ac(p, s)	2175.8	0.6
	Techtron AA-4[e] HTV R 106 N.S.	BOHCL(L)	A-Ac(p, s)	Σ	0.04
[65]	L.I. RCA 1P-28	150 W Xe-arc Eimac	A-H(t) Zeiss	2311.5	100

[a] S : N = 1.
[b] Dual-element As-Sb EDL.
[c] In MIBK.
[d] Unresolved lines.
[e] Without monochromator.

Abbreviations: see p. 187.

Non-resonance fluorescence lines can be used to eliminate flame and particle light-scatter but the detection limit deteriorates by about one order of magnitude.

5.3. ARSENIC

5.3.1. GENERAL CHARACTERISTICS AND FLUORESCENCE SPECTRUM

The only thorough study of the atomic fluorescence of arsenic was performed by Dagnall, Thompson, and West [111], who used an electrodeless discharge lamp (prepared from the iodide) to excite arsenic atomic vapour, and cool nitrogen-hydrogen or argon-hydrogen diffusion flames as well as normal premixed low-temperature flames as atom reservoirs. Most other workers have also used microwave-excited electrodeless discharge lamps as excitation sources because of the simplicity of their preparation, long lifetime, and intense radiation output [114, 150, 190]. Better detection limits have been obtained with single-element electrodeless discharge sources than with dual-element lamps, presumably because the power applied is shared in the latter lamps [97, 179]. Hell and Ricchio [190] reported a surprisingly good detection limit with a continuum source. Since the arsenic resonance lines which are responsible for all fluorescence transitions of arsenic lie in the far ultraviolet region, close proximity of the excitation source, flame, and monochromator entrance slit should be achieved to avoid considerable absorption of exciting radiation by the atmosphere. The use of lenses in entrance optics is equally undesirable. Detectors used should ideally be specially sensitized to have a good spectral response below 2000 Å.

The atomic fluorescence spectrum of arsenic is very complex. Spectral characteristics of all 15 lines which were found to exhibit atomic fluorescence [111] are listed in Table 5.5. In addition to resonance fluorescence at 1890.4, 1937.6 and 1972.6 Å, 12 non-resonance fluorescence signals were observed. The 2381.2, 2456.5, 2437.2, 2492.9, 3032.8 and 3119.6 Å lines are due to direct-line fluorescence, while the 2288.1, 2780.2, 2745.0, 2349.8 and 2860.4 Å lines are all examples of thermally-assisted fluorescence processes (called thermally-assisted direct-line fluorescence by the original authors [111]). The relatively high fluorescence intensity of the 2349.8 Å line is due to the low energy difference between the $^4P_{5/2}$ and the $^2P_{1/2}$ states (0.027 eV) and the favoured $^2P_{1/2} - {}^2D^0_{3/2}$ spin-allowed transition. The fluorescence signals from the $^2P_{3/2}$ state, which lies 0.21 eV above the $^4P_{5/2}$ state are much weaker than those from the $^2P_{1/2}$ state (cf. Table 5.5). These fluorescence signals (from the $^2P_{3/2}$ state) are very weak in the cool diffusion flame, but are much stronger in the hotter air-acetylene flame, as would be expected. The quenching effect of nitrogen and argon was found to be practically the same for the lines emanating from the $^4P_{5/2}$ state, whilst those from the $^4P_{3/2}$, $^4P_{1/2}$ and $^2P_{1/2}$ states are increased by *ca.* 33, 40, and 80% respectively, when argon is substituted for nitrogen (cf. Table 5.5).

TABLE 5.5. Atomic fluorescence spectrum of arsenic [111]

Wavelength (Å)	Energy levels (eV)	Transition	Relative fluorescence intensity[a, b]		
			A-Ac	N-H(d)	Ar-H(d)
1890.4	0—6.557	$^4S^0_{3/2} - {}^4P_{5/2}$	26	32	33
1937.6	0—6.398	$^4S^0_{3/2} - {}^4P_{3/2}$	40	34	46
1972.6	0—6.285	$^4S^0_{3/2} - {}^4P_{1/2}$	40	37	53
2288.1	1.353—6.770	$^2D^0_{5/2} - {}^2P_{3/2}$	58	3	5
2349.8	1.313—6.588	$^2D^0_{3/2} - {}^2P_{1/2}$	168	116	208
2381.2	1.353—6.557	$^2D^0_{5/2} - {}^4P_{5/2}$	33	52	56
2437.2	1.313—6.398	$^2D^0_{3/2} - {}^4P_{3/2}$	9	12	17
2456.5	1.353—6.398	$^2D^0_{5/2} - {}^4P_{3/2}$	45	48	67
2492.9	1.313—6.285	$^2D^0_{3/2} - {}^4P_{1/2}$	44	56	80
2745.0	2.255—6.770	$^2P^0_{1/2} - {}^2P_{3/2}$	N.M.	< 0.5	~ 0.5
2780.2	2.312—6.770	$^2P^0_{3/2} - {}^2P_{3/2}$	N.M.	0.5	1.0
2860.4	2.255—6.588	$^2P^0_{1/2} - {}^2P_{1/2}$	N.M.	34	63
2898.7	2.312—6.588	$^2P^0_{3/2} - {}^2P_{1/2}$	N.M.	5	85
3032.8	2.312—6.398	$^2P^0_{3/2} - {}^4P_{3/2}$	N.M.	11	15
3119.6	2.312—6.285	$^2P^0_{3/2} - {}^4P_{1/2}$	N.M.	N.M.	N.M.

[a] Values not corrected for detector spectral response.
[b] Relative emission intensity of EDL: 2349.8 > 2288.1 > 1972.6 > 1937.6 > 1890.4.

5.3.2. EXPERIMENTAL CONDITIONS AND DETECTION LIMITS

Papers concerned with the determination of arsenic by atomic fluorescence flame spectrometry are summarized in Table 5.6. This table indicates that all resonance lines as well as the 2349.8 Å line yield comparable detection limits in a given type of flame. Dagnall *et al.* [111] reported that the resonance lines gave the best response in a diffusion flame at low arsenic concentrations (< 200 μg/ml), but suffered serious self-absorption of emitted resonance radiation at high concentrations, which resulted in excessive curvature of calibration curves. The air-acetylene flame gave better linearity and sensitivity at high arsenic concentrations for a given

TABLE 5.6. Papers contributing to the atomic fluorescence flame spectrometry of arsenic

Ref.	Spectrometer Detector Band-width (Å)	Excitation source	Flame (burner)	Wavelength (Å)	Detection limit (μg/ml) S:N = 2
[111, 115]	Unicam SP900A EMI 9601 B 6	EDL	N-H(p)	1890.4 1937.6 1972.6 2349.8	0.20[a] 0.25[a] 0.50[a] 0.15[a]
			A-Ac(p)	1890.4 1937.6 1972.6 2349.8	1.0[a] 1.0[a] 2.0[a] 2.0[a]
[114]	L.I. EMI 6256 B 17	EDL	Ar-H(p)	1890.4 1937.6	0.60[a] 0.40[a]
[399]	S.A. A 3000 60 HTV R-213	EDL(mod.)	A-Ac(p)	1937.6	0.2
[190]	Beckman 440 RCA 1P-28	EDL	Ar-H(p)	1972.6	0.50
		450 W Xe-arc	Ar-H(p)	1972.6	2.0
		EDL[b]	A-H(p)[c]	1937.6	0.15
[179]	Techtron AA-4 HTV R-106 10	EDL[d]	A-Ac(p)	2349.8	25[a]
[97]	S.A. A 1740 HTV R-213	EDL[e]	O-H(t) (Beck- man)	1937.6	25
[264]	Perkin—Elmer 303 EMI 9592 B 20	BOHCL (S-W)	O-Ar-H(p)	1937.6	1.0
[145, 146]	Technicon AFS-6 filter	EDL	A-Ac(p)	1972.6	0.07
[32]	L.I. HTV R-166 N.S.	EDL(mod.)	A-Ac(p)	Σ	0.8
[416]	L.I. HTV R-166 N.S.	EDL	Ar-H(p, s) Alke- made burner	Σ	0.3

[a] S : N = 1
[b] EDL source positioned horizontally along the flame.
[c] Long flame.
[d] Dual-element As-Sb EDL.
Abbreviations: see p. 187.

line than the diffusion flame. When measurements are made at a non-resonance fluorescence line, much wider linear ranges for the calibration curves should be obtained, as self-absorption cannot occur. This was found to be true for fluorescence emission at 2349.8 Å. On this basis, and because of possible chemical interferences and scatter, measurement at 2349.8 Å (with a filter interposed between the excitation source and the flame, to transmit only radiation below *ca.* 2000 Å) in a separated air-acetylene flame should offer a sensitive, interference and scatter-free method for the determination of arsenic. Measurement with a non-dispersive system should lead to further improvement in detection limits by avoiding light losses which occur in the monochromator and by utilizing all fluorescence lines [32, 416].

5.3.3. INTERFERENCES

In diffusion flames, interference was observed from 10^{-2}M concentrations of various cations in the analysis of 10^{-3}M solution of arsenic. Up to 10% decrease in the fluorescence intensity was found with Al, Cu(II), Mg, Zn, Na and NH_4^+ (in decreasing order of interference). Anions such as Cl^-, NO_3^-, PO_4^{3-} and SO_4^{2-} had little effect. The decreases were considered to be due to incomplete dissociation of the metal arsenites in the relatively cool diffusion flame [111]. Thompson and Wildy [399] found a slight increase (*ca.* 10%) from a 100-fold weight ratio of aluminium in an air-acetylene flame, probably due to light-scattering.

5.4. BARIUM

Practically no information exists on the determination of barium by atomic fluorescence spectrometry. This fact is mainly due to the lack of the sufficiently intense excitation source which is necessary to get a favourable signal-to-noise ratio at the main resonance wavelength at 5535.5 Å, where the nitrous oxide-acetylene flame exhibits intense background emission. Demers [142] observed very weak atomic fluorescence at the 5535.5 Å line excited with a 450-W xenon arc source in the turbulent hydrogen-entrained air flame. Denton and Malmstadt [148] reported atomic fluorescence of barium at 5535.5 Å, excited by a laser-pumped tunable organic dye laser. Ultrasonic nebulization of the analysed solution was used to reduce scatter. Linear calibration curves suitable for quantitative analysis were obtained.

A detection limit of 0.2 μg/ml has been obtained by using a Technicon AFS-6 instrument with a pulsed hollow-cathode lamp source and an air-acetylene flame [145, 146].

5.5. BERYLLIUM

5.5.1. GENERAL CHARACTERISTICS AND FLUORESCENCE SPECTRUM

The low background emission around 2350 Å and the improved reducing properties of the separated nitrous oxide-acetylene flame allow a sensitive flame atomic fluorescence determination of this element by using boosted-output hollow-cathode lamps of the Sullivan-Walsh [203, 345] or Löwe [259] designs or electrodeless discharge lamps [62, 191, 226] as excitation sources. The fluorescence intensity was found to be critically dependent on the flame stoichiometry, the optimum intensity being obtained in a slightly fuel-rich flame. Robinson and Hsu [203, 345] observed beryllium fluorescence in a fuel-rich oxy-acetylene flame, but not in oxy-hydrogen, air-hydrogen, and air-acetylene flames. It is, therefore, not surprising that the fluorescence of this element reported in a turbulent air-hydrogen flame [483] was later shown [484] to be the result of scatter. Recently, Larkins and Willis [269] detected beryllium by atomic fluorescence in the fuel-rich nitrogen-shielded air-acetylene flame with fairly good sensitivity even though the metal is scarcely detectable by atomic absorption under these conditions.

The atomic fluorescence spectrum of beryllium in the nitrous oxide-acetylene flame exhibits only one line, that at 2348.6 Å, corresponding to the ${}^1P_1^0 - {}^1S_0$ resonance fluorescence transition.

5.5.2. EXPERIMENTAL CONDITIONS AND DETECTION LIMITS

Instrumental systems used for the flame atomic fluorescence determination of beryllium and the detection limits obtained are summarized in Table 5.7.

The shape of the analytical curve published by Larkins and Willis [259] (which gave an upper linear limit of *ca.* 5 μg/ml) differs from that published by Robinson and Hsu [345] in that the curve does not start to rise again at concentrations above 100 μg/ml.

5.5.3. INTERFERENCES

Hingle *et al.* [191] carried out an extensive interference study and concluded that atomic fluorescence was more selective than atomic emission for the determination of beryllium. They observed interference only from

TABLE 5.7. Papers contributing to the atomic fluorescence flame spectrometry of beryllium at 2348.6 Å

Ref.	Spectrometer Detector Band-width (Å)	Excitation source	Flame (burner)	Detection limit (μg/ml) S : N = 2
[191, 226]	Techtron AA-4 HTV R-213 10	EDL	No-Ac(p, s)	0.01[a, b]
[203, 345]	Techtron AA-3 HTV R-213 10	BOHCL (S-W)	No-Ac(p)	0.5
			O-Ac(t) Beckman 4020	10
[62]	L.I. RCA 1P-28	EDL	No-Ac (p, d.i.) Zeiss	0.04[a]
[259]	Techtron AA-4 HTV R-106 10	BOHCL(L)	No-Ac(p, s)	0.02
			A-Ac(p, s)	50
	Techtron AA-4[c] HTV R-166 N.S.	BOHCL(L)	No-Ac(p, s)	0.03
			A-Ac(p, s)	2
			A-H(p)	10
	Techtron AA-4[c] HTV R-166 with Zeiss filter (502280) $\lambda_{max} = 2050$ $\Delta\lambda_{(1/2)} = 350$	BOHCL(L)	No-Ac(p, s) A-Ac(p, s) A-H(p) No-H(p)	0.008 1 5 20

[a] S : N = 1.
[b] R.S.D. 3% at the 0.5 μg/ml level.
[c] Without monochromator.
Abbreviations: See p. 187.

1000-fold ratios (but not from 100-fold ratios) of aluminium, calcium, and molybdenum(VI). These results agree with those of Robinson and Hsu [345] who also concluded that, in general, interferences were greater in the oxy-acetylene flame than in the nitrous oxide-acetylene flame, and showed that interferences from anions in the oxygen-acetylene flame could be eliminated by adding EDTA to the solution analysed. The same authors also investigated the relative fluorescence intensities of beryllium in a wide range of different organic solvents [345].

5.6. BISMUTH

5.6.1. GENERAL CHARACTERISTICS AND FLUORESCENCE SPECTRUM

Thorough studies of the atomic fluorescence of bismuth were made by Dagnall, Thompson, and West [108] and later by Hobbs, Kirkbright, and West [193]. These authors showed that bismuth fluorescence could be easily excited either with an iodine electrodeless discharge lamp (owing to spectral overlap of the bismuth resonance line at 2061.7 Å and the non-resonance iodine line at 2061.6 Å) or with a bismuth electrodeless discharge lamp. Several other excitation sources have also been used (see Table 5.9).

In addition to two strong lines at which analytical measurements may be carried out, there are a number of weaker lines of minor analytical importance but of considerable theoretical interest in the complex bismuth spectrum. All bismuth fluorescence lines excited with an iodine electrodeless discharge lamp by overlap of the iodine line at 2061.6 Å and the 2061.7 Å ($^4S^0_{3/2}$ – $^4P_{5/2}$) bismuth resonance line are shown in Table 5.8, which indicates that only a small fraction of radiation absorbed at 2061.7 Å is re-emitted as resonance fluorescence. The most intense signals are observed at 3024.6 and 3067.7 Å, as expected, because of the favourable transition probabilities at these wavelengths. The fluorescence at 3024.6 Å is direct-line fluorescence, while the fluorescence signal at 3067.7 Å must be stepwise-line fluorescence, i.e. due to deactivation of the excited $^4P_{5/2}$ state by radiationless transition to the $7s^4P_{1/2}$ excited state (cf. Table 5.8). The high fluorescence intensity at 3067.7 Å arises because the $^4P_{1/2}$ excited state can be populated by radiationless transfer from other excited states. Additional weak fluorescence signals are due to direct-line fluorescence (at 2696.7 Å), stepwise-line fluorescence (at 2228.3, 2230.6, 2276.6, 2780.5, 2898.0, 2989.0, 2993.3, 3397.2, 3510.8, 4121.5 and 4722.2 Å) and thermally-assisted stepwise-line fluorescence (at 2627.9 and 2938.3 Å). In this last instance there must be a radiationless promotion from the $^4P_{5/2}$ to the $^2P_{3/2}$ state (energy difference of only 0.12 eV).

Excitation of bismuth fluorescence with a bismuth electrodeless discharge lamp which emits predominantly the 3067.7 Å resonance line resulted practically in only resonance fluorescence at this wavelength [193].

TABLE 5.8. Atomic fluorescence spectrum of bismuth [108][a]

Wavelength (Å)	Energy levels (eV)	Transition	Relative fluorescence intensity[b]	
			N-H	Ar-H
2061.7	0—6.012	${}^4S^0_{3/2} - {}^4P_{5/2}$	6.5	12
2228.3	0—5.562	${}^4S^0_{3/2} - {}^4P_{3/2}$		
2230.6	0—5.556	${}^4S^0_{3/2} - {}^2D_{5/2}$		
2276.6	0—5.444	${}^4S^0_{3/2} - {}^2D_{3/2}$		
2627.9	1.416—6.132	${}^2D^0_{3/2} - {}^2P_{3/2}$		
2696.7	1.416—6.012	${}^2D^0_{3/2} - {}^4P_{5/2}$	3	6
2780.5	1.416—5.873	${}^2D^0_{3/2} - 8s\,{}^4P_{1/2}$		
2898.0	1.416—5.692	${}^2D^0_{3/2} - {}^2P_{1/2}$		
2938.3	1.914—6.132	${}^2D^0_{5/2} - {}^2P_{3/2}$		
2989.0	1.416—5.562	${}^2D^0_{3/2} - {}^4P_{3/2}$		
2993.3	1.416—5.556	${}^2D^0_{3/2} - {}^2D_{5/2}$		
3024.6	1.914—6.012	${}^2D^0_{5/2} - {}^4P_{5/2}$	50	100
3067.7	0—4.040	${}^4S^0_{3/2} - 7s\,{}^4P_{1/2}$	28	54
3397.2	1.914—5.562	${}^2D^0_{5/2} - {}^4P_{3/2}$		
3510.8	1.914—5.444	${}^2D^0_{5/2} - {}^2D_{3/2}$		
4121.5	2.685—5.692	${}^2P^0_{1/2} - {}^2P_{1/2}$		
4722.2	1.416—4.040	${}^2D^0_{3/2} - 7s\,{}^4P_{1/2}$		

[a] Excited with an iodine EDL (the only line observed with this source is at 2061.63 Å).
[b] Relative fluorescence intensities of other lines are much less than 5 relative to 100 for Ar-H(d) flame at 3024.6 Å.

It is also apparent from Table 5.8 that the argon-containing diffusion flame gave twice the fluorescence signal obtained from the nitrogen-containing flame for lines originating directly from the ${}^4P_{5/2}$ state. Although the absolute intensities of these lines varied in the two flames examined, their ratio was constant. This is to be expected, because quenching of an excited state should not alter the ratio of intensities of lines emitted directly from that state.

5.6.2. EXPERIMENTAL CONDITIONS AND DETECTION LIMITS

Table 5.9 indicates that the most important analytical lines are at 3024.6 and 3067.7 Å. In the earlier study by Dagnall *et al.* [108], an unmodulated source was employed. The inability, with this system, to discriminate against the high flame background from OH species, and the relatively low

TABLE 5.9. Basic papers contributing to the atomic fluorescence flame spectrometry of bismuth

Ref.	Spectrometer Detector Band-width /Å)	Excitation source	Flame (burner)	Wave-length (Å)	Detection limit (μg/ml) S : N = 2
[411, 412]	L.I. RCA 1P-28	150 W Xe-arc	O-H(t) Beckman 4020	3067.7	2.0
[108, 115]	Unicam 900A EMI 9601B *ca.* 14	EDL(iodine)	Ar-H(p, d)	3024.6	0.05[a] (0.025)[a,b]
[58, 267, 268, 483]	L.I. RCA 1P-28 *ca.* 8	EDL(bismuth)	A-H(t) Beckman 4020	3067.7	0.7
[152]	L.I. RCA 1P-28	HHCL	A-H(t) Beckman 4020	3067.7	0.1
[265]	Perkin—Elmer 303 EMI 9592B 20	150 W Xe-arc	A-H(p)	3067.7	200
		HCL	A-H(p)	3067.7	10
[190]	Beckman 440 RCA 1P-28	150 W Xe-arc	A-H(p)	3067.7	0.9
			Ar-H(p, d)	3067.7	0.3
		450 W Xe-arc	A-H(p)	3067.7	0.5
			Ar-H(p, d)	3067.7	0.17
		EDL(bismuth)	Ar-H(p, d)	3067.7	0.19
		EDL(bismuth)[c]	A-H(p)[d]	3067.7	0.004 7
[399]	S.A. A 3000 HTV R-213 60	EDL(iodine) (mod.)	A-Ac(p)	2061.7	0.4
				3024.6	1.0[f]
[97]	S.A. A 1740 HTV R-213	EDL[g]	O-H(t)	3024.6	10
[193]	Unicam SP 900A EMI 9601B	EDL (bismuth) (mod.)	A-Ac(p, s)	3067.7	0.3[a]
				2230.6	0.6[a]
		EDL (iodine) (mod.)	A-Ac(p, s)	3024.6	0.09[a]
[258]	Techtron AA-4 HTV R-106 10	HCL	A-Ac(p, s)	3067.7	5.0
	Techtron AA-4[h] HTV R 166 N.S.	HCL	A-Ac(p, s)	Σ	4.0
		BOHCL(S-W)	A-Ac(p, s)	Σ	0.25

[a] S : N = 1.
[b] Cf. [115].
[c] EDL source positioned horizontally along the flame.
[d] Long flame (Autolam burner, hot mode).
[e] Excited with 2061.6Å iodine line from arsenic EDL.
[f] Band-width 12 Å.
[g] Dual-element Sb-Bi source.
[h] Without monochromator.
Abbreviations: see p. 187.

intensity and stability of the available bismuth electrodeless discharge lamp, resulted in the choice of a cool flame and measurement of direct-line fluorescence at 3024.6 Å excited with the iodine 2061.6 Å line rather than measurement of the resonance fluorescence excited at the bismuth resonance line at 3067.7 Å by a bismuth electrodeless discharge lamp source. The use of an electronically modulated bismuth electrodeless discharge lamp source and a separated air-acetylene flame permitted the determination of bismuth by measurement of the resonance fluorescence at 3067.7 Å [193]. Inert gas shielding of the air-acetylene flame was shown to result in improvement in detection limit for bismuth via suppression of flame background from OH species even when a modulated source and an a.c. detector were used, and the use of this flame minimizes chemical interferences. Scatter of source radiation may be avoided by measurement of the direct-line fluorescence at 3024.6 Å excited with an electronically modulated iodine electrodeless discharge lamp source. Linear calibration curves were obtained over the range *ca.* 0.1 – 200 μg/ml of bismuth at this wavelength.

The best detection limit reported was obtained by Hell and Ricchio [190], using a bismuth electrodeless discharge lamp source positioned horizontally and parallel to a long premixed air-hydrogen flame burning on an Autolam burner in its hot-operation mode. Relatively poor detection limits were reported for the non-dispersive fluorescence system [258]. This may be explained by the fact that the strongest bismuth fluorescence lines lie virtually outside the sensitivity range of the R-166 photomultiplier.

5.6.3. INTERFERENCES

Hundred-fold molar ratios of As(III), Cd, Fe(II), Fe(III), K, Na, NH_4^+, Sb(III), Zn, Cl^-, NO_3^-, PO_4^{3-}, and SO_4^{2-} did not produce a variation of more than 5% in the fluorescence signal from 10^{-4}M bismuth nitrate. Greater than 100-fold molar ratio of sulphuric acid produced a positive interference because of the strong S_2 emission in the diffusion flame (a d.c. detection system was used). Aluminium and magnesium at 100-fold molar ratio to Bi produced, as expected, significant decreases in the fluorescence signal (40 and 20%, respectively). This interference may be attributed to the formation of oxide particles which are not efficiently dissociated in the relatively cool diffusion flame [108]. These interferences and many other matrix effects may be readily overcome by using hotter flames. Of the ions studied by Hobbs *et al.* [193], only calcium and zirconium interfered (16 and 19% decrease, respectively) at 500-fold weight ratio to bismuth in the shielded air-acetylene flame. Aluminium and magnesium, which were reported to suppress bismuth fluorescence in the argon-hydrogen diffusion flame, caused only slight positive enhancement (*ca.* 5%) at 500-fold ratio under the conditions employed in that study.

5.7. CADMIUM

5.7.1. GENERAL CHARACTERISTICS AND FLUORESCENCE SPECTRUM

Cadmium was one of the first [444, 445, 446] elements determined by atomic fluorescence spectrometry because it is readily vaporized in all types of analytical flames, and there are several types of sufficiently intense excitation source for this element. The wavelength of its main resonance line is also favourable. Cadmium appears in almost every general study on atomic fluorescence and is often used to demonstrate the advantages of this technique over other flame techniques. The first systematic survey of experimental conditions for the determination of cadmium was completed by Dagnall *et al.* [102]; since then, more than sixty papers have appeared in which at least detection limits and optimum analytical conditions are presented.

Various excitation sources have been used successfully. Vapour discharge lamps yield high fluorescence signals despite considerable self-reversal [46, 102, 226, 253, 266]. An optimum operating current of 500 – 600 mA is usually recommended. Boosted-output hollow-cathode lamps exhibit approximately the same intensity as these sources, and have also been used by several authors [46, 244, 245]. Early fluorescence studies with cadmium showed the usefulness of electrodeless discharge lamps (prepared from the metal) as excitation sources [58, 267, 268, 483]. Since then, a great deal of attention has been paid to the preparation and resulting spectral characteristics of sources of this type [91, 97, 114, 126, 165, 270, 356, 358, 360, 398, 399]. Silvester and McCarthy [358, 360] reported a thorough study of the effect of filler-gas pressure, weight of cadmium introduced into the discharge tube and applied microwave power. Their results indicated that the optimum conditions seemed to be about 1 torr of helium as filler-gas with about 5 mg or less of cadmium metal and the highest power obtainable up to about 200 W. Cooke *et al.* [91] have shown that the use of the minimal microgram amounts of material is preferable because this largely obviates the problem of instability caused by movement of material within the lamp during operation, and maintains a totally gaseous discharge. Recent cadmium electrodeless discharge lamps (both single and multielement) exhibit a sufficiently good short-term and long-term stability and high intensity, and have lifetimes longer than 200 hours [91, 97, 270]. Xenon arc lamp sources are much less intense at 2288.0 Å than the line sources mentioned above [46, 103, 190, 411] (see also Section 3.2).

Cadmium fluorescence has been investigated in practically all types of flame employed in flame spectrometry. Although hydrogen flames exhibit very low background emission in the region of 2300 Å, many authors have found that the separated premixed air-acetylene flame gives the optimum signal-to-noise ratio, and allows the determination to be made free from most interferences [179, 244, 245, 258, 398, 399]. The enhancement of the fluorescence signal at 2288.0 Å due to the use of an argon-oxygen-hydrogen flame instead of an air-hydrogen flame was found to result in a more favourable signal-to-noise ratio in the former flame.

The atomic fluorescence spectrum of cadmium consists of two lines. Resonance fluorescence at the 2288.0 Å line corresponding to the $^1S_0 - {}^1P_1$ resonance transition is almost always measured. The resonance line at 3261.1 Å has a very low oscillator strength and the fluorescence signal at this wavelength is about three orders of magnitude weaker than that at 2288.0 Å. Experiments and calculations carried out in our laboratory [244, 245] for various premixed flames indicate that the fluorescence signal at 3261.1 Å is predominantly due to resonance fluorescence rather than to stepwise-line fluorescence (after excitation at 2288.0 Å). Only in the separated air-acetylene flame is the experimental ratio of fluorescence intensities at 2288.0 and 3261.1 Å lower than the theoretical one owing to the contribution of stepwise-line fluorescence. A more pronounced contribution from stepwise-line fluorescence to the fluorescence emission at 3261.1 Å has been observed in some non-flame cells [3, 245].

5.7.2. EXPERIMENTAL CONDITIONS AND DETECTION LIMITS

The various modified atomic absorption/flame emission spectrometers and laboratory instrumental systems used for the flame atomic fluorescence determination of cadmium, and the corresponding limits of detection are listed in Table 5.10. The detection limits at 2288.0 Å range over seven decades of concentration, depending particularly on the type of excitation source, the best being reported with electrodeless discharge lamps (the value of 0.001 pg/ml is the lowest detection limit which has ever been attained in flame spectrometry) and the worst with continuum sources. Results from our laboratory indicate [244, 245] that cadmium can be detected down to

TABLE 5.10. Basic papers contributing to the atomic fluorescence flame spectrometry of cadmium

Ref.	Spectrometer Detector Band-width (Å)	Excitation source	Flame (burner)	Wave-length (Å)	Detection limit (μg/ml) S : N = 2
[446]	L.I. RCA 1P-28	MVDL(Osram)	O-H(t) Beckman	2288.0	0.1[a]
[445]	L.I. RCA 1P-28	MVDL(Osram)	O-Ac(t)	2288.0	0.05[a]
[266]	L.I. RCA 1P-28 32	MVDL(Philips)	O-H(t) Beckman	2288.0	0.000 2[a]
[102] [424]	Unicam SP 900A EMI 9601 B *ca.* 60	MVDL(Osram)	A-Ac(p)	2288.0	0.002
[103]	Unicam SP 900A EMI 9601B *ca.* 60	150 W Xe-arc	A-P(p) A-H(p)	2288.0	0.25[b]
[186]	L.I. RCA 1P-28	MVDL(Philips)	A-H(t) Beckman 4020	2288.0	0.002
[58] [483]	L.I. RCA 1P-28 8	EDL	O-H(t) Zeiss	2288.0	0.000 001[b]
[267] [268]	L.I. RCA 1P-28 8	EDL	O-H(t) Beckman	2288.0	0.000 08[b]
[411] [412]	L.I. RCA 1P-28 29	150 W Xe-arc	O-H(t) Beckman 4020	2288.0	0.08[b]
[312]	Jarrell—Ash (Versat.) RCA 1P-28 64	MVDL(Philips)	A-H(t) HETCO	2288.0	0.005
[114]	L.I. EMI 6256B 17	EDL	A-H(p)	2288.0	0.000 1[b]
[192, 226, 227]	Unicam SP 900A EMI 9601B *ca.* 60	MVDL(Osram)	A-Ac(p, s)	2288.0 3261.1	0.000 5 1.0
[265]	Perkin—Elmer 303 EMI 9592B 20	HCL(Intensitron)	A-H(p)	2288.0	0.1
		150 W Xe-arc	A-H(p)	2288.0	2.0

TABLE 5.10. (continued)

Ref.	Spectrometer Detector Band-width (Å)	Excitation source	Flame (burner)	Wave-length (Å)	Detection limit (μg/ml) S : N = 2
[46]	Techtron AA-4 HTV R-106 10	EDL	A-P(p)	2288.0	0.001
		BOHCL(S-W)	A-P(p)	2288.0	0.001 5
		MVDL(Wotan)	A-P(p)	2288.0	0.002
				3261.1	300
		HCL(A.S.L.)	A-P(p)	2288.0	0.1
		150 W Xe-arc (Eimac)	A-P(p)	2288.0	1.0
[206]	Unicam SP 900A EMI 9601B	R.F. plasma	A-P(p) No-H(p)	2288.0 2288.0	0.08 3
[398] [399]	S.A. A 3000 HTV R-213 30	EDL(mod.)	A-Ac(p)	2288.0	0.000 3[c]
[179]	S.A. A 1740 HTV R-213	EDL[d]	A-Ac(p)	2288.0	10[b]
		EDL	A-Ac(p)	2288.0	0.001 7
[65]	L.I. RCA 1P-28	150 W Xe-arc Eimac	A-H(t) Zeiss	2288.0	0.08
[415]	L.I. HTV R-166 N.S.	MVDL(Philips)	A-H(t) Beckman 4020	2288.0	0.001
[190]	Beckman 440 RCA 1P-28	450 W Xe-arc	A-H(p) Autolam	2288.0	0.015
		150 W Xe-arc	A-H(p) Autolam	2288.0	0.033
[165]	L.I. HTV R-166 M.S.; filter	EDL	Ar-H-e.a.(t) HETCO	2288.0	0.000 1
[270] [430]	S.A. A 1740 HTV R-213 3.3	EDL[e]	A-P(p)	2288.0	0.000 3
[96] [433]	S.A. A 1740 HTV R-213 60	500 W Xe-arc	A-Ac(p)	2288.0	10
[97]	S.A. A 1740 HTV R-213	EDL	O-H(t) Beckman	2288.0	0.000 04[f] 0.000 01[g] 0.000 08[h]
[143] [144] [145] [146]	Technicon AFS-6 filter, $\lambda_{max} = 2175$ $\Delta\lambda_{(1/2)} = 225$	HCL(p)	A-Ac(p)	2288.0	0.03[i] 0.001[j]

TABLE 5.10. (continued)

Ref.	Spectrometer Detector Band-width (Å)	Excitation source	Flame (burner)	Wave-length (Å)	Detection limit (μg/ml) S : N = 2
[253]	L.I. RCA 1P-28 4	MVDL(Osram)	A-H(t) Beckman 4020	2288.0	0.000 1[k]
[32]	Perkin—Elmer 403 HTV R-136 20	EDL(mod.)	A-Ac(p)	2288.0	0.005
	L.I. HTV R-166 N.S.	EDL(mod.)	A-Ac(p)	2288.0	0.007
[356]	L.I. RCA 1P-28 8	EDL[l]	A-H(p, d.i.)	2288.0	0.000 002[m,n] 0.000 02
[258] [443]	Techtron AA-4 HTV R-106 10	HCL	A-Ac(p, s)	2288.0	0.02
		MVDL	A-Ac(p, s)	2288.0	0.001 5
			A-Ac(p)	2288.0	0.002
	Techtron AA-4[o] HTV R-166 N.S.	HCL	A-Ac(p, s)	2288.0	0.004
		MVDL	A-Ac(p, s)	2288.0	0.000 2[p]
			A-Ac(p)	2288.0	0.001 5
[244, 245]	Techtron AA-4 HTV R-106 10	BOHCL(S-W)	A-Ac(p, s)	2288.0	0.001
				3261.1	0.15
			A-H(p)	2288.0	0.000 4
				3261.1	0.25
			O-Ar-H(p)	2288.0	0.000 04
				3261.1	0.25
			A-Ac(p, s)	2288.0	0.000 01[r]

[a] S : N = 0.75.
[b] S : N = 1
[c] Linear range 0.001—2 μg/ml.
[d] Dual-element As-Sb source.
[e] Dual-element Cd-Zn source.
[f] Dual-element Cd-Ag source.
[g] Dual-element Cd-Pb source.
[h] Dual-element Cd-Cr source.
[i] Cf. [143, 144]
[j] Linear range 0—8 μg/ml.
[k] Photon-counting detection.
[l] With A-antenna.
[m] With aluminium ellipse.
[n] Linear calibration curve from about 0.000 01 to 1.0 μg/ml.
[o] Without monochromator.
[p] Upper linear limit 1.0 μg/ml.
[r] With the sampling cup technique (See Section 3.3), sample volume of 100 μl, i.e. absolute detection limit 1×10^{-12} g.

Abbreviations: See p. 187.

0.15 μg/ml at 3261.1 Å in a separated air-acetylene flame. However, poor linearity of the calibration curve makes this less sensitive line useless at cadmium concentrations above 10 μg/ml. A contradictory result was published by Bennett [46], who obtained a linear calibration curve from 300 up to 5000 μg/ml at 3261.1 Å with a Wotan vapour discharge lamp source operating at maximum current. This fact cannot be satisfactorily explained without detailed knowledge of the experimental conditions used in both studies, such as the method of illumination of the flame cell, the source spectral line-width, etc.

Since the main fluorescence line at 2288.0 Å lies in the region of maximum spectral response of the solar-blind photomultiplier, the use of non-dispersive systems seems to be very attractive [32, 165, 415]. Larkins [258] reported that the detection limit was ten times better when the monochromator was omitted and only an HTV R-166 photomultiplier used.

The use of the sampling cup technique [245, 393] enables the determination of cadmium to be performed with an absolute detection limit of 1 pg, using a sample volume of 100 μl. A linear calibration curve was obtained over more than three decades of concentration, and the relative standard deviation was 5% at the 10^{-10} g level.

5.7.3. INTERFERENCES

The effects of 100-fold molar ratios of 41 cations and 18 anions and of some complexing agents on 10^{-5}M cadmium were examined by Dagnall *et al.* [102]. None of the ions and compounds investigated caused variation of the fluorescence signal by more than ±5% in a premixed air-acetylene flame. Bratzel *et al.* [59] studied the influence of hydrochloric and nitric acid concentration on cadmium fluorescence in a turbulent oxy-hydrogen flame. They did not explain the unusual fact that the effect of the acids depends upon the cadmium concentration. The effect of 1000 μg/ml of Al, Cu, Li, and Sn on 10 μg/ml of cadmium in two different turbulent air-hydrogen mixtures was investigated by Goodfellow [185, 186]. In the fuel-rich flame, only tin was found to suppress the fluorescence signal (by about 10%). It is probable that compound formation is responsible. In the fuel-lean flame, the interference of tin was more pronounced, and other cations investigated also appeared to suppress the fluorescence signal by about 10%.

Bratzel, Dagnall and Winefordner compared various premixed laminar and unpremixed turbulent flames [61] as well as various types of premixed and unpremixed flames produced by using a direct-injection nebulizer-burner [62], with respect to "chemical" interferences and light-scatter affecting the atomic fluorescence of cadmium. They found that 1000-fold ratio of aluminium gave three times as much scatter in the unpremixed turbulent air-hydrogen flame (critically dependent on the flame height) as in the premixed laminar air-hydrogen flame. They also showed that similar interferences occur in both premixed and unpremixed air-hydrogen and nitrous oxide-hydrogen flames burning on a direct-injection Zeiss burner, and recommended a premixed nitrous oxide-acetylene flame of that type to overcome the effect of ions known to cause interferences in low-temperature flames.

Our experience with the sheathed premixed air-acetylene flame [245, 388] indicates that no scatter signal occurs at 2288.0 Å from 1000 μg/ml of aluminium and is in agreement with the results of Thompson and Wildy [399].

Nickel was found to affect the determination of cadmium at 2288.0 Å owing to spectral interference when atomic fluorescence was excited with a d.c. 500-W xenon arc lamp, using a wide spectral band-pass (*ca.* 60 Å). Nickel lines at 2290.0, 2311.0, 2312.3, 2314.0, 2325.8 and 2345.5 Å are responsible [96].

5.7.4. ORGANIC SOLVENTS AND EXTRACTION

Addition of 10% v/v concentrations of 8 water-miscible solvents to 10^{-5}M cadmium iodide was found to increase the fluorescence response by about 15–35%. The presence of these small amounts of organic solvents caused no appreciable change in flame conditions (premixed air-acetylene flame). Measurements of MIBK extracts of a cadmium iodide complex resulted in a scale reading *ca.* 4 times greater than that of an aqueous solution of the same concentration. A detection limit of *ca.* 10^{-9}M for the MIBK extracts at 2288.0 Å was reported with a non-optimized atomic fluorescence apparatus [102].

5.8. CALCIUM

5.8.1. GENERAL CHARACTERISTICS AND FLUORESCENCE SPECTRUM

The lack of suitable excitation sources and the relatively good atomic emission and absorption characteristics of this element appear to be the main reasons that the atomic fluorescence determination of calcium has not been satisfactorily developed. The manufacture of electrodeless discharge lamps for this element (as well as for other alkaline earth elements) presents numerous problems. The lamp walls are subject to attack by calcium metal, resulting in short lifetimes and instability of output. Special glasses and methods of construction have been suggested for the preparation of calcium and other alkaline-earth tubes having lifetimes longer than 100 hours [6]. In most early work on calcium, continuous excitation sources have therefore been used [65, 140, 142, 190, 265, 328, 433]. Their use, however, does not offer any marked advantages for calcium determination by atomic fluorescence over the other flame techniques. The recent application of pulsed hollow-cathode lamps [32, 143, 144, 145, 146] and of a laser [177] to excite atomic fluorescence seems to overcome the problem with sources.

At present, conventional low-temperature flames have been employed in all papers dealing with calcium fluorescence. This has resulted in the observation of interferences similar to those encountered in atomic emission and absorption spectrometry. Efforts to avoid these interferences by the use of a separated nitrous oxide-acetylene flame will probably be of limited value, because of the poor performance of this flame for making atomic fluorescence measurements in the region of 4200 Å. The fluorescence spectrum of calcium has not been investigated in detail, but it is not expected that lines other than that at 4226.7 Å ($^1P_1^0 - {}^1S_0$ resonance transition) would fluoresce.

5.8.2. EXPERIMENTAL CONDITIONS AND DETECTION LIMITS

It can be seen from Table 5.11 that the most sensitive atomic fluorescence determination of calcium was carried out with a Technicon AFS-6 filter instrument. A modified Perkin–Elmer 403 atomic absorption spectrometer, which utilized the same type of excitation source as the Technicon AFS-6, yielded a much poorer detection limit [32]. Demers *et al.* [143, 144] reported an upper linear calibration curve limit of 6 μg/ml and a relative standard deviation of 0.67% at the 10 μg/ml level. Detection limits of analytical significance were also obtained by Fraser and Winefordner [177] who excited calcium fluorescence with Model 1000 N_2 Laser and Dial-a-line Dye Laser (AVCO Everett Research Laboratory, Everett, Mass. 02149), and by Demers and Ellis [140, 142].

5.8.3. INTERFERENCES

An extensive examination of factors affecting the atomic fluorescence of calcium was undertaken by Demers and Ellis [140, 142, 164] using an

TABLE 5.11. Papers contributing to the atomic fluorescence flame spectrometry of calcium at 4226.7 Å

Ref.	Spectrometer Detector Band-width (Å)	Excitation source	Flame (burner)	Detection limit (μg/ml) S : N = 2
[142]	L.I. EMI 6255B 27	450 W Xe-arc	H-e.a.(t) HETCO	0.02[a]
	L.I. RCA 1P-28 27	450 W Xe-arc	A-H(p)[b] Tri-flame	0.2
[265]	Perkin—Elmer 303 RCA 1P-28 20	150 W Xe-arc	A-H(p)	2.0
		HCL(Intensitron)	A-H(p)	0.2
[483]	L.I. RCA 1P-28 16	EDL	A-H(t)	0.02
[253]	L.I. RCA 1P-28 40	150 W Xe-arc	A-H(t) Beckman	0.05[e]
[190]	Beckman 440 RCA 1P-28	450 W Xe-arc 150 W Xe-arc	A-H(p) Autolam	0.2
[65]	L.I. RCA 1P-28	150 W Xe-arc Eimac	A-H(t) Zeiss	10
[144, 145, 146, 147]	Technicon AFS-6 filter, $\lambda_{max} = 4232$ $\Delta\lambda_{(1/2)} = 40$	HCL(p)	A-Ac(p)	0.003 0.001[c] 0.000 3[d]
[132]	L.I. EMI 9592 A	HCL	A-Ac(p)	0.8[e]
[206]	Unicam SP 900A EMI 9601 B	R.F. plasma	A-Ac(p) A-P(p)	0.1
[32]	Perkin—Elmer 403 HTV R-106 20	HCL(p) (Intensitron)	A-Ac(p, s)	0.05
[176]	L.I. RCA 1P-28 24	L	A-H(p) (Tri-flame)	0.01

[a] S : N = 1.
[b] With "thermal concentrator" producing dry aerosol.
[c] Reference [144].
[d] [144, 145] linear range 0.30 μg/ml.
[e] Photon-counting detection system.
Abbreviations: See p. 187.

a.c. continuum excitation source, a hydrogen/entrained air flame and a direct-injection burner. No spectral interferences other than scatter were observed because a monochromator of good resulution was employed. Interferences due to light-scatter were easily compensated for by means of the zero suppression circuit of the lock-in amplifier and by scanning the fluorescence line. As expected, calcium was found to be subject to numerous and serious chemical interferences in the flame used. Among the ions studied, Ba, Cu(II), Co(II), K, La, Mn(II), Ni, Zn, ClO_4^-, and Cl^- did not interfere at concentrations up to 1000 μg/ml. Al, Cr, Fe(III), Li, Mg, Pb, CO_3^{2-}, HCO_3^-, I^-, F^-, S^{2-}, SO_4^{2-}, NO_3^-, acetate, and citrate caused serious suppression, and phosphate, silicate, and oxalate depressed the signal completely. The interferences described were attributed to relatively inert compound formation, and could be removed by the prior addition of strontium to the sample solution (to a final concentration of 0.1% with respect to $SrCl_2 . 6\,H_2O$) in nearly all instances, the sulphide and silicate interferences being exceptions.

Chromium was found to interfere in the determination of calcium at 4226.7 Å in both thermal and fluorescence emission when a continuous excitation source was used for the fluorescence determination with a wide spectral slit-width [96].

5.9. CHROMIUM

5.9.1. GENERAL CHARACTERISTICS AND FLUORESCENCE SPECTRUM

The atomic fluorescence of chromium has been observed by many authors but little more than optimum operating conditions and detection limits have been reported. Xenon-arc lamps [96, 264, 433], demountable hot hollow-cathode lamps [152, 349], and electrodeless discharge lamps (both single and multielement) [45, 97, 114, 398, 483] were found to be insufficiently intense light sources for this element. Only very recently have satisfactory analytical results been obtained by exciting the atomic fluorescence of chromium with pulse-operated hollow-cathode lamps [143, 144, 145, 146, 147, 495], a boosted-output hollow-cathode lamp of the Löwe design [258, 260], and a tunable organic dye laser [176]. A slightly fuel-rich separated air-acetylene flame seems to be the most satisfactory atomic cell [143, 144, 145, 146, 258].

The atomic fluorescence spectrum of chromium has not been studied in detail. Atomic fluorescence is usually measured at the 3578.7, 3593.5 and 3605.3 Å resonance multiplet ($a^7S_3 - y^7P^\circ$ transitions); of these lines, the 3578.7 Å line is the strongest. Less often, measurements are made at less sensitive resonance lines at 4254.4, 4274.8 and 4289.7 Å ($a^7S_3 - z^7P^0$ transitions).

TABLE 5.12. Atomic absorption and atomic fluorescence data for chromium
Reprinted from Ref. [316] by permission of Pergamon Press Ltd.

Atomic absorption			Atomic fluorescence		
	Experimental ratios	Theoretical ratios		Experimental ratios	Theoretical ratios
$\frac{A_{3578.7}}{A_{3593.5}}$	1.18	1.14	$\frac{I_F\ 3578.7}{I_F\ 3593.5}$	1.215	0.995
$\frac{A_{3593.5}}{A_{3605.3}}$	1.4	1.4	$\frac{I_F\ 3593.5}{I_F\ 3605.3}$	1.46	1.915
$\frac{A_{3578.7}}{A_{3605.3}}$	1.65	1.6	$\frac{I_F\ 3578.7}{I_F\ 3605.3}$	1.77	1.9

Omenetto and Rossi [314, 316] investigated excitation and de-excitation processes governing the chromium resonance triplet at 3578.7, 3593.5 and 3605.3 Å. They compared theoretical and experimental atomic absorption and atomic fluorescence intensity ratios and demonstrated that while atomic absorption ratios agree very well with theoretical absorption oscillator strength ratios, for fluorescence poor agreement is obtained (see Table 5.12), indicating that the quantum efficiencies must be different. From experimental evidence it seems that owing to collisional mixing of the three states, the $y^7P_3^0$ level contributes to the higher $y^7P_4^0$ (thermally-assisted anti-Stokes fluorescence) and, to a greater extent, to the lower $y^7P_2^0$ level (stepwise-line fluorescence). The experimental results could be interpreted from a detailed understanding of the role played by the thermal energy of the flame. Attempts to isolate the exciting line from the source and to excite the fluorescence at 3593.49 Å with overlapping neon and vanadium lines at 3593.52 and 3593.33 Å, respectively, were not successful because of the low intensity of the sources used.

5.9.2. EXPERIMENTAL CONDITIONS AND DETECTION LIMITS

It can be seen from Table 5.13 that only a few of the detection limits reported for chromium have a real analytical value. Results obtained with

TABLE 5.13. Papers contributing to the atomic fluorescence flame spectrometry of chromium

Ref.	Spectrometer Detector Band-width (Å)	Excitation source	Flame (burner)	Wave-length (Å)	Detection limit (μg/ml) S : N = 2
[483]	L.I. RCA 1P-28 48	EDL	A-H(t) Zeiss	3578.7, 3593.5, 3605.3 } a	10
[114]	L.I. EMI 6256 B	EDL	A-H(p)	3578.7	0.05[b]
[264]	Perkin—Elmer 303 EMI 9592 B 20	150 W Xe-arc	A-H(p)	3578.7	0.3
[265]	Perkin—Elmer 303 EMI 9592 B 20	HCL Intensitron	A-H(p)	3578.7	20
		150 W Xe-arc	A-H(p)	3578.7	10
[312]	Jarrell—Ash (Versat.) RCA 1 P-28 64	MVDL (Hg-90 W) Philips	A-H(t) HETCO	3593.5	5.0
[349]	Jarrell—Ash (Versat.) RCA 1P-28 64	HHCL	A-H(t) HETCO	3578.7	1.0
[398]	S.A. A3000 HTV R-213	EDL(mod.)	A-Ac(p)	3578.7	0.05
[96, 433]	S.A. A 1740 HTV R-213 60	500 W Xe-arc	A-Ac(p)	3578.7, 3593.5, 3605.3 } a	3.0
				4258.7, 4274.8, 4289.7 }	20
[97]	S.A. A 1740 HTV R-213	EDL[c]	O-H(t) Beckman	3593.5	20
[143, 144 145, 146]	Technicon AFS-6 filter, $\lambda_{max} = 3605$ $\Delta\lambda_{(1/2)} = 89$	HCL(p)	A-Ac(p)	3578.7, 3593.5, 3605.3 } a	0.02[d] 0.000 3
[258]	Techtron AA-4[e] RCA-931 A UG 1 Schott filter $\lambda_{max} = 3600$, $\Delta\lambda_{(1/2)} = 600$	BOHCL(L)	A-Ac(p, s)	Σ	0.001 5

TABLE 5.13 (continued)

Ref.	Spectrometer Detector Band-width (A)	Excitation source	Flame (burner)	Wave-length (A)	Detection limit (μg/ml) S : N = 2
[45]	L.I. RCA 1P-28 64	EDL	A-H(t) HETCO	3578.7	0.1
[32]	Perkin—Elmer 403 HTV R-136 20	HCL(p) Intensitron	A-Ac(p, s)	3578.7 } a 3593.5 }	0.5
[176]	L.I. RCA 1P-28 40	L	A-H(p) (Tri-flame)	3578.7 } 3593.5 } a 3605.3 }	0.03

[a] Unresolved lines.
[b] S : N = 1.
[c] Dual-element Cr—Ni and Cr—Mn EDL sources.
[d] Reference [143, 144].
[e] Without monochromator.
Abbreviations: See p. 187.

continuous excitation sources are practically useless, taking into account the fact that the value of 0.3 μg/ml reported for a 150-W xenon arc lamp [264] was not later reproduced [265]. Omenetto and Rossi [312] were able to detect 5 μg/ml of chromium at the 3593.5 Å line with a mercury vapour discharge lamp source, owing to overlap with the mercury 3593.48 Å line. A detection limit of 0.0003 μg/ml resulted from the use of a Technicon AFS-6 multichannel filter atomic fluorescence spectrometer [145, 146] which detects all three components of the chromium resonance triplet. The manufacturers report a linear calibration over the range 0 – 30 μg/ml. Very good detection limits have been achieved by Larkins [258] with a non-monochromator atomic fluorescence system using a Schott UG 1 filter. Fraser and Winefordner [176] reported a very promising detection limit, using a tunable organic dye laser as an excitation source, which could be improved (according to the authors) from 10 to 100-fold by a more extensive optimization of experimental conditions.

5.9.3. INTERFERENCES

No systematic study of interferences affecting the determination of chromium by atomic fluorescence has been reported so far. Cresser and West [96] investigated the spectral interferences associated with the use of a d.c. 500-W xenon arc lamp source in an air-acetylene flame, and found the fluorescence determination to be free from the iron and cobalt interference encountered in flame emission spectrometry.

5.10. COBALT

5.10.1. GENERAL CHARACTERISTICS AND FLUORESCENCE SPECTRUM

A large number of papers have been published on the atomic fluorescence of cobalt, but most contain only limits of detection. Fluorescence may be generated in nearly all types of conventional premixed and turbulent flames. Boosted-output hollow-cathode lamps [173, 258, 264, 276, 388], pulse-excited hollow-cathode lamps [143, 144, 145], and microwave-excited electrodeless discharge lamps [58, 114, 173, 267, 268, 306, 398, 483] have all been used as excitation sources with considerable success. Demountable hot hollow-cathode lamps were less satisfactory [152, 349]. Continuous excitation sources also gave results of analytical utility [96, 103, 142, 162, 164, 190, 265].

Detailed studies of the atomic fluorescence characteristics of cobalt were conducted practically simultaneously in our laboratory [276] and by Fleet *et al.* [173] using a modified Techtron AA-4 atomic absorption spectrophotometer. The atomic fluorescence spectra of cobalt were excited with a high-intensity hollow-cathode lamp [173, 276] and an electrodeless discharge lamp [173]. Fleet and co-workers compared the emission intensities of the two sources and showed a ratio of *ca.* 20 : 1 in favour of the discharge lamp (cobalt-cobalt chloride). They evaluated air-hydrogen, air-propane, and air-acetylene flames and concluded that the air-propane and air-hydrogen flames were most suitable with the Techtron instrument used in an unmodulated mode (i.e. with an unmodulated discharge lamp) whereas all three flames should be suitable for use with a modulated source and synchronous detection. Results from our laboratory indicated that there is only a small enhancement of the fluorescence signal when argon-oxygen is used as oxidant in place of air, and the separated air-acetylene flame was the best fluorescence flame cell for the sensitive and interference-free atomic fluorescence determination of cobalt. Results by Larkins [258] supported this statement.

Details of the atomic fluorescence spectrum of cobalt are given in Table 5.14. The most sensitive fluorescence line is at 2407.3 Å. Fluorescence signals obtained in the region 3400 – 3600 Å are too weak and ill-defined to be of analytical value. If a wide band-pass is used, the lines at 2407.3,

TABLE 5.14. Atomic fluorescence spectrum of cobalt [276]

Wavelength (Å)	Energy levels (eV)	Transition	Relative source emission intensity[a, b]	Relative fluorescence intensity[a, b, c]
2407.3	0—5.149	$a^4F_{9/2} - x^4G^0_{11/2}$	91	100
2411.6	0.101—5.240	$a^4F_{7/2} - x^4G^0_{9/2}$	59	38
2414.5[d]	0.174—5.308	$a^4F_{5/2} - x^4G^0_{7/2}$	54	15
2415.3[d]	0.224—5.356	$a^4F_{3/2} - x^4G^0_{5/2}$	39	
2424.9	0—5.111	$a^4F_{9/2} - x^4F^0_{9/2}$	61	50
2432.2	0.101—5.197	$a^4F_{7/2} - x^4F^0_{7/2}$	42	16
2521.4	0—4.916	$a^4F_{9/2} - x^4D^0_{7/2}$	100	45
2529.0	0.101—5.002	$a^4F_{7/2} - x^4D^0_{5/2}$	58	7

[a] Excited with a BOHCL(S-W).
[b] All intensities corrected for detector spectral response.
[c] In an A-H flame.
[d] Lines unresolved in fluorescence spectrum.

2411.6, 2414.5 and 2415.3 Å appear as one broad analytical line, with the 2424.9 Å line usually being resolved. The fact that all main fluorescence lines occur close together is very advantageous when non-monochromator fluorescence measurements are made [258].

5.10.2. EXPERIMENTAL CONDITIONS AND DETECTION LIMITS

Basic publications dealing with the determination of cobalt by atomic fluorescence spectrometry and the detection limits obtained are summarized in Table 5.15. Very good detection limits have been reported from use of electrodeless discharge lamps and boosted-output hollow-cathode lamp excitation sources. Hell and Ricchio [190] reported a surprisingly good detection limit (0.02 μg/ml) with a 450-W xenon arc, which showed the superior performance of the 450-W lamp over the 150-W lamp under the same experimental conditions. Comparative measurements from our study [276] have indicated that there is very little difference between the limits of detection obtained with turbulent and with premixed flames or those obtained with air-hydrogen and argon-oxygen-hydrogen flames. The unshielded air-acetylene flame gives a poorer detection limit but the shielded flame should give approximately the same limit of detection as the air-hydrogen flame. The calibration curve at 2407.3 Å in the premixed air-acetylene flame is linear up to 10 μg/ml. The striking difference between detection limits in our study and those reported by Fleet *et al.* [193], which were obtained with practically the same equipment and excitation source, could only be explained in terms of the different entrance optics used in the two studies and/or the different radiative output of the individual lamps. The best detection limit achieved at present is that of 1.5 ng/ml reported by Larkins [258] for a non-dispersive atomic fluorescence system.

5.10.3. INTERFERENCES

Matoušek and Sychra [276, 277] found appreciable interferences from 100-fold ratios of a number of elements that form refractory oxides and from sulphuric, phosphoric, and hydrofluoric acids (even at low concentrations) in the air-hydrogen flame. In all cases, the fluorescence signal was depressed. The degree of interference was reduced in the argon-oxygen-hydrogen flame, and interferences were virtually eliminated in the hotter air-acetylene flame. No effect was observed from 100-fold amounts of 11 foreign ions on the fluorescence signal of cobalt in the air-propane flame [173].

Cresser and West [96] investigated spectral interferences affecting the fluorescence signal of cobalt excited with a d.c. 500-W xenon arc continuum source in the air-acetylene flame. They found the fluorescence determination to be more selective than the flame emission determination at a longer wavelength.

5.10.4. ORGANIC SOLVENTS AND EXTRACTION

It was found that 80% of methanol in water improves the detection limit in the air-hydrogen flame by a factor of approximately 2.5, i.e. to 0.008 μg/ml. Following an extraction procedure with the APDC-MIBK system, as little as 0.0004 μg/ml of cobalt can be detected with an air-hydrogen flame at 2407.3 Å [276].

5.11. COPPER

5.11.1. GENERAL CHARACTERISTICS AND FLUORESCENCE SPECTRUM

The atomic fluorescence of copper has been reported by several authors, but only Smith *et al.* [368] and Kolihová and Sychra [245, 249] have made a more complete investigation. The results obtained by these authors indicate

TABLE 5.15. Basic papers contributing to the atomic fluorescence flame spectrometry of cobalt

Ref.	Spectrometer Detector Band-width (Å)	Excitation source	Flame (burner)	Wavelength (Å)	Detection limit (μg/ml) S : N = 2
[142, 162, 164]	L.I. EMI 6255 B[b] 27	450 W Xe-arc	H-e.a.(t) HETCO	2407.3	0.5 0.18[a,c]
[142]	L.I. RCA 1P-28 27	450 W Xe-arc	A-H(p)[d] Tri-flame	2407.3	0.5
[58, 267, 268, 483]	L.I. RCA 1P-28 8	EDL	Ar-H e.a.(t) Zeiss	2407.3	0.1[a]
[114]	L.I. EMI 6256 B ca. 20	EDL	A-Ac(p)	2407.3	0.04[a]
[264]	Perkin – Elmer 303 EMI 9592B 20	BOHCL(S-W)	O-Ar-H(p)	2407.3	0.04
[265]	Perkin – Elmer 303 EMI 9592B 20	150 W Xe-arc HCL	A-H(p) A-H(p)	2407.3 2407.3	2.0 0.2
[398]	S.A. A. 3000 HTV R-213	EDL(mod.)	A-Ac(p)	2407.3	0.005
[349]	Jarrell – Ash 82500 RCA 1P-28 64	HHCL	A-H(t) HETCO	2407.3	0.3
[276, 388]	Techtron AA-4 HTV R-106 10	BOHCL(S-W)	O-Ar-H(p) O-Ar-H(t)	2407.3	0.01
			A-H(p)	2407.3 2424.9 2521.4	0.02 0.1 0.2
			A-H(t)	2407.3	0.01
			A-Ac(p)	2407.3	0.05
[173]	Techtron AA-4 HTV R-106 10	EDL	A-P(p) A-H(p)	2407.3	0.005
			A-Ac(p)	2407.3	0.01
		BOHCL(S-W)	A-P(p)	2407.3	0.20
			A-H(p)	2407.3	0.20
			A-Ac(p)	2407.3	0.33
[190]	Beckman 440 RCA 1P-28	450 W Xe-arc 150 W Xe-arc	A-H(p) A-H(p)	2407.3 2407.3	0.021 0.14
[145, 146]	Technicon AFS 6 filter, $\lambda_{max} = 2250$ $\Delta\lambda_{(1/2)} = 525$	HCL(p)	A-Ac(p)	2407.3	0.002
[306]	S.A. A. 1740 HTV R-213 30	EDL	A-Ac(p, s)	2407.3	0.02
[258]	Techtron AA-4 HTV R 106 10	BOHCL(L) HCL	A-Ac(p, s) A-Ac(p, s)	2407.3 2407.3	0.015 0.15
	Techtron AA-4 HTV R-166 N.S.	BOHCL(L) HCL	A-Ac(p, s) A-Ac(p, s)	Σ Σ	0.0015[g] 0.03

[a] S : N = 1.
[b] Photomultiplier cooled at −15 °C.
[c] Reference [164].
[d] With "thermal concentrator" producing dry aerosol.
[e] Dual element (nickel – cobalt).
[f] Without monochromator.
[g] Upper linear limit 2.5 μg/ml.
Abbreviations: See p. 187.

that copper fluorescence is easily excited in all conventional flames, the best sensitivity being obtained in turbulent argon/hydrogen/entrained air [368] and premixed fuel-lean oxygen/argon/hydrogen and air-hydrogen [249] flames. Hollow-cathode lamps, both of Sullivan-Walsh [244, 245, 249, 264, 268] and Löwe [258, 260] design, as well as hot [151, 152] and pulsed [143, 144, 145, 146] hollow-cathode lamps have proved satisfactory as excitation sources. Copper electrodeless discharge lamps were found to be unsuitable sources because of their short lifetime and unsatisfactory operating characteristics [268, 483].

Most authors have reported only copper resonance fluorescence at the two main resonance lines at 3247.5 and 3274.0 Å. A more detailed examination of the atomic fluorescence spectrum of copper excited with a boosted-output copper hollow-cathode lamp of the Sullivan-Walsh design performed in our laboratory [249] indicated that the fluorescence spectrum of copper consists of nine lines. Spectral characteristics of these lines are listed in Table 5.16. All the lines, except for those at 5105.5 and 5782.1 Å, are due to resonance fluorescence; direct-line fluorescence observed at the latter lines is extremely weak. Data in Table 5.16 for relative source and fluorescence intensities of the 3247.5 and 3274.0 Å lines suggest a difference in the

TABLE 5.16. Atomic fluorescence spectrum of copper

Wave-length (Å)	Energy levels (eV)	Transitions	Relative source emission intensity[a, b]	Relative fluorescence intensity[a, b]	
				Ar-O-H	A-Ac
2165.1	0—5.725	$4s\,^2S_{1/2} - 4s\,4p\,^2D^0_{3/2}$	9	0.34	0.45
2178.9[c]	0—5.688	$4s\,^2S_{1/2} - 4s\,4p\,^2P^0_{3/2}$	13 }	1.25	1.5
2181.7[c]	0—5.681	$4s\,^2S_{1/2} - 4s\,4p\,^2P^0_{1/2}$	7 }		
2225.7	0—5.569	$4s\,^2S_{1/2} - 4s\,4p\,^4D^0_{1/2}$	12	0.24	0.30
2492.2	0—4.973	$4s\,^2S_{1/2} - 4s\,4p\,^4P^0_{3/2}$	23	N.C.[d]	N.C.[d]
3247.5	0—3.817	$4s\,^2S_{1/2} - 4p\,^2P^0_{3/2}$	100	100	100
3274.0	0—3.786	$4s\,^2S_{1/2} - 4p\,^2P^0_{1/2}$	64	50.2	50.7
5105.5	1.389—3.817	$4s^2\,^2D_{5/2} - 4p\,^2P^0_{3/2}$	N.M.	N.C.[d]	N.C.[d]
5782.1	1.642—3.786	$4s^2\,^2D_{3/2} - 4p\,^2P^0_{1/2}$	N.M.	N.C.[d]	N.C.[d]

[a] Excited with BOHCL (S-W).
[b] All intensities corrected for detector spectral response.
[c] Lines unresolved in fluorescence spectrum.
[d] N.C. = Not calculated (intensities too weak to be compared with the most sensitive line).

fluorescence quantum efficiencies for the two transitions. This is in agreement with the observations of Smith *et al.* [368]. Collisional mixing of the $^2P_{1/2}$ and $^2P_{3/2}$ states and a contribution from thermally assisted anti-Stokes fluorescence (i.e. excitation with the 3274.0 Å line and subsequent emission of the 3247.5 Å line) cannot be excluded.

5.11.2. EXPERIMENTAL CONDITIONS AND DETECTION LIMITS

Basic instrumental data of systems used for the atomic fluorescence determination of copper, and the detection limits attained are summarized in Table 5.17. The lowest detection limit of 0.3 ng/ml was obtained by the use of a pulsed hollow-cathode lamp source and an improved Technicon AFS-6 instrument [145, 146]. Very low limits of detection were also obtained with a 450-W xenon arc continuum source [142, 164, 190].

For both of the main resonance lines the linear part of the calibration curves obtained with a premixed oxygen-argon-hydrogen flame covers more than four decades of concentration (from the detection limit up to a concentration of *ca.* 15 μg/ml). To avoid dilution problems, less sensitive lines can be used, particularly the 2225.7 Å line, which gives linear calibration curves up to *ca.* 200 μg/ml. The surprisingly high sensitivity of the lines lying between 2100 and 2300 Å could be used to advantage with non-dispersive systems coupled to solar-blind photomultipliers. The detection limits shown in Table 5.17 [244, 245, 249] at this wavelength could thus be improved by *ca.* one order of magnitude.

5.11.3. INTERFERENCES

Goodfellow [185] examined the effect of large quantities of potassium on the atomic fluorescence of copper in the turbulent air-hydrogen flame and found practically no interference except at the highest ratio of potassium to copper (500-fold) when there was a slight depression (about 10%). Smith *et al.* [368] investigated the effect of 1000-fold ratio of concomitant elements in the turbulent argon-hydrogen-entrained air flame. Chloride was found to give an enhancement of about 20% in the copper signal when present at greater than fivefold molar ratio to copper. However, the fluorescence intensity was independent of the chloride concentration beyond this point. In the presence of 0.01M hydrochloric acid, the following cations did not interfere: Ag, Ca, Cd, Co, Cr, Fe, Ga, Ge, Hg, In, K, Li, Mg, Mn, Mo, Na, NH_4^+, Ni, Sr, Te, Tl and Zn. Bromide, carbonate, EDTA, iodide,

TABLE 5.17. Basic papers contributing to the atomic fluorescence flame spectrometry of copper

Ref.	Spectrometer Detector Band-width (Å)	Excitation source	Flame (burner)	Wave-length (Å)	Detection limit (μg/ml) S : N = 2
[103]	Unicam SP 900A EMI 9601B ca. 30	150 W Xe-arc	A-H(p)	3247.5	0.4[a]
			A-P(p)	3247.5	1.0[a]
[186]	L.I. RCA 1P-28	EDL	A-H(t) Beckman 4020	3247.5	0.02
[58, 483]	L.I. RCA 1P-28 8	EDL[b]	A-H(t) Zeiss	3247.5	0.005[a]
[267, 268]	L.I. RCA 1P-28 8	EDL[c]	O-H(t) Beckman 4020	3247.5	0.05[a]
[411, 412]	L.I. RCA 1P-28 8	150 W Xe-arc	O-H(t) Beckman 4020	3247.5	0.35[a]
[142, 164]	L.I. EMI 6255B[d] 2.7	450 W Xe-arc (Osram)	O-H(t) HETCO	3247.5	0.1[a]
			H-e.a.(t) HETCO	3247.5	0.02[a]
[142]	L.I. RCA 1P-28 2.7	450 W Xe-arc (Osram)	A-H(p)s Tri-flame	3247.5	0.2
				3274.0	0.3
			A-H(p) Tri-flame	3247.5	1.0
				3274.0	2.0
[151, 152]	L.I. RCA 1P-28 32	HHCL	A-H(t) Beckman 4020	3247.5	0.001
[312]	Jarrell—Ash(Max. Versat.) RCA 1P-28 6.4	MVDL Hg 90 W continuum	A-H(t) HETCO	3247.5	0.1
[264]	Perkin—Elmer 303 EMI 9592B(9529P) 20	BOHCL(S-W)	A-H(p)	3247.5	0.001
[265]	Perkin—Elmer 303 EMI 9592B(9529P) 20	150 W Xe-arc	A-H(p)	3247.5	0.3
		HCL(Intensitron)	A-H(p)	3247.5	0.01

TABLE 5.17 (continued)

Ref.	Spectrometer Detector Band-width (Å)	Excitation source	Flame (burner)	Wave-length (Å)	Detection limit (μg/ml) S : N = 2
[114]	L.I. EMI 6256B 17	EDL	A-H(p)	3247.5	0.03[a]
[206]	Unicam SP 900A EMI 9601B	R.F. plasma	A-P(p) A-Ac(p) No-H(p)	3247.5 3247.5 3247.5	0.05 0.3 5.0
[398, 399]	S.A. A3000 HTV R-213	EDL(mod.)	A-Ac(p)	3247.5	0.006
[349]	Jarrell—Ash 82500 RCA 1P-28 6.4	HHCL	A-H(t) HETCO	3247.5	0.1
[368]	L.I. RCA 1P-28 35—64	BOHCL(S-W)	Ar-H(t) HETCO	3247.5	0.003
			A-H(t)	3247.5	0.02
			No-H(t)	3247.5	0.03
			O-H(t)	3247.5	0.1
			H-e.a.(t)	3247.5	0.5
[65]	L.I. RCA 1P-28	150 W Xe-arc Eimac	A-H(t) Zeiss	3247.5	0.2
[96, 433]	S.A. A 1740 HTV R-213 60	500 W Xe-arc	A-Ac(p)	3247.5	3.0
				3274.0	2.0
[190]	Beckman 440 RCA 1P-28	450 W Xe-arc	A-H(p) Autolam	3247.5	0.015
		150 W Xe-arc	A-H(p) Autolam	3247.5	0.083
[291, 292, 293]	L.I. RCA 1P-28 filter, $\lambda_{max} = 3225$ $\Delta\lambda_{(1/2)} = 60$	HCL(p)	A-H(p)	3247.5	0.003
[143, 144, 145, 146, 147, 475]	Technicon AFS-6 filter, $\lambda_{max} = 3264$ $\Delta\lambda_{(1/2)} = 40$	HCL(p)	A-Ac(p)	3247.5	0.03[f] 0.000 3[g]
[253]	L.I. RCA 1P-28 4.0	150 W Xe-arc	A-H(t) Beckman 4020	3247.5	0.05[h]

TABLE 5.17 (continued)

Ref.	Spectrometer Detector Band-width (Å)	Excitation source	Flame (burner)	Wave-length (Å)	Detection limit (μg/ml) S : N = 2
[32]	Perkin—Elmer 403 HTV R-136 20	HCL(p)	A-Ac(p)	3247.5	0.4
	L.I. HTV R-166 N.S.	HCL(p)	A-Ac(p)	3247.5	0.01
[258, 443]	Techtron AA-4[i] RCA-931 A UG 1 Schott filter λ_{max} = = 3600, $\Delta\lambda_{(1/2)}$ = 600	BOHCL(L)	A-Ac(p, s)	Σ	0.001
[244, 245, 249]	Techtron AA-4 HTV R-106 10	BOHCL(S-W)	O-Ar-H(p)	2165.1	0.3
				2178.9, 2181.7 [j]	0.06
				2225.7	0.5
				3247.5	0.001 5
				3274.0	0.002 5
			A-H(p)	3247.5	0.003
				3274.0	0.005
			A-Ac(p, s)	3247.5	0.006
				3274.0	0.01

[a] S : N = 1.
[b] With A-antenna.
[c] With tapered rectangular cavity.
[d] Photomultipler cooled at —15 °C.
[e] With thermal concentrator producing dry aerosol.
[f] Reference [143], [144].
[g] Linear range 0—9 μg/ml.
[h] Photon-counting detection.
[i] Without monochromator.
[j] Unresolved lines.
Abbreviations: See p. 187.

nitrate, and sulphate did not interfere provided the hydrochloric acid molarity was more than five times the foreign anion molarity. Aluminium reduced the fluorescence signal of copper by 42%, phosphate by 56% and silicate by 20%. Depressions in signal were also noted when the concentration of other anions, such as carbonate, nitrate and sulphate, exceeded one fifth of the hydrochloric acid molarity. This effect could easily be overcome by ensuring a large excess of hydrochloric acid at all times or by chelating the copper with 0.001M EDTA. The latter authors also showed that in favourable instances, nearly 1% total dissolved solids did not cause significant scattering of the exciting radiation in low-temperature hydrogen-supported flames (see also Section 4.2). Results from our laboratory [245] indicate that the virtually interference-free fluorescence determination of copper can be carried out in a separated air-acetylene flame.

Cresser and West [96] showed that the atomic fluorescence signal of copper excited in the air-acetylene flame with a d.c. 500-W xenon arc source is subject to spectral interferences (in both thermal emission and fluorescence) from nickel and silver, when measured at 3247.5 and 3274.0 Å, respectively.

5.12. GALLIUM

5.12.1. GENERAL CHARACTERISTICS AND FLUORESCENCE SPECTRUM

Although the atomic fluorescence of gallium has been reported by several authors [45, 114, 190, 266, 313, 349, 483, etc.], it cannot be considered to be completely examined. In most cases, only detection limits have been reported. Various types of hydrogen-supported flames have been used almost exclusively as atom cells for measuring gallium fluorescence and gave similar results. There is no quantitative description of the complete atomic fluorescence spectrum. Mansfield *et al.* [266] reported decreasing fluorescence intensity of lines in the order 4172.1 > 4033.0 > 2874.2 Å. However, much attention has been paid to the explanation and evaluation of excitation processes at these basic fluorescence wavelengths. Omenetto and Rossi [45, 313, 316] showed that thermally-assisted excitation processes play a significant role in the fluorescence transitions at these wavelengths and confirmed this experimentally. Because the metastable $^2P_{3/2}$ state (see Fig. 5.1) is considerably thermally populated (the ratio of the number of atoms populating the $^2P_{3/2}$ metastable state and the $^2P_{1/2}$ ground state is 1.2 in an air-hydrogen flame), direct-line fluorescence at 4172.1 Å (excited with the 4033.0 Å line) is always accompanied with resonance fluorescence at the same wavelength (31% of the resonance fluorescence at 4172.1 Å in the air-hydrogen flame [316]). Resonance fluorescence at 4033.0 Å is accompanied by thermally-assisted anti-Stokes fluorescence (excitation at 4172.1 Å and emission at 4033.0 Å). Omenetto and Rossi [316] showed that 29.2% of the fluorescence signal at 4033.0 Å is due to anti-Stokes

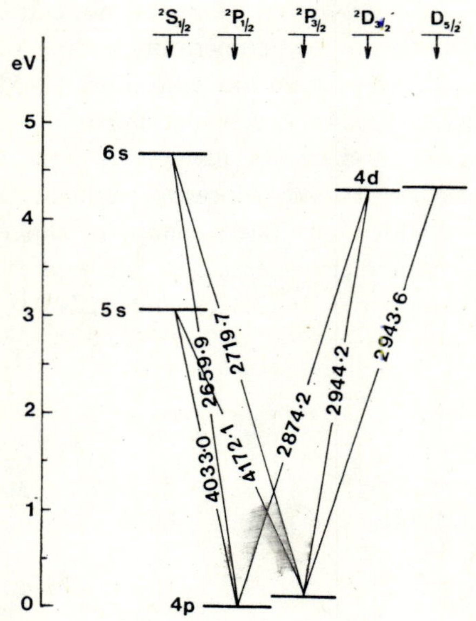

FIG. 5.1. Simplified atomic term diagram for gallium.

fluorescence. This type of fluorescence can be used for scatter-free fluorescence measurements when the 4033.0 Å line is cut off from the primary excitation source. With very efficient entrance optics, Benetti *et al.* [45] could detect as little as 0.02 μg/ml of gallium by measuring "pure" anti-Stokes fluorescence at 4033.0 Å in the turbulent air-hydrogen flame. Resonance fluorescence at 2659.9, 2719.7, 2944.2, 2943.6 Å was also predicted. Anti-Stokes fluorescence at 2874.2 Å, induced by the absorption of the 2944.2 Å line and the "thermally-assisted" stepwise-line fluorescence at the 2943.6 Å line are also likely to occur.

5.12.2. EXPERIMENTAL CONDITIONS AND DETECTION LIMITS

Detection limits attained with various experimental systems are summarized in Table 5.18. All the types of excitation source used give analytically

TABLE 5.18. Basic papers contributing to the atomic fluorescence flame spectrometry of gallium

Ref.	Spectrometer Detector Band-width /Å)	Excitation source	Flame (burner)	Wave-length (Å)	Detection limit (μg/ml) S : N = 2
[266]	L.I. RCA 1P-28 16	EDL	O-H(t) Beckman	4172.1	10[a]
[186]	L.I. RCA 1P-28	MVDL	A-H(t) Beckman 4020	4172.1	0.2
[151] [152]	L.I. RCA 1P-28	HHCL	A-H(t) Beckman 4020	4172.1	0.5
[265]	Perkin—Elmer 303 EMI 9592 B 20	150 W Xe-arc	A-H(p)	4172.1	5.0
		HCL(Intensitron)	A-H(p)	4172.1	0.5
[58, 267, 268, 483]	L.I. RCA 1P-28 16	EDL	A-H(t) Zeiss (Beckman)	4172.1	1.0[b]
[114]	L.I. EMI 6256 B 17	EDL	A-H(p)	4172.1	0.30[b]
[349]	Jarrell—Ash (Versat) RCA 1P-28 6.4	MVDL	A-H(t) HETCO	4172.1	2.0
		HHCL	A-H(t) HETCO	4172.1	1.0
[190]	Beckman 440 RCA 1P-28	450 W Xe-arc	Ar-H e.a.(p) Autolam	4172.1	0.14
		150 W Xe-arc	Ar-H e.a.(p) Autolam	4172.1	0.16
		EDL	Ar-H e.a.(p) Autolam	4172.1	0.007 2
[270, 433]	S.A. A 1740 HTV R-213 3.3	EDL[c]	A-P(p)	4172.1	1.0
				4033.0	3.0
[45]	Jarrell—Ash (Versat) RCA 1P-28 6.4	MVDL Philips	A-H(t) HETCO	4033.0	0.02
[258]	Techtron AA-4 HTV R-106 1.0	MVDL	A-Ac(p, s)	4172.1	0.5
				2943.6	10
	Techtron AA-4[d] HTV R-166 N.S.	MVDL	A-Ac(p, s)	Σ	6
[176]	L.I. RCA 1P-28 14.4	L	A-H(p) Tri-flame	4033.0	0.3

[a] S : N = 0.75.
[b] S : N = 1.
[c] Ga-In dual EDL source.
[d] Without monochromator.

Abbreviations: See p. 187.

useful results; surprisingly low detection limits were obtained with metal vapour discharge lamp sources, which were very often considered to be too weak and self-reversed to give satisfactory fluorescence signals. Both the 4033.0 and 4172.1 Å lines can be used for the measurement, the 4172.1 Å line being slightly more sensitive [270]. For practical applications, the 4033 Å line and anti-Stokes fluorescence measurement should be preferred for the reasons mentioned above. Non-dispersive systems with solar-blind photomultipliers are not advantageous because of the low intensity of the ultraviolet gallium fluorescence lines [258].

5.12.3. INTERFERENCES

No information exists about "chemical" interferences which could affect the determination of gallium; these interferences are expected to occur particularly in cool flames. When using continuum excitation sources and the 4033.0 Å line, serious spectral interferences can occur from the manganese fluorescence triplet at 4030.8, 4033.1 and 4034.5 Å and, if a wide band-pass is used, from potassium fluorescence lines at 4044.1 and 4047.2 Å [96]. The 4172.1 Å line is relatively free from spectral interferences.

5.13. GERMANIUM

5.13.1. GENERAL CHARACTERISTICS AND FLUORESCENCE SPECTRUM

The earliest fluorescence result reported for germanium appears to be that of Zacha *et al.* [483] who reported an atomic fluorescence signal for germanium in a turbulent air-hydrogen flame, which was later stated to be due to scatter rather than fluorescence [484]. Dagnall *et al.* [113] observed weak fluorescence emission in an air-acetylene flame enriched with oxygen, which appeared to be chemiluminescent in nature. Analytically significant fluorescence signals for this refractory element were obtained, as expected, in a nitrous oxide-acetylene flame [119, 190, 259]. The separated slightly fuel-rich nitrous oxide-acetylene flame is sufficiently hot and exhibits sufficient reducing properties to ensure efficient reduction or dissociation of the germanium oxides formed within the flame, resulting in the maintenance of a substantial population of free ground-state atoms of germanium. The principal atomic fluorescence lines of germanium (see Table 5.19) lie in the wavelength region where the nitrous oxide-acetylene flame exhibits

TABLE 5.19. Atomic fluorescence spectrum of germanium [259]

Wavelength (Å)	Energy levels (eV)	Transition	Relative source emission intensity[a, b]	Relative fluorescence intensity[a, b, c]
2498.0	0—4.962	$^3P_0 - {}^1P_1^0$	8	1
2592.5	0.069—4.850	$^3P_1 - {}^3P_2^0$	48	34
2651.2[d]	0.171—4.850	$^3P_2 - {}^3P_2^0$	100	100
2651.6[d]	0—4.674	$^3P_0 - {}^3P_1^0$		
2691.3	0.069—4.674	$^3P_1 - {}^3P_1^0$	28	19
2709.6	0.069—4.643	$^3P_1 - {}^3P_0^0$	37	31
2754.6	0.171—4.674	$^3P_2 - {}^3P_1^0$	40	35
3039.1	0.883—4.962	$^1D_2 - {}^1P_1^0$	69	35
3269.5	0.883—4.674	$^1D_2 - {}^3P_1^0$	13	12

[a] Excited with a BOHCL(L).
[b] All intensities uncorrected for detector spectral response.
[c] In a No-Ac(p, s) flame.
[d] Lines unresolved both in emission and fluorescence spectrum.

relatively low background emission. Results published recently by Larkins and Willis [259] show surprisingly that relatively strong fluorescence signals can be obtained even in fuel-rich air-hydrogen and shielded air-acetylene flames, while the nitrous oxide-hydrogen flame is of limited use. The fact that the emission signal observed in the air-acetylene flame is actually due to atomic fluorescence was checked experimentally by means of scatter tests.

Germanium lines which exhibit atomic fluorescence are listed in Table 5.19. All fluorescence lines observed are probably due to resonance fluorescence; at some non-resonance wavelengths, contributions from direct-line fluorescence cannot be excluded, particularly at 3039.1 and 3269.5 Å. The main fluorescence wavelength is the unresolved doublet at 2651 Å.

5.13.2. EXPERIMENTAL CONDITIONS AND DETECTION LIMITS

Analytical data from papers dealing with the atomic fluorescence flame spectrometry of germanium are listed in Table 5.20. Both continuum and line sources have been used successfully to excite germanium fluorescence. The most intense line source seems to be the electrodeless discharge lamp employed by Dagnall *et al.* [119], the use of which results in the best reported

TABLE 5.20. Papers contributing to the atomic fluorescence flame spectrometry of germanium

Ref.	Spectrometer Detector Band-width (Å)	Excitation source	Flame (burner)	Wave-length (Å)	Detection limit (μg/ml) S : N = 2
[190]	Beckman 440 RCA 1P-28	450 W Xe-arc	No-Ac(p, s) Autolam	2651.2 2651.6 } [a]	1.8
		EDL	No-Ac(p, s) Autolam	2651.2 2651.6 } [a]	2.8
[119]	Techtron AA-4 HTV R-106 10	EDL	No-Ac(p, s)	2651.2 2651.6 } [a]	0.2[b, c] 0.10[b, d] 0.066[b, e]
[259]	Techtron AA-4 HTV R-106 10	BOHCL(L)	No-Ac(p, s)	2651.2 2651.6 } [a] 3039.1	2.0[d] 20[d]
			A-Ac(p, s) A-H(p)	2651.2 2651.6 } [a]	60[d]
	Techtron AA-4[f] HTV R-166 N.S.	BOHCL(L)	No-Ac(p, s)	Σ	3.5[d]
			A-Ac(p, s)	Σ	25[d]
			A-H(p)	Σ	7
	Techtron AA-4[f] HTV R-166 filter, $\lambda_{max} = 2660$, $\Delta\lambda_{(1/2)} = 20$	BOHCL(L)	No-Ac(p, s)	Σ	0.6[d]
			A-Ac(p, s)	Σ	15[d]

[a] Unresolved lines.
[b] S : N = 1.
[c] Unshielded flame.
[d] Nitrogen-shielded flame.
[e] Argon-shielded flame.
[f] Without monochromator.
Abbreviations: See p. 187.

detection limit. Comparison of detection limits for the nitrous oxide-acetylene flame with and without separation shows that a threefold improvement can be obtained when argon is used as a sheathing gas. The last-mentioned authors reported a linear calibration over the range 1 – 1000 μg/ml and a relative standard deviation of 1.25% for a 10 μg/ml germanium solution. Very good detection limits can be obtained with a non-dispersive atomic fluorescence system, particularly when it is used in conjunction with a narrow-band interference filter [259].

5.13.3. INTERFERENCES

No effects were found from 100-fold weight ratios of 22 cations and anions commonly associated with germanium in ores, on the atomic fluorescence intensity produced by 10 μg/ml of germanium in the nitrous oxide-acetylene flame [119].

5.14. GOLD

5.14.1. GENERAL CHARACTERISTICS AND FLUORESCENCE SPECTRUM

The atomic fluorescence of gold can be easily generated in conventional low-temperature flames. Various excitation sources have been used, such as 450-W and 150-W xenon arcs [142, 265, 411, 412], shielded hollow-cathode lamps [145, 146, 258, 265], demountable hot hollow-cathode lamp [152], electrodeless discharge lamps [58, 267, 268, 483], and boosted-output hollow-cathode lamps [1, 258, 277, 278, 388], the last of which seems to be the most suitable. Gold electrodeless discharge lamps are at least as intense as boosted-output hollow-cathode lamps but are very difficult to operate owing to rapid plating-out of the metallic gold on the discharge-tube walls, which results in poor stability of the radiative output and reduction in lamp life. This problem can be partly minimized by the use of the metal chloride [5] in preference to the iodide [58, 267, 268].

Matoušek and Sychra [277, 278, 388] have carried out a detailed investigation of the atomic fluorescence of gold, using premixed nitrogen-shielded air-acetylene, air-hydrogen, and argon-oxygen-hydrogen flames and a boosted-output hollow-cathode lamp as the excitation source. The relative intensities of lines emitted by the source and of the corresponding fluorescence lines are listed in Table 5.21. Four lines exhibit atomic fluorescence, the two resonance lines being the most intense. For the resonance doublet it can be seen that the quantum efficiency of the fluorescence process for the 2428.0 Å line is practically twice the quantum efficiency for the 2676.0 Å line. The 3122.8 Å line is due to direct-line fluorescence. For the 3029.2 Å line, the upper energy limit lies 0.122 eV above the upper energy level of the 2428.0 Å line. In this case, thermally-assisted stepwise-line fluorescence must be assumed. This fluorescence signal, however, is too weak to have any practical value.

TABLE 5.21. Atomic fluorescence spectrum of gold

Wavelength Å	Energy levels (eV)	Transition	Relative source emission intensity[a, b]	Relative fluorescence intensity[a, c]
2428.0	0—5.105	$^2S_{1/2} - {}^2P^0_{3/2}$	100	100
2676.0	0—4.632	$^2S_{1/2} - {}^2P^0_{1/2}$	96	52
3029.2	1.136—5.227	$^2D_{5/2} - {}^4P^0_{5/2}$	42	1
3122.8	1.136—5.105	$^2D_{5/2} - {}^2P^0_{3/2}$	12	5

[a] Corrected for detector spectral response.
[b] Excited with a BOHCL (S-W).
[c] In an Ar-O-H flame.

5.14.2. EXPERIMENTAL CONDITIONS AND DETECTION LIMITS

Detection limits reported by different authors for the atomic fluorescence flame spectrometric determination of gold are summarized in Table 5.22. The most intense line at 2428.0 Å permits the atomic fluorescence determination to be made with a detection limit of 5 μg/ml in the argon-oxygen-hydrogen flame. In this flame, an approximately 2.5-fold increase in the fluorescence signal occurs (at the 2428.0 Å line) compared to the air-hydrogen flame, owing to lower quenching. Since the noise of the measurement for both flames is practically the same, the higher signal in the argon-oxygen-hydrogen flame results in a better detection limit. Although the fluorescence signal in the separated air-acetylene flame is higher than that in the air-hydrogen flame, the relatively higher noise for the former flame makes the detection limits identical for both resonance lines in both flames. The upper linear limit of the calibration curves for the 2428.0 Å line in all three flames investigated was approximately 20 μg/ml.

TABLE 5.22. Papers contributing to the atomic fluorescence flame spectrometry of gold

Ref.	Spectrometer Detector Band-width (Å)	Excitation source	Flame (burner)	Wave-length (Å)	Detection limit (μg/ml) S : N = 2
[152]	L.I. RCA 1P-28 32	HHCL	A-H(t) Beckman 4020	2676.0	0.05
[411, 412]	L.I. RCA 1P-28 8	150 W Xe-arc	O-H(t) Beckman 4020	2676.0	3.5
[58, 267 268, 483]	L.I. RCA 1P-28 32	EDL	O-H(t) Zeiss	2676.0	0.2
[265]	Perkin—Elmer 303 EMI 9592B 20	150 W Xe-arc	A-H(p)	2676.0	10
		HCL	A-H(p)	2676.0	10
[277, 278, 388]	Techtron AA-4 HTV R-106 10	BOHCL(S-W)	Ar-O-H(p)	2428.0 2676.0 3122.8	0.005 0.02 2.5
			A-Ac(p, s)	2428.0 2676.0 3122.8	0.015 0.07 2.0
[258]	Techtron AA-4 HTV R-106 10	BOHCL(L)	A-Ac(p, s)	2428.0	0.05
		HCL	A-Ac(p, s)	2428.0	1.0
	Techtron AA-4[a] HTV R 166 N.S.	BOHCL(L)	A-Ac(p, s)	2428.0 2676.0	0.005[b]
		HCL	A-Ac(p, s)	2428.0 2676.0	0.1
[145, 146]	Technicon AFS-6 filter	HCL(p)	A-Ac(p)	2428.0	0.009[c]

[a] Without monochromator.
[b] Upper linear limit 3.0 μg/ml.
[c] Linear range 0—10 μg/ml.
Abbreviations: See p. 187.

Recent results published by Larkins [258] have shown that with an R-166 non-dispersive fluorescence system and a shielded air-acetylene flame, the detection limit obtained was the same as the best reported for a system with a monochromator.

5.14.3. INTERFERENCES

Matoušek *et al.* [277, 278] investigated the effect of inorganic acids (concentrations 0.01 – 0.5M) and of 100-fold ratios of 18 cations (which could be present in practical analyses) on the fluorescence signal produced by 5 μg/ml of gold in air-hydrogen, argon-oxygen-hydrogen and air-acetylene flames.

The chemical interferences of sulphuric, phosphoric, and hydrofluoric acids were more pronounced in the cooler air-hydrogen flame than in the other two flames, and the interference of Zn, Bi, Rh, Al, and W in the air-hydrogen flame were completely eliminated in the hotter flames. In contrast, palladium strongly decreased the fluorescence signal in the air-acetylene flame but had no effect in the air-hydrogen and argon-oxygen-hydrogen flames. This interference was probably due to the spatial distribution of the sample within the flame, because it was not observed in atomic absorption with the long-path air-acetylene flame. The chemical interference of the acids and cations could be eliminated by the addition of lanthanum to give a final concentration of 0.1%.

Sensitivity of the fluorescence determination was further increased by the extraction procedure which is commonly used in atomic absorption spectrometry. Gold was extracted as its bromide complex from 1M hydrobromic acid or its chloride complex from 3M hydrochloric acid into MIBK [278].

5.15. INDIUM

5.15.1. GENERAL CHARACTERISTICS AND FLUORESCENCE SPECTRUM

Atomic fluorescence of indium was reported for the first time by Mansfield *et al.* [266] who observed fluorescence at the following lines given in order of relative intensities: 4101.8 > 3256.1 > 3258.6 > 3039.4 Å. Since then several authors have mentioned indium fluorescence, but except in the studies by Omenetto and Rossi [45, 313, 316] only limits of detection with various excitation source/flame atomizer systems have been published

TABLE 5.23. Papers contributing to the atomic fluorescence flame spectrometry of indium

Ref.	Spectrometer Detector Band-width (Å)	Excitation source	Flame (burner)	Wave-length (Å)	Detection limit (μg/ml) S : N = 2
[266]	L.I. RCA 1P-28 16	EDL	O-H(t) Beckman	4101.8	10[a]
[151, 152]	L.I. RCA 1P-28	HHCL	A-H(t) Beckman 4020	4101.8	50
[265]	Perkin—Elmer 303 EMI 9592 B 20	150 W Xe-arc	A-H(p)	4101.8	2.0
		HCL (Intensitron)	A-H(p)	4101.8	0.2
[58, 483]	L.I. RCA 1P-28 8.0	EDL	Ar-H e.a.(t) Zeiss	4101.8	0.1[b]
[267, 268]	L.I. RCA 1P-28 8.0	EDL	A-H(t) Beckman	4101.8	0.2[b]
[114]	L.I. EMI 6256 B 17	EDL	No-Ac(p)	4511.3	1.0[b]
[96]	S.A. 1740 HTV R-213 60	500 W Xe-arc	A-Ac(p)	4101.8	20
				4511.3	15
[349]	Jarrell—Ash(Versat) RCA 1P-28 6.4	MVDL	A-H(t) HETCO	4511.3	0.5
		HHCL	A-H(t) HETCO	4511.3	0.3
[270]	S.A. A 1740 HTV R-213 3.3	EDL[c]	A-P(p)	4511.3	0.2
				4101.8	0.7
[45]	Jarrell—Ash(Versat) RCA 1P-28 6.4	MVDL Philips	A-H(t) HETCO	4101.8	0.015
[258]	Techtron AA-4 HTV R-106 1.0	MVDL	A-Ac(p, s)	4511.3	0.4
				3039.4	2.0
	Techtron AA-4[d] HTV R-166 N.S.	MVDL	A-Ac(p, s)	Σ	1.0
[176]	L.I. RCA 1P-28 24	L	A-H(p) Tri-flame	4101.8	0.01

[a] S : N = 0.75.
[b] S : N = 1.
Abbreviations: See p. 187.

[c] Ga-In dual-element EDL source.
[d] Without monochromator.

(see Table 5.23). Electrodeless discharge lamps do not seem to be very intense sources at present, although the dual-element lamp for gallium and indium described by Marshall and West [270] proved adequate. Poor detection limits were obtained with xenon-arc continuum sources [96, 265] despite the fact that most sensitive indium fluorescence lines lie in the wavelength region where these sources exhibit relatively high radiance. The best limit of detection (0.01 μg/ml) was obtained with a laser source in the premixed air-hydrogen flame [176]. With the latter instrumental arrangement, the analytical curve (log-log plot) for the 4101.8 Å line was linear over four decades of concentration.

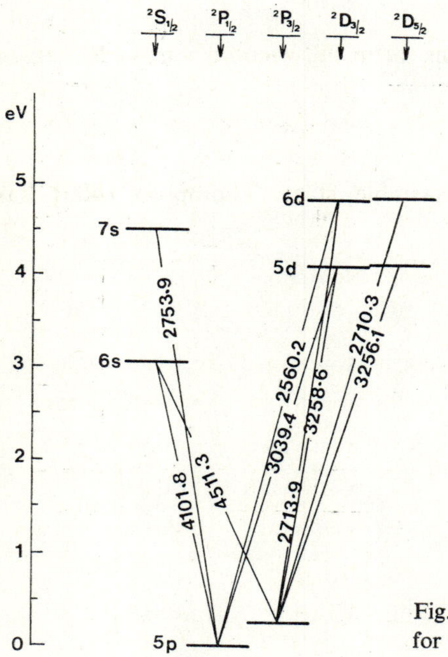

Fig. 5.2. Simplified atomic term diagram for indium.

As was the case for gallium (cf. Section 5.12), most attention has been paid to the investigation of the excitation processes at the basic fluorescence wavelengths. Although the metastable $^2P_{3/2}$ state lies slightly higher above the $^2P_{1/2}$ ground-state (compared to gallium), it is still appreciably thermally populated [313] and the fluorescence signal at 4511.3 Å is due to a mixture of direct line fluorescence (69.3%) and resonance fluorescence (30.7%) [316] (see Fig. 5.2). Moreover, the observed thermally-assisted anti-Stokes fluorescence at 4101.8 Å (27% of the total fluorescence signal at this wavelength [316]) is of analytical importance. By filtering off radiation at 4101.8 Å from the excitation source and using a very efficient optical system,

Benetti *et al.* [45] obtained an excellent detection limit (0.015 µg/ml) from "pure" anti-Stokes fluorescence, and eliminated any problems of scatter. Other lines at which fluorescence was measured or predicted are also shown in Fig. 5.2. Combinations of resonance fluorescence and direct-line fluorescence are also likely to occur.

Cresser and West [96] have examined spectral interferences after excitation of indium with a d.c. 500-W xenon arc lamp and found appreciable positive interference from cobalt.

5.16. IODINE

The main resonance lines of iodine lie in the vacuum ultraviolet region of the spectrum (1782.9 and 1830.4 Å) and cannot be used for the direct determination of this element with conventional instrumentation. Iodine has, however, a metastable state 0.941 eV above the ground state and a line at 2062.4 Å emanating from this metastable state. Thompson [400] has calculated the fraction of iodine atoms that are free in the metastable state at temperatures corresponding to analytical flames, and tried to generate iodine fluorescence in the air-acetylene flame with a microwave-excited arsenic electrodeless discharge lamp (containing iodine) as source. This lamp emitted at 2062.4 Å as strongly as a lamp containing only iodine, but gave a more stable output. The fluorescence signal observed at the 2062.4 Å line was very weak, and resulted in a detection limit of 1.6 mg/ml. This disappointing result is thought to be due mainly to the low oscillator strength of the 2062.4 Å line, and indicates that it is not feasible at present to determine iodine directly by atomic fluorescence spectrometry.

5.17. IRIDIUM

There is no information about the atomic fluorescence of iridium other than that reported by Manning and Heneage [265] and Larkins [258]. The former excited the atomic fluorescence of iridium with a 150-W xenon arc and a conventional shielded hollow-cathode lamp, and obtained detection limits of 100 and *ca.* 1000 µg/ml, respectively, in an air-hydrogen flame at 2544.0 Å. Larkins reported the relatively good detection limit (compared with other flame methods) of 4.0 µg/ml for the R-166 non-dispersive atomic fluorescence system coupled with a separated air-acetylene flame. Better detection limits are expected with more suitable excitation sources or with pulsed hollow-cathode lamps.

5.18. IRON

5.18.1. GENERAL CHARACTERISTICS AND FLUORESCENCE SPECTRUM

Although early measurements of the atomic fluorescence of iron did not look very promising (primarily owing to the use of excitation light-sources of low intensity [103, 152, 164, 264]), recent results are very promising [114, 145, 146, 157, 190, 258, 276]. Boosted-output hollow-cathode lamps

TABLE 5.24. Atomic fluorescence spectrum of iron [157, 276]

Wavelength (Å)	Energy levels (eV)	Transition	Relative lamp emission intensity		Relative fluorescence intensity	
			A[a]	B[b]	A[a, c]	B[b, d]
2483.3	0—4.991	$a^5D_4 - x^5F^0_5$	49		100	
2488.2	0.052—5.033	$a^5D_3 - x^5F^0_4$	29	100[e]	46	100[e]
2489.8[e]	0.121—5.099	$a^5D_0 - x^5F^0_1$	25[e]		23[e]	
2490,6[e]	0.087—5.064	$a^5D_2 - x^5F^0_3$				
2522.9	0—4.913	$a^5D_4 - x^5D^0_4$	33	35[e]	34	51.8[e]
2527.4	0.052—4.955	$a^5D_3 - x^5D^0_3$	12		12	
2719.0[e]	0—4.558	$a^5D_4 - y^5P^0_3$	36	31[e]	15	5.0[e]
2720.9[e]	0.052—4.607	$a^5D_3 - y^5P^0_3$	N.M.		N.D.	
2983.6	0—4.154	$a^5D_4 - y^5D^0_3$	N.M.	23	N.D.	7.0
3020.5[e]	0.087—4.191	$a^5D_2 - y^5D^0_2$				
3020.6[e]	0—4.103	$a^5D_4 - y^5D^0_4$	84[e]	24[e]	12[e]	8.5[e]
3021.1[e]	0.052—4.154	$a^5D_3 - y^5D^0_3$				
3565.4[e]	0.958—4.434	$a^5F_3 - z^3G^0_4$	N.M.	8[e]	N.D.	5.4[e]
3570.1[e]	0.915—4.386	$a^5F_4 - z^3G^0_5$	N.M.		N.D.	
3719.9	0—3.332	$a^5D_4 - z^5F^0_5$	N.M.	135	N.D.	32.8
3734.9	0.859—4.177	$a^5F_5 - y^5F^0_5$	62	N.M.	6	N.D.
3737.1	0.052—3.368	$a^5D_3 - z^5F^0_4$	100	N.M.	7	N.D.
3820.4[e]	0.859—4.103	$a^5F_5 - y^5D^0_4$	N.M.		N.D.	
3824.4[e]	0—3.241	$a^5D_4 - z^5D^0_3$	N.M.	127[e]	N.D.	19.3[e]
3825.9[e]	0.915—4.154	$a^5F_4 - y^5D^0_3$	N.M.		N.D.	
3859.9	0—3.211	$a^5D_4 - z^5D^0_4$	N.M.	50	N.D.	2.3

[a] Excited with a BOHCL(S-W); all intensities corrected for detector spectral response [276].

[b] Excited with an EDL in an A-Ac flame; all intensities uncorrected for detector spectral response [157].

[c] In an A-H flame.

[d] In an A-Ac flame.

[e] Lines unresolved both in emission and fluorescence spectrum.

[258, 276], pulsed hollow-cathode lamps [143, 145] and electrodeless discharge lamps [114, 157] have proved to be sufficiently intense excitation sources, and are capable of giving rise to detection limits of about 0.01 μg/ml or less. There was very little difference between the limits of detection obtained with turbulent and with premixed flames or with air-hydrogen and argon-oxygen-hydrogen flames [276]. The shielded air-acetylene flame gave slightly poorer limits of detection than hydrogen-supported flames, but eliminated several interferences found in the air-hydrogen flame.

The most extensive investigations of the atomic fluorescence characteristics of iron were carried out almost simultaneously by Matoušek and Sychra [276] and by Ebdon *et al.* [157]. In the former study, fluorescence was excited with a boosted-output hollow-cathode lamp of Sullivan-Walsh design, in the latter work with an electrodeless discharge lamp [prepared from iron(II) chloride with argon as filler-gas at 3 torr, and operated in a three-quarter wave resonant cavity at 50 W]. Details of the atomic fluorescence spectra obtained with the two sources are reproduced in Table 5.24. As expected, the atomic fluorescence spectrum of iron is very complex, and the mechanisms of the fluorescence transitions for most lines are practically impossible to evaluate. All lines listed in the table are probably due to resonance fluorescence originating either from the ground-state or from low-lying metastable states, except for the lines at 3565.4, 3570.1, 3734.9, 3820.4 and 3825.9 Å, for which direct-line fluorescence or stepwise-line fluorescence would be expected. The most intense fluorescence occurs at 2483.3 Å; other lines corresponding to $^5D - x^5F$ transitions are also very intense.

5.18.2. EXPERIMENTAL CONDITIONS AND DETECTION LIMITS

Basic information from papers dealing with the atomic fluorescence of iron is reviewed in Table 5.25. The 2483.3 Å line gives the best detection limits, but the 2522.9, 2719.0 and 3719.9 Å lines also yield detection limits lower than 1 μg/ml.

Calibration curves at 2483.3 Å in a premixed air-acetylene flame are approximately linear to 40 μg/ml, but for the 3719.9 and 3734.9 Å lines, the linearity is extended to 120 μg/ml.

5.18.3. INTERFERENCES

Solute volatilization interferences observed for iron in the air-hydrogen flame are eliminated or considerably suppressed in the air-acetylene flame.

TABLE 5.25. Papers contributing to the atomic fluorescence flame spectrometry of iron

Ref.	Spectrometer Detector Band-width (Å)	Excitation source	Flame (burner)	Wave-length (Å)	Detection limit (μg/ml) S : N = 2
[103]	Unicam SP 900A EMI 9601B 24.0	150 W Xe-arc	A-H(p) A-P(p)	2483.3	5.0[a]
[152]	L.I. RCA 1P-28	HHCL	A-H(t) Beckman 4020	2483.3	5.0
[264, 265]	Perkin—Elmer 303 EMI 9592 B 20	150 W Xe-arc	A-H(p)	2483.3	1.0
		BOHCL(S-W)	A-H(p)	2483.3	0.1
		HCL(Intensitron)	A-H(p)	2483.3	0.4[b]
[483]	L.I. RCA 1P-28 16.0	EDL	A-H(t) Zeiss	2483.3	0.25[a]
[164]	L.I. EMI 6255 B[c] 2.8	450 W Xe-arc	H-e.a.(t) HETCO	2483.3	1.8[a]
			O-H(t) HETCO	2483.3	25[a]
[312]	Jarrell—Ash(Versat.) RCA 1P-28 6.4	MVDL, Hg-90 W[d]	A-H(t) HETCO	2483.3	1.0
[114]	L.I. EMI 6256 B 17	EDL	A-H(p)	2483.3 3719.9	0.008[a] 0.50[a]
[349]	Jarrell—Ash(Versat.) RCA 1P-28 6.4	HHCL	A-H(t) HETCO	2483.3	2.0
[366]	L.I. RCA 1P-28 8.0	EDL	Ar-H-e.a.(t) HETCO	2483.3	0.3[a]
[157]	Unicam SP 900A EMI 9601 B 8.0	EDL	A-Ac(p)	2483.3	0.009[a]
				2527.4	0.015[a]
				3719.9	0.84[a]
				3820.4	1.2[a]
				2719.0	3.0[a]
[276, 277, 388]	Techtron AA-4 HTV R-106 9.9	BOHCL(S-W)	O-Ar-H(p)	2483.3	0.02
			A-Ar-H(t)	2483.3	0.02
			A-H(t)	2483.3	0.02
			A-Ac(p)	2483.3	0.08
			A-H(p)	2483.3	0.03
				2522.9	0.10
				2719.0	0.4

TABLE 5.25 (continued)

Ref.	Spectrometer Detector Band-width (Å)	Excitation source	Flame (burner)	Wave-length (Å)	Detection limit (μg/ml) S : N = 2
[398]	S.A. A 3000 HTV R-213 30	EDL(mod.)	A-Ac(p)	2483.3	0.005
[65]	L.I. RCA 1P-28	150 W Xe-arc Eimac	A-H(t) Zeiss	2483.3	5.0
[291, 292, 293, 294]	L.I. RCA 1P-28 filter, $\lambda_{max} = 2486$, $\Delta\lambda_{(1/2)} = 150$	HCL(p)	A-H(p) A-Ac(p)	2483.3	0.10
[143, 145, 146, 147]	Technicon AFS-6 filter, $\lambda_{max} = 2475$ $\Delta\lambda_{(1/2)} = 137$	HCL(p)	A-Ac(p)	2483.3	0.03[e] 0.000 6[f]
[190]	Beckman 440 RCA 1P-28	150 W Xe-arc	A-H(p) Autolam	2483.3	0.67
		450 W Xe-arc	A-H(p) Autolam	2483.3	0.035
[32]	Perkin—Elmer 403 HTV R-136 20	HCL(p) Intensitron	A-Ac(p, s)	2483.3	0.15
	L.I. HTV R-166 N.S.	HCL(p) Intensitron	A-Ac(p)	Σ	0.08
[176]	L.I. RCA 1P-28 24	L	A-H(p) (Tri-flame)	3719.9	0.3
[258]	Techtron AA-4 HTV R-106 9.9	BOHCL(L)	A-Ac(p, s)	2483.3	0.05
	Techtron AA-4[g] HTV R-166 N.S.	BOHCL(L)	A-Ac(p)	Σ	0.04
			A-Ac(p, s)	Σ	0.003

[a] S : N = 1.
[b] Reference [265].
[c] Photomultiplier cooled at —15 °C.
[d] Excitation with the overlapping Hg 2482.72 Å line.
[e] Reference [143].
[f] Linear range 0—19 μg/ml.
[g] Without monochromator.

Abbreviations: See p. 187.

With the latter flame Ebdon *et al.* [157] did not observe the interference from chromium, tungsten and cobalt found by the authors [277], but reported a slight depression (less than 10%) of the iron fluorescence signal caused by 1000-fold amounts of thorium, cadmium and phosphates, and a strong depression by the same amounts of vanadium, manganese and zinc. The effect of the last three elements is very surprising because none of them is known to interfere appreciably with iron determination by atomic absorption.

Spectral interference with iron by platinum can result as a direct consequence of overlap of the 2719.02 Å iron line and the platinum 2719.04 Å line.

5.18.4. EXTRACTION

By extraction of the pyrrolidinedithiocarbamate complex into methyl isobutyl ketone, iron can be determined down to 0.2 ng/ml in the air-hydrogen flame [276].

5.19. LEAD

5.19.1. GENERAL CHARACTERISTICS AND FLUORESCENCE SPECTRUM

The atomic fluorescence of lead in flames has been thoroughly examined. Comprehensive investigations were reported almost simultaneously from our laboratory [277, 388, 390] and by workers from Imperial College [74]. Results from these studies give a complete picture of the atomic fluorescence transitions involved in the lead spectrum, and indicate that fluorescence (rather than other flame techniques) is particularly suitable for lead determination. Excitation sources used in these studies (boosted-output hollow-cathode lamp of Sullivan-Walsh design and electrodeless discharge lamp prepared from lead iodide, respectively) proved to be sufficiently intense, and gave similar detection limits under comparable conditions.

The basic spectral characteristics of all fluorescence lines of lead are summarized in Table 5.26. As expected from atomic absorption measurements, the resonance lines at 2170.0 and 2833.1 Å give relatively high fluorescence signals, owing to resonance fluorescence. The additional intense signals at 4057.8, 3639.6, 7229.0 Å, and 2613.7 Å must be almost exclusively due to direct-line fluorescence, because of the great energy difference

TABLE 5.26. Atomic fluorescence spectrum of lead [74, 390]

Wavelength (Å)	Energy levels (eV)	Transition	Relative lamp emission intensity[a]		Relative fluorescence intensity[a, b]	
2170.0	0–5.712	$^3P_0 - {}^3D_1^0$	0.4[c]	15[d]	3.2[c]	16[d]
2613.7 } [e]	0.969–5.712	$^3P_1 - {}^3D_1^0$ }	18.8	49	4.0	13
2614.2 }	0.969–5.711	$^3P_1 - {}^3D_2^0$ }				
2802.0	1.320–5.744	$^3P_2 - {}^3F_3^0$	17.2	55	2.6	8
2823.2	1.320–5.711	$^3P_2 - {}^3D_2^0$	32.8	7	1.4	N.D.
2833.1	0–4.375	$^3P_0 - {}^3P_1^0$	34.3	79	52.5	68
2873.3	1.320–5.612	$^3P_2 - {}^3F_2^0$	4.4	26	1.3	N.D.
3639.6	0.969–4.375	$^3P_1 - {}^3P_1^0$	14.7	47	16.1	33
3683.5	0.969–4.334	$^3P_1 - {}^3P_0^0$	70.5	65	7.6	25
4019.6	2.660–5.744	$^1D_2 - {}^3F_3^0$	4.8	N.M.	0.7	N.D.
4057.8	1.320–4.375	$^3P_2 - {}^3P_1^0$	100	100	100	100
7229.0	2.660–4.375	$^1D_2 - {}^3P_1^0$	77.1	N.M.	65.2	N.M.

[a] All intensities corrected for detector spectral response.
[b] In an Ar-O-H flame.
[c] Excited with an EDL.
[d] Excited with a BOHCL(S-W).
[e] Lines unresolved both in emission and fluorescence spectrum.

between the ground-state of the atom and the two other components of the triplet $6p^2\,{}^3P$ state, and also because of the very low atomic absorption sensitivity at these lines, which indicates a very low thermal population of the $^3P_1$ and $^3P_2$ states. For the fluorescence lines at 4057.8 and 3683.5 Å this mechanism was confirmed experimentally by filtering out the resonance lines from the source with a glass filter. (Results published earlier from our laboratory [390], which indicated considerable contributions of resonance fluorescence to both lines, were not correct.) The remaining fluorescence lines of lead are due to stepwise-line fluorescence. The energy which the excited atoms must lose by a radiationless deactivation or acquire by thermally-assisted activation is very small. The most intense fluorescence occurs at the 4057.8 Å line, as expected from theoretical calculations [27].

Since lead is readily atomized, even in low-temperature flames, hydrogen-supported flames should result in better signal-to-noise ratios than acetylene-supported flames. Comparative measurements made in our laboratory with various premixed flames have shown that although the fluorescence signal in the sheathed air-acetylene flame is slightly higher than that in the air-hydrogen flame, signal-to-noise ratios for all lines except for the 2170.0 Å line are better in the latter flame. The highest fluorescence signal was obtained in the oxy-hydrogen flame diluted with argon, as might be expected. By using argon, a 4.0-fold increase in the fluorescence signal was observed for the 4057.8 Å line, which compares well with the results published by Browner *et al.* [74]. The use of the nitrogen sheath makes both air-hydrogen and argon-oxygen-hydrogen flames steadier, and produces a further improvement (approximately 2-fold) in the detection limits. The fluorescence signal in the hydrogen-oxygen-argon flame at a given flow-rate is critically dependent on the hydrogen flow-rate, whereas the fluorescence signal in the air-hydrogen flame is practically independent of the fuel-flow.

5.19.2. EXPERIMENTAL CONDITIONS AND DETECTION LIMITS

Basic information (including limits of detection) from papers contributing to the atomic fluorescence flame spectrometry of lead are listed in Table 5.27. The lowest limits of detection have been reported at 4057.8 Å, using line excitation sources. The value listed for the 450-W xenon arc source should be lowered by at least a factor of 5 if the optimized flame conditions and fluorescence wavelength are used.

Calibration curves at 4057.8 Å for various premixed flames are linear over a range of three orders of magnitude, the upper linearity limit being at about 150 μg/ml. Wavelengths corresponding to resonance fluorescence give shorter linear ranges.

The high volatility of lead at about 1000 °C allows the sampling-cup technique (cf. Section 3.3.3) to be used for small sample volumes. As little as 1.2 ng of lead can be determined by this technique [393]. Calibration curves bear the same linear relationship at low concentrations (1.0 – 1000 μg of lead) as those obtained in atomic fluorescence flame spectrometry. Some applications of the sampling-cup technique for lead are discussed in Chapter 6.

5.19.3. INTERFERENCES

Only fluoride, molybdate and aluminium (in 100-fold ratio to Pb) were found to cause serious depression (80, 97 and 12%, respectively) of the fluorescence signal of lead in the argon-oxygen-hydrogen flame [277] (cf. also Table 4.1). The interference of fluoride can be lowered by the addition of EDTA (to a final concentration of 0.005M). The effect of cations can be completely eliminated in the same manner. Browner *et al.* [74] reported (for the same flame) interference from 100-fold ratio of aluminium

TABLE 5.27. Papers contributing to the atomic fluorescence flame spectrometry of lead

Ref.	Spectrometer Detector Band-width (Å)	Excitation source	Flame (burner)	Wave-length (Å)	Detection limit (μg/ml)
[103]	Unicam SP 900A EMI 9601 B *ca.* 9.0	150 W Xe-arc	A-H(p)	2833.1	10[a]
			A-P(p)	2833.1	20[a]
[411, 412]	L.I. RCA 1P-28 4.8	150 W Xe-arc	Ar-H, e.a.(t) Beckman	4057.8	7.5[a]
[152]	L.I. RCA 1P-28	HHCL	A-H(t) Beckman 4020	4057.8	1.0
[264, 265]	Perkin—Elmer 303 EMI 9592 B 20	150 W Xe-arc	A-H(p)	4057.8	3.0
		BOHCL(S-W)	A-H(p)	4057.8	0.02[b]
		HCL(Intensi-tron)	A-H(p)	4057.8	1.0
[58, 267, 268, 483]	L.I. RCA 1P-28 32	EDL	Ar-H, e.a.(t) Beckman, Zeiss	4057.8	0.5[a, c]
			A-H(t) Beckman	4057.8	0.5[a, c]
[114]	L.I. EMI 6256 B 17	EDL	A-H(p)	4057.8	0.06[a]
				2833.1	0.40[a]
[349]	Jarrell—Ash (Versat.) RCA 1P-28 6.4	HHCL	A-H(t) HETCO	4057.8	1.0
[366]	L.I. RCA 1P-28 *ca.* 13	EDL	Ar-H, e.a.(t) HETCO	4057.8	0.2[a]
[277, 388, 390]	Techtron AA-4 HTV R-106 9.9	BOHCL(S-W)	Ar-O-H(p, s)	4057.8	0.02
				2833.1	0.05
				3639.6	0.05
				2170.0	0.06
			A-H(p)	4057.8	0.10
				2833.1	0.12
				2170.0	0.20
			A-Ac(p, s)	4057.8	0.20
				2833.1	0.15
				2170.0	0.12

TABLE 5.27 (continued)

Ref.	Spectrometer Detector Band-width (Å)	Excitation source	Flame (burner)	Wave-length (Å)	Detection limit (μg/ml)
[74]	Perkin—Elmer 290 RCA 1P-28 20	EDL(mod.)	Ar-O-H(p, s)	4057.8	0.01[a]
[65]	L.I. RCA 1P-28	150 W Xe-arc Eimac	Ar-H, e.a.(t) Zeiss	4057.8	20.0
[97]	S.A. A 1740 HTV R-213	EDL[d]	O-H(t) Beckman	4057.8	10.0[d]
[143, 145, 146, 147]	Technicon AFS-6 filter, $\lambda_{max} = 4053$, $\Delta\lambda_{(1/2)} = 34$	HCL(p)	No-Ac(p, s)	4057.8	0.07[e]
		EDL(p)	A-Ac(p)	4057.8	0.01[f]
[190]	Beckman 440 RCA 1P-28	150 W Xe-arc	A-H(p) Autolam	2833.1	3.3
		450 W Xe-arc	A-H(p) Autolam	2833.1	0.61
[258]	Techtron AA-4[g] HTV R-166 N.S.	HCL	A-Ac(p, s)	2833.1	1.0
[245, 393]	Techtron AA-4 HTV R-106 9.9	BOHCL(S-W)	A-Ac(p, s)	4057.8	0.012[h]

[a] S : N = 1.
[b] Reference [264].
[c] Reference [268].
[d] Dual-element Pb-Cd source.
[e] Reference [143].
[f] Linear range 0–160 μg/ml.
[g] Without monochromator.
[h] With the "sampling cup" technique; sample volume 100 μl, i.e. absolute detection limit 1.2×10^{-9} g.

Abbreviations: See p. 187.

which could be removed by the use of a leaner flame, but only at the expense of introducing interference from calcium and magnesium.

5.20. MAGNESIUM

5.20.1. GENERAL CHARACTERISTICS AND FLUORESCENCE SPECTRUM

The atomic fluorescence spectrum of magnesium is very simple and consists only of the main resonance line at 2852.1 Å ($^1P_1^0 - {}^1S_0$ transition). The resonance lines at 2025.8 and 4571.1 Å have too low *gf*-values to be observed in fluorescence.

Magnesium fluorescence has been the subject of many investigations, the most complete and important of which appear to be those of Ellis and Demers [140, 142], West and Williams [429], Larkins and Willis [259] and Kolihová [244, 245]. Except for the first-cited authors, all the others excited the atomic fluorescence of magnesium with a boosted-output hollow-cathode lamp of Sullivan-Walsh design, which appears to be the most intense excitation source for this element. Continuum excitation sources such as the 450-W xenon arc [140, 164, 190] yield detection limits only about one order of magnitude higher. A quite low limit of detection was obtained by using an electrodeless discharge lamp source [483], However, magnesium electrodeless discharge lamps are subject to attack on the quartz by magnesium metal, and have very short lifetimes. Although this problem can be overcome to some extent (using special glass with quartz windows, cf. Section 3.2.2.3.4), magnesium electrodeless discharge lamps do not seem to be promising as excitation sources.

Magnesium fluorescence can be easily generated in all types of commonly used analytical flames. Of the premixed flames, hydrocarbon flames yield higher fluorescence signals and signal-to-noise ratios than hydrogen flames. For the latter flames, richer flames generally result in higher sensitivity. For the air-hydrogen flame especially, the hydrogen flow-rate is critical (with respect to sensitivity). An approximately twofold higher signal and signal-to-noise ratio result from the use of an argon-oxygen-hydrogen flame instead of an air-hydrogen flame [245].

Nitrous oxide supported flames result in lower fluorescence signals and signal-to-noise ratios [259, 429], but the nitrous oxide-acetylene flame should always be used when serious solute volatilization interferences are expected [429].

5.20.2. EXPERIMENTAL CONDITIONS AND DETECTION LIMITS

Papers contributing to the atomic fluorescence flame spectrometry of magnesium are briefly reviewed in Table 5.28. Data from this table clearly demonstrate the facts partly summed up in the previous paragraph. The hollow-cathode lamp sources and the shielded air-acetylene flame usually result in the best sensitivity. Note that lowest limits of detection can be obtained when the monochromator is replaced with a suitable interference filter [143, 145, 146, 259]. Note also that with the non-monochromator system and with complete non-dispersive systems, lowest limits of detection occur in the air-hydrogen flame, owing to the more favourable flame background emission of this flame. A very interesting result was obtained with a radiofrequency plasma source [206].

The upper limit of linearity of the calibration curve for the air-acetylene flame is 2 μg/ml [245] and for the nitrous oxide-acetylene flame about 100 μg/ml [429].

5.20.3. INTERFERENCES

Serious solute volatilization interferences occur, as expected, in a cool hydrogen-entrained air flame [62, 140]. These have already been discussed in Section 4.2.4. Bratzel, Dagnall, and Winefordner [62] studied the effect of several typical interferents on magnesium in premixed and unpremixed oxygen-hydrogen, argon/hydrogen/entrained air, nitrous oxide-hydrogen, and nitrous oxide-acetylene flames on a direct-injection burner, and found the premixed oxygen-hydrogen and particularly the nitrous oxide-acetylene flames to be the most efficient for the suppression of these effects. Silicate still interfered in the nitrous oxide-acetylene flame, but the interference could be reduced by the addition of a small amount of EDTA at pH 12.

West and Williams [429] observed no cationic interferences (at a 1000-fold ratio) with magnesium in a premixed nitrous oxide-acetylene flame. Aluminium, molybdenum and titanium all caused a depressive interference in the air-propane and air-acetylene flames; with up to 800 μg/ml of aluminium, for example, the signal for 1 μg/ml of magnesium decreased with increase of aluminium concentration, but remained constant at higher concentrations. Similar behaviour was found for the other elements which form refractory oxides. Hence, magnesium can be reliably determined at reduced sensitivity in the presence of such elements by ensuring that enough interferent is present. In the presence of aluminium, for example, magnesium can be determined in the range 0.1 – 2.0 μg/ml with only a 50% loss in sensitivity.

TABLE 5.28. Papers contributing to the atomic fluorescence flame spectrometry of magnesium at 2852.1 Å

Ref.	Spectrometer Detector Band-width (Å)	Excitation source	Flame (burner)	Detection limit (μg/ml) S : N = 2
[103]	Unicam SP 900A EMI 9601 B 6.0	150 W Xe-arc	A-P(p) A-H(p)	2.0[a]
[411, 412]	L.I. RCA 1P-28 6.4	150 W Xe-arc	Ar-H, e.a.(t) Beckman	0.2[a]
[264, 265]	Perkin—Elmer 303 RCA 1P-28 20	150 W Xe-arc	A-H(p)	0.1
		HCL(Intensitron)	A-H(p)	0.01[b]
[58, 483]	L.I. RCA 1P-28 4.8	EDL	Ar-H, e.a.(t) Zeiss	0.008[a]
[58, 267, 268]	L.I. RCA 1P-28 8.0	EDL	Ar-H, e.a.(t) O-H(t) Beckman	0.2[a]
[140, 164]	L.I. EMI 6255 B[c] 2.8	450 W Xe-arc	H-e.a.(t) HETCO	0.004[a]
			O-H(t) HETCO	0.18[a]
[349]	Jarrell—Ash (Versat.) RCA 1P-28 6.4	HHCL	A-H(t) HETCO	0.03
[312]	Jarrell—Ash (Versat.) RCA 1P-28 4.8	MVDL, Hg-90 W[d]	A-H(t) HETCO	0.5
[206]	Unicam SP 900 A EMI 9601 B	R.F. plasma	A-P(p)	0.005
[429]	Techtron AA-4 HTV R-106 9.9	BOHCL(S-W)	A-P(p) A-Ac(p)	0.001
			No-Ac(p)	0.05
[366]	L.I. RCA 1P-28 3.2	EDL	Ar-H, e.a.(t) HETCO	0.01[a]
[65]	L.I. RCA 1P-28	150 W Xe-arc Eimac	Ar-H, e.a.(t) Zeiss	0.04
[291, 292, 293, 294]	L.I. RCA 1P-28 filter, $\lambda_{max} = 2875$ $\Delta\lambda_{(1/2)} = 115$	HCL(p)	A-H(p)	0.001
			A-Ac(p)	0.02[e]

TABLE 5.28 (continued)

Ref.	Spectrometer Detector Band-width (Å)	Excitation source	Flame (burner)	Detection limit (μg/ml) S : N = 2
[143, 145, 146, 147]	Technicon AFS-6 filter, $\lambda_{max} = 2851$ $\Delta\lambda_{(1/2)} = 35$	HCL(p)	A-Ac(p)	0.005[f] 0.000 1[g]
[190]	Beckman 440 RCA 1P-28	150 W Xe-arc	A-H(p) Autolam	0.038
		450 W Xe-arc	A-H(p) Autolam	0.002 4
[253]	L.I. HTV R-166 4.0	150 W Xe-arc	A-H(t) Beckman 4020	0.2[h]
[259]	Techtron AA-4 HTV R-106 9.9	BOHCL(S-W)	A-Ac(p, s)	0.000 5
			A-H(p)	0.000 8
			No-Ac(p, s)	0.004
			No-H(p)	0.03
	Techtron AA-4[i] HTV R-166 filter, $\lambda_{max} = 2850$ $\Delta\lambda_{(1/2)} = 40$	BOHCL(S-W)	A-Ac(p, s) A-H(p) No-Ac(p, s) No-H(p)	0.000 09 0.000 05 0.003 0.01
	Techtron AA-4[i] HTV R-166 N.S.	BOHCL(S-W)	A-Ac(p, s) A-H(p) No-Ac(p, s) No-H(p)	0.000 5 0.000 1 0.08 0.07
[258]	Techtron AA-4 HTV R-106 9.9	HCL	A-Ac(p)	0.005
			A-Ac(p, s)	0.000 8
		BOHCL(S-W)	A-Ac(p)	0.000 6
			A-Ac(p, s)	0.000 15
	Techtron AA-4[i] HTV R-166 N.S.	HCL	A-Ac(p)	0.009
			A-Ac(p, s)	0.001 5
		BOHCL(S-W)	A-Ac(p)	0.000 9
			A-Ac(p, s)	0.000 2
[244, 245]	Techtron AA-4 HTV R-106 9.9	BOHCL(S-W)	A-Ac(p, s)	0.000 7
			A-H(p)	0.000 7
			Ar-O-H(p, s)	0.000 3

[a] S : N = 1.
[b] Reference [265].
[c] Photomultiplier cooled at —15 °C.
[d] Excitation with the overlapping Hg 2852.42 Å.
[e] Reference [294].
[f] Reference [143].
[g] Linear range 0—8 μg/ml.
[h] Photon counting detection.
[i] Without monochromator.

Abbreviations: See p. 187.

5.21. MANGANESE

5.21.1. GENERAL CHARACTERISTICS AND FLUORESCENCE SPECTRUM

Although atomic fluorescence of manganese has been reported by many authors [e.g. 58, 65, 103, 114, 164, 176, 483], in most cases only limits of detection for manganese resonance fluorescence lines have been reported. The most extensive study was carried out by Ebdon, Kirkbright, and West [158] with an electrodeless discharge lamp source. Argon-fllled lamps have been found to be more stable and have a longer operating life than those containing helium [360]. The most satisfactory results were obtained for lamps which contained *ca.* 1 mg of manganese(II) chloride and argon at a pressure of 4 torr, operated in the three-quarter wave resonant cavity at a power input of 50 W.

The atomic fluorescence spectrum of manganese obtained with this source showed that the most intense fluorescence occurred at the 2800 and 4030 Å resonance triplets. The individual lines of these triplets were unresolved with the spectral band-pass available. The use of an optical wide-band filter to prevent irradiation of the flame by 2800 Å radiation from the source revealed that the fluorescence emission signal at 4030 Å originated from both resonance and stepwise-line fluorescence.

Surprising, however, were the observations of atomic fluorescence at the 3044.6, 3577.9 and 3834.4 Å non-resonance lines and particularly of ionic

TABLE 5.29. Atomic fluorescence spectrum of manganese [245]

Wavelength (Å)	Energy levels (eV)	Transitions	Relative source emission intensity[a, b]	Relative fluorescence intensity[a, b, c]
2794.8	0—4.433	$a^6S_{5/2} - y^6P^0_{7/2}$	13.9	100
2798.3	0—4.428	$a^6S_{5/2} - y^6P^0_{5/2}$	12.4	73.6
2801.1	0—4.423	$a^6S_{5/2} - y^6P^0_{3/2}$	11.0	47.7
4030.8	0—3.075	$a^6S_{5/2} - z^6P^0_{7/2}$	100	22.1
4033.1	0—3.073	$a^6S_{5/2} - z^6P^0_{5/2}$	84.8	19.5
4034.5	0—3.072	$a^6S_{5/2} - z^6P^0_{3/2}$	61.5	14.6

[a] HCL-source.
[b] All intensities corrected for detector spectral response.
[c] In A-Ac(p, s) flame.

resonance fluorescence lines of considerable relative fluorescence intensities at 2576.1, 2593.7, and 2605.7 Å. Sufficient population of the a^6D manganese states is highly improbable even at the relatively high temperature of the air-acetylene flame as is high ionization of manganese atoms which could give rise to ionic fluorescence of the relative intensities reported. The atomic fluorescence spectrum of manganese investigated in our laboratory with an excitation source which is probably of comparable absolute intensity (a shielded A.S.L. manganese hollow-cathode lamp overrun at 30 mA) and of very similar relative emission intensities at both atomic and ionic lines, and measured in the same flame as used by Ebdon *et al.* [158], consisted only of resonance fluorescence at both resonance manganese triplets (see Table 5.29). No ionic fluorescence was observed. The interesting results reported by Ebdon *et al.* would need independent verification with a similar source.

5.21.2. EXPERIMENTAL CONDITIONS AND DETECTION LIMITS

Detection limits reported for manganese determination by atomic fluorescence are summarized in Table 5.30. The 2794.8 Å line (usually as part of the unresolved ultraviolet resonance triplet) exhibits the highest signal-to-noise ratio. The lowest limit of detection (0.5 ng/ml) so far reported was obtained with an improved prototype Technicon AFS-6 multichannel atomic fluorescence spectrometer [145, 146]. High sensitivity has also been attained with 450-W xenon arc continuum sources [164, 190].

The shielded air-acetylene flame seems to be the best flame for manganese with regard to sensitivity and freedom from interferences. At the 2800 Å triplet, the atomic fluorescence signal was reported to be about 30% greater in the shielded flame and the signal to noise level increased by a factor of two on shielding. A linear calibration curve was obtained up to 10 μg/ml [158].

5.21.3. INTERFERENCES

Of 26 ions studied at 1000-fold weight ratio in the shielded air-acetylene flame at 2800 Å, only four produced serious interferences. Three of the most refractory elements – thorium, silicon, and vanadium – gave a positive interference (about 4, 5 and 16%, respectively), probably by scattering of incident radiation from the source. Magnesium seriously reduced the fluorescence signal (−37%) even when only a 100-fold amount was present (−11%).

TABLE 5.30. Papers contributing to the atomic fluorescence flame spectrometry of manganese

Ref.	Spectrometer Detector Band-width (Å)	Excitation source	Flame (burner)	Wave-length (Å)	Detection limit (μg/ml) S : N = 2
[103]	Unicam SP 900A EMI 9601 B *ca.* 8.0	150 W Xe-arc	A-H(p)	2794.8	0.15[a]
			A-P(p)	2794.8	0.30[a]
[152]	L.I. RCA 1P-28	HHCL	A-H(t) Beckman 4020	2794.8	5.0
[264, 265]	Perkin—Elmer 303 EMI 9592 B 20	150 W Xe-arc	A-H(p)	2794.8	1.0; 0.5[b]
		HCL (Intensitron)	A-H(p)	2794.8	0.05[b]
[58, 483]	L.I. RCA 1P-28 8.0	EDL	Ar-H, e.a.(t) Zeiss	2794.8	0.006[a]
[58, 267, 268]	L.I. RCA 1P-28 8.0	EDL	A-H(t) O-H(t) Beckman 4020	2794.8	1.0[a]
[164]	L.I. EMI 6255 B[c]	450 W Xe-arc	H-e.a.(t) HETCO	2794.8	0.04[a]
			O-H(t) HETCO	2794.8	35
[312]	Jarrell—Ash (Versat.) RCA 1P-28 6.4	MVDL, Hg-90 W[d]	A-H(t) HETCO	2794.8	0.5
[114]	L.I. EMI 6256 B 17	EDL	A-H(p)	2794.8	0.014[a]
[349]	Jarrell—Ash (Versat.) RCA 1P-28 6.4	HHCL	A-H(t) HETCO	2794.8	0.05
[206]	Unicam SP 900A EMI 9601 B	R.F. plasma	A-P(p)	2794.8	0.1
[158]	Unicam SP 900A EMI 9601 B 16.0	EDL	A-Ac(p, s)	2794.8[e] 2798.3[e] 2801.1[e]	0.001
				4030.8[e] 4033.1[e] 4034.5[e]	0.02

TABLE 5.30 (continued)

Ref.	Spectrometer Detector Band-width (Å)	Excitation source	Flame (burner)	Wave-length (Å)	Detection limit (μg/ml)
[65]	L.I. RCA 1P-28	150 W Xe-arc Eimac	Ar-H, e.a.(t) Zeiss	2794.8	0.2
[97]	S.A. A 1740 HTV R-213	EDL[f]	O-H(t) Beckman	2794.8	0.1
[143, 145, 146, 147]	Technicon AFS-6 filter, $\lambda_{max} = 2803$, $\Delta\lambda_{(1/2)} = 32.5$	HCL(p)	A-Ac(p)	2794.8	0.003[g] 0.000 5[h]
[190]	Beckman 440 RCA 1P-28	150 W Xe-arc	A-H(p) Autolam	2794.8	0.030
		450 W Xe-arc	A-H(p) Autolam	2794.8	0.0025
[253]	L.I. HTV R-166 12.0	150 W Xe-arc	A-H(t) Beckman 4020	2794.8	0.01[j]
[176]	L.I. RCA 1P-28 14.4	L	A-H(p) (Tri-flame)	4030.8[e] 4033.1[e] 4034.5[e]	0.3
[258]	Techtron AA-4[j] HTV R166 N.S.	HCL	A-Ac(p)	Σ	0.09
			A-Ac(p, s)	Σ	0.01[k]

[a] S : N = 1
[b] Reference [265].
[c] Photomultiplier cooled at —15 °C.
[d] Excitation with Hg lamp continuum.
[e] Unresolved lines.
[f] Dual-element Mn-Cr source.
[g] Reference [143].
[h] Linear range 0—8 μg/ml.
[i] Photon counting detection.
[j] Without monochromator.
[k] Upper linear limit 1.0 μg/ml.
Abbreviations: See p. 187.

Interference studies with a continuum source [96] showed that the manganese triplet at 4030 Å was susceptible to spectral interference from the 4044.1 Å potassium resonance line (when a wide spectral band-pass was used) and from the gallium 4033.0 Å resonance line. Potassium (1000 μg/ml) was found to give a 12% enhancement in the determination of 20 μg/ml of manganese, but the same concentration of gallium gave less than 5% enhancement.

5.22. MERCURY

5.22.1. GENERAL CHARACTERISTICS AND FLUORESCENCE SPECTRUM

The main resonance line of mercury lies at 1849 Å and is not accessible to conventional instrumentation because of appreciable atmospheric absorption at this wavelength. Resonance fluorescence at the weaker intercombination line at 2536.5 Å ($^3P_1 - {}^1S_0$ transition) has, therefore, been generally used so far in all atomic fluorescence studies on mercury. Mercury metal vapour discharge lamps used in early atomic fluorescence studies [445] exhibited a high degree of self-reversal so that poor detection limits were obtained. Much lower limits of detection have resulted from using electrodeless discharge lamps (simply prepared from the metal) [73, 90, 356, 483] and particularly mercury pen lamps (cf. Section 3.2.2.1) [414–416], germicidal lamps [258, 417] and a specially constructed mercury discharge tube [420].

5.22.2. EXPERIMENTAL CONDITIONS AND DETECTION LIMITS

Atomization of mercury does not present any problem, because mercury exhibits a relatively high atomic vapour pressure even at room temperature. Detection limits attained in various flames with various sources are listed in Table 5.31. The possibility of using low-temperature flames with low background emission and the favourable wavelength of the mercury fluorescence line are good conditions for non-dispersive fluorescence systems to be used [258, 361, 415]. The lowest detection limit reported so far for the atomic flame-fluorescence spectrometric determination of mercury was obtained with such a system and an HTV R-166 solar-blind photomultiplier. The calibration curve was found to be linear from the detection limit (0.25 μg/ml) up to a concentration of 1000 μg/ml [420].

TABLE 5.31. Papers contributing to the atomic fluorescence flame spectrometry of mercury at 2536.5 Å

Ref.	Spectrometer Detector Band-width (Å)	Excitation source	Flame (burner)	Detection limit (μg/ml) S : N = 2
[445]	L.I. RCA 1P-28 8.0	MVDL Hanovia	O-H(t) Beckman	5.0[a]
			O-Ac(t) Beckman	1.0[a]
[446]	L.I. RCA 1P-28 6.4	EDL	O-H(t) Beckman	0.1[a]
			O-Ac(t) Beckman	0.1[a]
[266]	L.I. RCA 1P-28 16	EDL	O-H(t) Beckman	0.1[a]
[58, 483]	L.I. RCA 1P-28 16	EDL	A-H(t) Zeiss	0.1[b]
[267, 268]	L.I. RCA 1P-28 16	EDL	O-H(t) Beckman	0.1[b]
[114]	L.I. EMI 6256 B 17	EDL	A-H(p)	0.10[b]
[414]	L.I. RCA 1P-28	Hg-pen	O-H(t) HETCO	0.002[c]
[415]	L.I. HTV R-166 N.S.	Hg-pen	A-H(t) Beckman 4020	1.0[d]
[73]	Perkin—Elmer 290 RCA 1P-28 20	EDL	A-H(p)	0.08[b]
[417]	L.I. FEU-39A	Hg-g. lamp	A-P(p)	0.002
[420]	L.I. RCA 1P-28 32	MVDL[e]	A-TG(p)	0.02
	L.I. RCA 1P-28 filter[f]	MVDL[e]	A-TG(p)	0.002
	L.I. HTV R-166 N.S.	MVDL[e]	A-TG(p)	0.000 25

TABLE 5.31 (continued)

Ref.	Spectrometer Detector Band-width (Å)	Excitation source	Flame (burner)	Detection limit (μg/ml) S : N = 2
[270, 433]	S.A. A 1740 HTV R-213 60	EDL[g]	A-P(p)	0.6
[399]	S.A. A 3000 HTV R-213 12.0	EDL(mod.)	A-Ac(p)	0.2[h]
[65]	L.I. RCA 1P-28	150 W Xe-arc Eimac	A-H(z) Zeiss	100
[97]	S.A. A 1740 HTV R-213	EDL[i,j]	O-H(t) Beckman	12.0[i] 16.0[j]
[145, 146]	Technicon AFS-6 filter	EDL(p)	A-Ac(p)	0.4[k]
[356]	L.I. RCA 1P-28 8.0	EDL	A-H(p, d.i.)	0.03[l]
[165]	L.I. HTV R-166 filter	EDL	Ar-H-e.a.(t) HETCO	2.5
[416]	L.I. HTV R-166 Chlorine filter	Hg-pen	Ar-O-H(p)	1.0
[90]	L.I. EMI 9601B *ca.* 10	EDL	A-P(p)	0.25[m] 0.05[n]
[258]	Techtron AA-4 HTV R-106 9.9	Hg-g. lamp	A-Ac(p) A-Ac(p, s)	2.0 0.5
	Techtron AA-4[o] HTV R-166 N.S.	Hg-g. lamp	A-Ac(p) A-Ac(p, s)	0.7 0.07

[a] S : N = 0.75.
[b] S : N = 1.
[c] In MIBK
[d] Upper linear limit 10 μg/ml.
[e] Special construction.
[f] Baird-Atomic prominent ultraviolet line filter A.1.
[g] Multielement Bi—Hg—Se—Te source.
[h] Linearity range 0.5—80 μg/ml.
Abbreviations: See p. 187.

[i] Dual-element Hg—Ag source.
[j] Dual-element Hg—Zn source.
[k] Linear range 0—50 μg/ml.
[l] With aluminium ellipse.
[m] D.C. measurement, linear range 0 to 300 μg/ml.
[n] Photon counting, band-width 20 Å, linear range 0—50 μg/ml.
[o] Without monochromator.

The use of reducing agents in the analyte significantly increases the atomic fluorescence signal of mercury. Browner *et al.* [73]. found that the addition of excess of ascorbic acid to mercury(II) solutions produced about a 4-fold increase in the fluorescence signal. A very sensitive flame method for mercury has been developed by Vitkun *et al.* [417]. A solution containing mercury and a solution of stannous chloride were simultaneously nebulized

TABLE 5.32. Precision of the determination of mercury by the reduction of mercury to metal during atomization [417]

Mercury concentration (μg/ml)	Relative fluorescence signal	Absolute standard deviation of the signal	Relative standard deviation, %
0.01	3.35	0.76	23.1
0.025	8.4	1.06	12.6
0.05	17.2	1.06	5.8
0.10	34.5	1.48	4.3
0.20	70.0	1.16	1.65

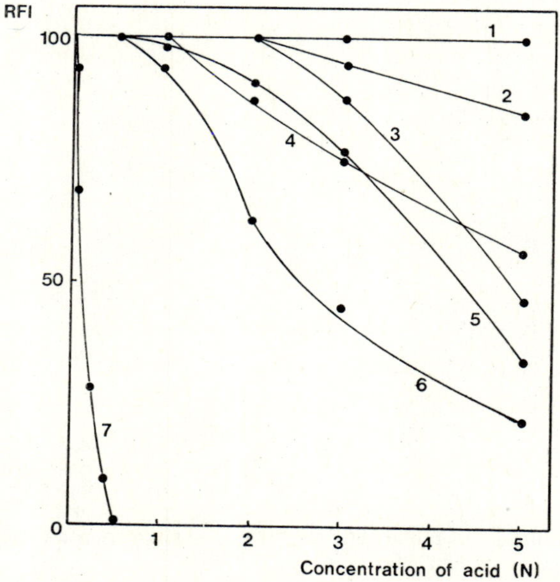

FIG. 5.3. Effects of acid on relative fluorescence intensity of mercury in a premixed air-propane flame [417]: (1) H_3PO_4; (2) H_2SO_4; (3) $HClO_4$; (4) CH_3COOH; (5) HNO_3; (6) HCl; (7) HBr.

into a chamber-type nebulizer-burner. Mercury was very efficiently reduced to the metal under these conditions and reproducibly supplied into the flame gases. A detection limit of 2 ng/ml was reported. Precision data on the 1 μg/ml level are shown in Table 5.32. Without use of the reducing agent, a drastic drop in sensitivity (about two orders of magnitude) resulted.

Vitkun and co-workers also performed a detailed investigation of interferences encountered with this system. Ions which form complex anions with mercury (such as halides), sulphides, and nitrates interfere owing to the decreased rate of mercury reduction. The effect of some anions is shown in Fig. 5.3. Note e.g. the more pronounced effect of hydrobromic acid than of hydrochloric acid, which compares well with the stability of $HgCl_4^{2-}$ and $HgBr_4^{2-}$ anions. Of a range of cations examined, only those capable of reduction with stannous chloride (silver, gold, platinum, selenium, tellurium) interfere, by the formation of metal amalgams or mercury selenide (or telluride).

5.23. MOLYBDENUM

5.23.1. GENERAL CHARACTERISTICS AND FLUORESCENCE SPECTRUM

Attempts to generate the atomic fluorescence of molybdenum in hydrogen/entrained air and hydrogen-oxygen flames were unsuccessful [483, 484]. The first detailed atomic fluorescence study on molybdenum was carried out in our laboratory [277, 388]. Fluorescence excited with a boosted-output hollow-cathode lamp in a nitrogen-shielded fuel-rich air-acetylene flame was measured at a carefully optimized fuel-to-oxidant ratio. Although the fluorescence signal-to-noise ratios were not too satisfactory from the analytical point of view, primarily owing to the insufficiently intense excitation source and relatively strong background emission of the fuel-rich air acetylene flame (despite the separation of the reaction zones) in the region 3000 – 4000 Å, the results obtained gave a satisfactory picture of the atomic fluorescence spectrum of this element, and some interferences were investigated. A more detailed study was later made by Dagnall *et al.* [118], who described the atomic fluorescence characteristics of molybdenum in a separated nitrous oxide-acetylene flame, and used an electrodeless discharge lamp source.

Details of the fluorescence spectrum of molybdenum obtained with both excitation sources are shown in Table 5.33. All fluorescence lines observed

TABLE 5.33. Atomic fluorescence spectrum of molybdenum [118, 388]

Wavelength (Å)	Energy levels (eV)	Transition	Relative source emission intensity		Relative fluorescence intensity	
			A[a]	B[b]	A[a,c]	B[b,d]
3132.6	0—3.957	$a^7S_3 - y^7P_4^0$	49	100	96	100
3158.2	0—3.925	$a^7S_3 - z^7D_3^0$	18	26	10	13
3170.3	0—3.909	$a^7S_3 - y^7P_3^0$	31	63	39	59
3194.0	0—3.881	$a^7S_3 - y^7P_2^0$	25	58	32	46
3798.3	0—3.263	$a^7S_3 - z^7P_4^0$	92	80	100	35
3864.1	0—3.208	$a^7S_3 - z^7P_3^0$	100	67	83	25
3903.0	0—3.176	$a^7S_3 - z^7P_2^0$	67	54	46	7

[a] Excited with a BOHCL(S-W); all intensities corrected for detector spectral response [388].
[b] Excited with an EDL; all intensities uncorrected for detector spectral response [118].
[c] In a A-Ac(p, s) flame.
[d] In a No-Ac(p, s) flame.

are due to resonance fluorescence. The marked differences in relative fluorescence intensities are caused mainly by different spectral distributions of emitted power of the two sources, but partly by the fact that measurements were made in different flames.

5.23.2. EXPERIMENTAL CONDITIONS AND DETECTION LIMITS

Detection limits obtained in our study (2.0, 9.0 and 10.0 μg/ml for the 3132.6, 3798.3 and 3864.1 Å lines, respectively) with a modified Techtron AA-4 instrument have little analytical significance. The poor detection limits at 3798.3 and 3964.1 Å are caused by the very unfavourable signal-to-noise ratio in consequence of strong background emission from the fuel-rich air-acetylene flame at these wavelengths. Demers and Mitchell [143] achieved a surprisingly good detection limit (0.2 μg/ml) in the air-acetylene flame at 3132.6 Å with a Technicon AFS-6 atomic fluorescence instrument. This value is only twice that obtained with the same equipment using an argon-shielded nitrous oxide-acetylene flame.

When an electronically modulated microwave-excited electrodeless discharge lamp source was used [118, 433], the most intense fluorescence line at 3132.6 Å yielded detection limits of 0.75 and 0.46 μg/ml (signal-to-noise ratio = 1) for the nitrogen- and argon-shielded fuel-rich nitrous oxide-acetylene flames, respectively. The range of linearity of the calibration curve was *ca.* 5 – 50 μg/ml, with a marked curvature toward the concentration axis between 100 and 1000 μg/ml.

It should be pointed out that detection limits in both flames are appreciably affected by fuel-to-oxidant ratios and changes in observation height in the flame.

5.23.3. INTERFERENCES

There are a considerable number of interferences in the molybdenum determination by atomic fluorescence (see Table 5.34). The extent of these interferences is not unexpected because the atomic absorption of molybdenum is also subject to considerable interferences, both in air-acetylene and nitrous oxide-acetylene flames. The results listed in Table 5.34 are very difficult to compare and evaluate, because different excesses of

TABLE 5.34. Effect of diverse ions on the determination of molybdenum by atomic fluorescence spectrometry [118, 277]

Interferent	Change in fluorescence signal, %		Interferent	Change in fluorescence signal, %	
	No-Ac[a]	A-Ac[b]		No-Ac[a]	A-Ac[b]
Al	+34	—5	Ni	+10	+11
As	+25	N.M.	Pb	+29	N.I.
Ca	N.M.	—95	Ta	+27	N.M.
Cd	+18	N.M.	V (V)	+27	+5
Co	N.I.	+5	W (VI)	N.I.	+5
Cr	+31	—11	Zn	+26	N.I.
Cu	+33	N.I.	Cl^-	N.I.	N.I.
Fe (III)	N.I.	+7	NO_3^-	N.I.	N.I.
K	+63	N.M.	SO_4^{2-}	N.I.	N.I.
Mg	+20	—78	PO_4^{3-}	N.I.	N.I.
Mn	N.M.	—39	F^-	N.I.	—57
Na	—44	N.I.			

[a] 25 μg/ml Mo added (ammonium molybdate); 100-fold excess of diverse ions.
[b] 50 μg/ml Mo added (ammonium molybdate); 10-fold excess of diverse ions.
[c] 200-fold excess.

diverse ions and different experimental conditions were utilized to get the values.

More work is required to understand more fully the mechanisms of these interferences and to find means for their elimination.

5.24. NICKEL

5.24.1. GENERAL CHARACTERISTICS AND FLUORESCENCE SPECTRUM

The first investigation of the atomic fluorescence of this element was made by Armentrout [23]. He used a boosted-output hollow-cathode lamp

TABLE 5.35. Atomic fluorescence spectrum of nickel [276]

Wavelength (Å)	Energy levels (eV)	Transition	Relative lamp emission intensity[a, b]	Relative fluorescence intensity[a, b, c]
2290.0	0—5.412	$a^3F_4 - x^3F_3^0$	5	5
2311.0	0—5.363	$a^3F_4 - w^3F_4^0$	19	52
2312.3	0.165—5.525	$a^3F_3 - w^3F_3^0$	8	12
2314.0	0.275—5.631	$a^3F_2 - w^3F_2^0$	13	7
2320.0	0—5.342	$a^3F_4 - y^3G_5^0$	32	100
2325.8	0.165—5.494	$a^3F_3 - y^3G_4^0$	12	6
2345.5	0—5.284	$a^3F_4 - x^3D_3^0$	14	13
3002.5	0.025—4.153	$a^3D_3 - y^3D_3^0$	59	13[d]
3003.6	0.109—4.235	$a^3D_2 - y^3D_2^0$	22	
3050.8	0.025—4.088	$a^3D_3 - y^3F_4^0$	74	12
3101.6[e]	0.109—4.105	$a^1D_2 - y^3F_3^0$	57[e]	9[e]
3101.9[e]	0.423—4.418	$a^1D_2 - y^1F_3^0$		
3134.1	0.212—4.167	$a^3D_1 - y^3F_2^0$	25	7
3414.8	0.025—3.655	$a^3D_3 - z^3F_4^0$	95	13
3461.7	0.025—3.606	$a^3D_3 - z^5F_4^0$	99	4
3515.1	0.109—3.635	$a^3D_2 - z^3F_3^0$	58	6
3524.5	0.025—3.542	$a^3D_3 - z^3P_2^0$	100	20

[a] Excited with a BOHCL(S-W).
[b] All intensities corrected for detector spectral response.
[c] In an A-H flame.
[d] Lines unresolved in fluorescence spectrum.
[e] Lines unresolved both in emission and fluorescence spectrum.

of Sullivan-Walsh design as an excitation source and observed strong fluorescence at 3524.5 and 2320.0 Å.

The same type of lamp was used as a source in a more complete study carried out by Matoušek and Sychra [276], who described in detail the atomic fluorescence spectrum, the shapes of calibration curves for all lines observed, and limits of detection in various premixed and turbulent flames.

Relative intensities of nickel lines exhibiting atomic fluorescence are listed in Table 5.35. The complexity of the atomic spectrum of nickel does not permit complete evaluation of the mechanism of the fluorescence transitions. Since nickel has several low-lying metastable states which can be thermally populated, all fluorescence lines listed in Table 5.35 are probably due largely to resonance fluorescence, but the possibility of contributions from stepwise-line fluorescence cannot be excluded. Very weak fluorescence was also observed at several other lines (above 3000 Å) not included in Table 5.35.

5.24.2. EXPERIMENTAL CONDITIONS AND DETECTION LIMITS

Boosted-output hollow-cathode lamps of Sullivan-Walsh design seem to be the most suitable excitation source for nickel at present. Further improvement in the sensitivity of nickel determination by atomic fluorescence can be expected with lamps of Löwe design, which have been reported to be approximately ten times more intense [260] than the earlier type.

Electrodeless discharge lamps (both of single and dual-element type) exhibit comparable radiative output but lower stability. Lamps prepared from nickel chloride gave the best results [114, 306]. Detection limits obtained with continuous excitation sources are considerably inferior to those obtained with line sources because of the low radiance of xenon arc lamps below 2500 Å.

The study made in our laboratory showed that sheathed premixed flames give the best signal-to-noise ratios [266, 267] (cf. also Table 3.10). Although the highest signal-to-noise ratio was obtained in the oxy-hydrogen flame diluted with argon, from the practical point of view (freedom from chemical interferences and scatter), the separated air-acetylene flame should be preferred in spite of the fourfold drop in sensitivity in this flame.

The 2320.0 Å line gave the lowest limits of detection. Eight other lines yielded detection limits below 1 μg/ml [276] (see Table 5.36). The calibration curve at 2320.0 Å in the argon-oxygen-hydrogen flame was linear up to 10 μg/ml. Less sensitive lines (e.g. 3414.8, 3461.6 Å) exhibited wider linear ranges, and could be used for practical analytical measurements to avoid dilution problems. Since the four most sensitive lines of nickel lie in the

TABLE 5.36. Papers contributing to the atomic fluorescence flame spectrometry of nickel

Ref.	Spectrometer Detector Band-width (Å)	Excitation source	Flame (burner)	Wave-length (Å)	Detection limit (μg/ml) S:N=2
[23]	L.I. RCA 1P-28 20	BOHCL(S-W)	A-H(t) Beckman 4020	2320.0	0.1[a]
[152]	L.I. RCA 1P-28	HHCL	A-H(t) Beckman 4020	2320.0	0.5
[264, 265]	Perkin—Elmer 303 EMI 9592 B 20	150 W Xe-arc	A-H(p)	2320.0	3.0
		BOHCL(S-W)	Ar-O-H(p)	2320.0	0.003[b]
		HCL(Intensitron)	A-H(p)	2320.0	0.1
[483]	L.I. RCA 1P-28 16.0	EDL	A-H(t) Zeiss	2320.0	0.04[c]
[164]	L.I. EMI 6255 B[d] 2.8	450 W Xe-arc	H-e.a.(t) HETCO	2320.0	1.0[c]
			O-H(t) HETCO	2320.0	5.0[c]
[312]	Jarrell—Ash(Versat.) RCA 1P-28 6.4	MVDL, Hg-90 W[e]	A-H(t) HETCO	2320.0	3.0
[114]	L.I. EMI 6256 B 17	EDL	A-H(p)	2320.0	0.006[c]
				3414.8	0.10[c]
				2345.5	0.20[c]
[349]	Jarrell—Ash(Versat.) RCA 1P-28 6.4	HHCL	A-H(t) HETCO	3414.8	1.0
[366]	L.I. RCA 1P-28 20.0	EDL	Ar-H-e.a.(t) HETCO	2320.0	0.04[c]
[276, 277, 388]	Techtron AA-4 HTV R-106 9.9	BOHCL(S-W)	Ar-O-H(p)	2320.0	0.003
			Ar-O-H(t)	2320.0	0.006
			A-Ac(p)	2320.0	0.03
			A-H(t)	2320.0	0.005
			A-H(p)	2320.0	0.006

TABLE 5.36 (continued)

Ref.	Spectrometer Band-width (Å) Detector	Excitation source	Flame (burner)	Wave-(Å) length	Detection limit (μg/ml) S:N=2
			A-H(p)	2311.0	0.1[f]
				3414.8	0.2
				2290.0	0.3
				2345.5	0.3
				3002.5[g] 3003.6[g]	0.3
				3524.5	0.35
				3050.8	0.5
				3461.6	0.6
[65]	L.I. RCA 1P-28	150 W Xe-arc Eimac	A-H(t) Zeiss	2320.0	10.0
[97]	S.A. A 1740 HTV R-213	EDL[h]	O-H(t) Beckman	2320.0	0.8
[143, 145, 146, 147]	Technicon AFS-6 filter, $\lambda_{max}=2250$ $\Delta\lambda_{(1/2)}=525$	HCL(p)	A-Ac(p)	2320.0	0.08[i] 0.001[j]
[306]	S.A. A 1740 HTV R-213 30	EDL[k]	A-Ac(p, s)	2320.0	0.02
[258]	Techtron AA-4[l] HTV R-166 N.S.	BOHCL(L)	A-Ac(p)	Σ	0.01
			A-Ac(p, s)	Σ	0.002[m]

[a] S : N = 0.75.
[b] Reference [264].
[c] S : N = 1.
[d] Photomultiplier cooled at —15 °C.
[e] Excitation with Hg lamp continuum.
[f] Band-width 6.6 Å.
[g] Lines unresolved in fluorescence.
[h] Dual-element Ni-Cr source.
[i] Reference [143].
[j] Linear range 0—21 μg/ml.
[k] Dual-element Ni–Co source
[l] Without monochromator.
[m] Upper linear limit 5 μg/ml.
Abbreviations: See p. 187.

optimum spectral response region of the solar-blind photomultiplier, the application of simple non-dispersive systems resulted in very low detection limits [258].

5.24.3. INTERFERENCES

Serious depressive effects of some refractory elements and mineral acids on nickel were observed in the premixed air-hydrogen flame [277]. In the hotter air-acetylene flame, these solute volatilization interferences were practically eliminated. Norris and West [306] also reported no interferences from 100-fold ratios of 24 cations and 1000-fold ratio of iron in the shielded air-acetylene flame. Serious spectral interferences were observed from the group of cobalt lines between 3502.3 and 3533.4 Å with the nickel line at 3524.5 Å excited with a continuum source [96].

Larkins [258] demonstrated that, in the case of nickel (an element with a rich atomic emission spectrum), the use of a non-dispersive system with an R-166 solar-blind photomultiplier could lead to much higher scatter-to-fluorescence ratio than use of a monochromator, since all fluorescence and non-fluorescence lines from the lamp were scattered in the flame and detected by the photomultiplier. Thus scatter, which is negligible with a dispersive system, could be a problem when a non-dispersive system is used (under otherwise identical conditions).

5.25. PALLADIUM

5.25.1. GENERAL CHARACTERISTICS AND FLUORESCENCE SPECTRUM

The literature on the atomic fluorescence of palladium is relatively sparse. The only detailed study was carried out in our laboratory [391]. A palladium boosted-output hollow-cathode lamp of Sullivan-Walsh design proved much more satisfactory as an excitation source than an electrodeless discharge lamp source prepared according to the method of Aldous *et al.* [5]. The latter source (prepared from the chloride) operated both in the 1/4-wave and 3/4-wave cavities was rather unstable and less intense than expected.

The atomic fluorescence spectrum of palladium is very complex. Spectral characteristics of lines which exhibit atomic fluorescence are listed in Table 5.37. The highest fluorescence intensities correspond to the lines having lower electronic states at 0.814 and 0.961 eV ($^3D_3$ and $^3D_2$ multiplet levels, respectively). Considering the high population of the ^3D multiplet levels (see characteristic concentrations in atomic absorption for corresponding lines [391]), the measured fluorescence signals for the lines originating from these levels are mostly due to resonance fluorescence. In those cases

TABLE 5.37. Atomic fluorescence spectrum of palladium [391]

Wavelength (Å)	Energy level (eV)	Transition	Relative lamp emission intensity[a, b]	Relative fluorescence intensity[b, c]
2447.9	0–5.063	$^1S_0 - {}^1P_1^0$	9	10
2476.4	0–5.005	$^1S_0 - {}^3D_1^0$	8	5
2763.1	0–4.486	$^1S_0 - {}^3P_1^0$	7	2
3242.7	0.814–4.636	$^3D_3 - {}^3D_3^0$	49	32
3251.6	1.251–5.063	$^3D_1 - {}^1P_1^0$	6	4
3258.8	1.251–5.055	$^3D_1 - {}^1D_2^0$	9	3
3302.1	1.251–5.005	$^3D_1 - {}^3D_1^0$	12	7
3373.0	0.961–4.636	$^3D_2 - {}^3D_3^0$	18	13
3404.6	0.814–4.454	$^3D_3 - {}^3F_4^0$	100	100
3421.2	0.961–4.584	$^3D_2 - {}^3D_2^0$	43	19
3433.4	1.453–5.063	$^1D_2 - {}^1P_1^0$	12	7
3441.4	1.453–5.055	$^1D_2 - {}^1D_2^0$	19	4
3460.8	0.814–4.395	$^3D_3 - {}^3F_3^0$	36	13
3481.2	1.251–4.812	$^3D_1 - {}^3F_2^0$	33	5
3489.8	1.453–5.005	$^1D_2 - {}^3D_1^0$	6	2
3516.9	0.961–4.486	$^3D_2 - {}^3P_1^0$	40	17
3553.1	1.453–4.946	$^1D_2 - {}^1F_3^0$	49	8
3609.6	0.961–4.395	$^3D_2 - {}^3F_3^0$	61	40
3634.7	0.814–4.224	$^3D_3 - {}^3P_2^0$	90	98

[a] Excited with a BOHCL(S-W).
[b] All intensities corrected for detector spectral response.
[c] In an Ar-O-H flame.

where the upper level of the transition concerned is also optically connected with the ground state of the atom, contributions from direct-line fluorescence are also likely to occur (e.g. for the lines at 3251.6, 3302.1, 3516.9 Å, etc.). Stepwise-line fluorescence could also contribute in some cases, because of the relatively low intensities of fluorescence corresponding to resonance transitions to the $^1S_0$ level.

The highest fluorescence signals were obtained in the oxygen-hydrogen flame diluted with argon. Increases of 2.2, 2.8, 2.3, 2.6, 2.2, 2.0, 1.2, and 1.5-fold, resulting from using the argon-oxygen-hydrogen flame in place of the air-hydrogen flame, were found for the 3404.6, 3634.7, 3609.6, 3242.7, 3421.2, 3516.9, 2447.9, and 3302.1 Å lines, respectively [391]. It is very interesting to note that the ratio of relative fluorescence signals for the 3404.6 and 3634.7 Å lines varies significantly with the flame used (1.0, 1.25, and 1.6 in the argon-oxygen-hydrogen, air-hydrogen, and separated air-acetylene flames, respectively). This suggests differences in the quantum efficiencies of fluorescence at the two transitions in all three flames. High-temperature nitrous oxide-supported flames yield much lower fluorescence signals [259].

5.25.2. EXPERIMENTAL CONDITIONS AND DETECTION LIMITS

Basic experimental data of systems used for the determination of palladium by atomic fluorescence spectrometry and corresponding detection limits are listed in Table 5.38. Eight lines give detection limits better than 1.0 μg/ml in the argon-oxygen-hydrogen flame, the best limit of detection being at 3404.6 Å. As expected, this flame provided better detection limits for all lines than the two other flames, except for the 3242.7 Å line [391]. For this line, the high noise of the premixed argon-oxygen-hydrogen flame in this spectral region causes deterioration of the detection limit compared with the other two flames.

Since the most sensitive fluorescence lines of palladium lie above 3400 Å, the solar-blind photomultiplier in non-dispersive systems should be replaced by an RCA 931A photomultiplier equipped with a filter transmitting this wavelength. It is of interest to note that for this purpose a Schott UG 1 glass filter is almost as good as a high-quality interference filter, and the limits of detection achieved are practically the same as those with a monochromator [259].

The detection limit obtained with a continuous excitation source is surprisingly high [265]. A lower detection limit should result from the use of a more intense continuous source (e.g. 450-W xenon arc lamp) under optimized experimental conditions.

Calibration curves (log-log plots) at 3404.6, 3634.7, and 3609.6 Å for the argon-oxygen-hydrogen flame are linear up to almost 200 μg/ml.

5.25.3. INTERFERENCES

The effects of inorganic acids and of 100-fold ratios of several ions on the fluorescence signal produced by 5 μg/ml of palladium were investigated in our laboratory [391]. The depressive effect of acids (except for phosphoric

TABLE 5.38. Papers contributing to the atomic fluorescence flame spectrometry of palladium

Ref.	Spectrometer Detector Band-width (Å)	Excitation source	Flame (burner)	Wave-length	Detection limit (μg/ml) S : N = 2
[265]	Perkin—Elmer 303 EMI 9592 B 20	150 W Xe-arc	A-H(p)	3404.6	50
		HCL (Intensitron)	A-H(p)	3404.6	2.0
[312]	Jarrell—Ash(Versat) RCA 1P-28 6.4	MVDL(Cd) 3403.7 Å	A-H(t) HETCO	3404.6	10
[391]	Techtron AA-4 HTV R-106 9.9	BOHCL(S-W)	O-Ar-H(p)	3404.6 3634.7 3609.6 3421.2 2447.9 3516.9 3242.7 3302.1	0.04 0.05 0.10 0.20 0.20 0.30 0.50 0.50
			A-Ac(p, s)	3404.6 3634.7 3609.6	0.08 0.18 0.30
			A-H(p)	3404.6 3634.7 3609.6	0.06 0.12 0.20
[259]	Techtron AA-4 HTV R-106 9.9	BOHCL(S-W)	A-Ac(p, s)	3404.6 2476.4	0.15 5.0
			A-H(p)	3404.6 2476.4	0.3 7.0
			No-H(p)	3404.6 2476.4	0.7 40
			No-Ac(p, s)	3404.6 2476.4	1.0 13
	Techtron AA-4[a] HTV R-166 N.S.	BOHCL(S-W)	A-Ac(p, s)	Σ	0.2
			A-H(p)	Σ	0.4
			No-H(p)	Σ	30
			No-Ac(p, s)	Σ	50
	Techtron AA-4[a] RCA 931 A UG 1 Schott filter $\lambda_{max} = 3\,600$; $\Delta\lambda_{(1/2)} = 600$	BOHCL(S-W)	A-Ac(p, s)	Σ	0.09
			A-H(p)	Σ	0.06
			No-H(p)	Σ	0.9

[a] Without monochromator.

Abbreviations: See p. 187.

acid), nickel, and cobalt could be eliminated by the addition of strontium (chloride) or EDTA to a final concentration of 0.2% and 0.01M, respectively. The depressive effect of phosphoric acid could be eliminated only by the addition of EDTA. Practically the same interferences were found in the separated air-acetylene flame as have been reported for atomic absorption [391], i.e. aluminium, cobalt, nickel, and hydrofluoric acid. However, these interferences were more pronounced in atomic fluorescence, particularly for nickel and aluminium. All cationic interferences in all three flames were found to be strongly dependent on the height of measurement above the burner top and on the concentration of acids in the solution analysed. No interferences were found from 100-fold ratios of Na^+, K^+, Ca^{2+}, Mg^{2+}, Mn^{2+}, Cr^{3+}, Bi^{3+}, Zn^{2+}, Cu^{2+}, Rh^{3+}, MoO_4^{2-}, VO_3^-, OsO_4^{2-}, RuO_4^{2-}, SiO_4^{2-}, and HNO_3.

5.25.4. EXTRACTION

By spraying the MIBK extracts of the palladium chelate with APDC into an air-hydrogen flame, palladium can be determined down to 0.5 ng/ml. A linear calibration curve was obtained over the concentration range 0.001 – 5.0 µg/ml [391].

5.26. PLATINUM

5.26.1. GENERAL CHARACTERISTICS AND FLUORESCENCE SPECTRUM

There are, at the time of writing, only three reports dealing with the atomic fluorescence of platinum. The study performed in our laboratory [277, 388] presented the atomic fluorescence spectrum excited with a boosted-output hollow-cathode lamp of Sullivan-Walsh design (see Table 5.39) and some detection limits in various premixed flames (see Table 5.40).

As for palladium, the atomic fluorescence spectrum of platinum is very complex. Twenty-nine lines were found to exhibit atomic fluorescence; because of the existence of low-energy metastable states, most of the fluorescence transitions are probably due to a combination of resonance fluorescence and direct-line and stepwise-line fluorescence. The most intense fluorescence signals occurred at 2659.4, 3064.7, 2174.7 and 2702.4 Å. The best limit of detection (0.5 µg/ml) was obtained in an argon-oxygen-hydrogen flame at 2659.4 Å. The detection limit at 3064.7 Å is poor because of the

TABLE 5.39. Atomic fluorescence spectrum of platinum [388]

Wavelength (Å)	Energy levels (eV)	Transition	Relative lamp emission intensity[a, b, c]	Relative fluorescence intensity[a, b, c]
2049.4	0—6.045	$^3D_3 - 32^0_3$	3	9
2067.5	0—5.994	$^3D_3 - 30^0_4$	1	3
2084.6	0.102—6.045	$^3F_4 - 27^0_3$	2	6
2103.3	0.102—6.045	$3F_4 - 30^0_4$	5	8
2128.6	0.096—5.919	$^3D_2 - 29^0_1$	10	13
2144.2	0—5.780	$^3D_3 - 27^0_3$	11	6
2174.7	?	? ?	16	38
2274.4	0.096—5.546	$^3D_2 -$?	12	3
2357.1	0.096—5.354	$^3D_2 - 18^0_1$	8	3
2440.1	0—5.079	$^3D_3 - 15^0_3$	24	6
2467.4	0—5.023	$^3D_3 - 12^0_2$	17	4
2487.2	0—4.983	$^3D_3 - 11^0_4$	43	7
2498.5	0.096—5.057	$^3D_2 - 13^0_2$	16	3
2628.0	0.096—4.812	$^3D_2 - 10^0_2$	30	16
2646.9	0—4.683	$^3D_3 - 8^0_3$	12	13
2650.9	0.102—4.778	$^3F_4 - z^5F^0_5$	60	2
2659.4	0—4.660	$^3D_3 - 7^0_4$	76	100
2677.1	0—4.630	$^3D_3 - 6^0_2$	7	3
2702.4	0.096—4.682	$^3D_2 - 8^0_3$	31	30
2705.9	0.102—4.682	$^3F_4 - 8^0_3$	20	22
2719.0	0.102—4.660	$^3F_4 - 7^0_4$	17	18
2734.0	0.096—4.630	$^3D_2 - 6^0_2$	40	21
2830.3	0—4.379	$^3D_3 - 4^0_3$	57	13
2893.9	0.096—4.379	$^3D_2 - 4^0_3$	16	3
2929.8	0—4.230	$^3D_3 - 3^0_3$	40	13
2998.0	0.096—4.230	$^3D_2 - 3^0_3$	58	17
3042.6	0.102—4.176	$^3F_4 - 2^0_5$	100	6
3064.7	0—4.044	$^3D_3 - 1^0_2$	95	73
3408.1	0.102—3.739	$^3F_4 - z^5D^0_4$	63	3

[a] Excited with a BOHCL(S-W).
[b] All intensities corrected for detector spectral response.
[c] In a shielded A-Ac flame.

presence of strong OH emission at this wavelength, even in an argon-sheathed air-acetylene flame.

Since the strongest fluorescence lines lie below 3000 Å, the use of the R-166 solar-blind photomultiplier with a non-dispersive instrument seems worthwhile. Table 5.40 shows that a non-dispersive system yields detection limits at least as good as those obtained with a monochromator, but that

TABLE 5.40. Papers contributing to the atomic fluorescence flame spectrometry of platinum

Ref.	Spectrometer Detector Band-width (Å)	Excitation source	Flame (burner)	Wave-length (Å)	Detection limit (μg/ml) S : N = 2
[265]	Perkin—Elmer 303 EMI 9592 B 20	150 W Xe-arc	A-H(p)	2659.4	500
		HCL(Intensitron)	A-H(p)	2659.4	50
[277, 388]	Techtron AA-4 HTV R-106 9.9	BOHCL(S-W)	O-Ar-H(p)	2659.4	0.5
			A-H(p)	2659.4	0.7
			A-Ac(p, s)	2659.4	1.0
				2174.7	1.7
				2128.6	2.5
				2049.4	4.0
				2103.3	4.0
				2702.4	4.0
				2084.6	6.0
				2628.0	8.0
				2144.2	10
				2646.9	10
				2719.0	10
				2734.0	10
				3064.7	10
[259]	Techtron AA-4 HTV R-106 9.9	BOHCL(L)	A-Ac(p, s)	2659.4	1.0
			A-H(p)	2659.4	1.0
			No-Ac(p, s)	2659.4	4.0
			No-H(p)	2659.4	8.0
	Techtron AA-4[a] HTV R-166 N.S.	BOHCL(L)	A-Ac(p, s)	Σ	0.2
			A-H(p)	Σ	0.15
			No-Ac(p, s)	Σ	8
	Techtron AA-4[a] HTV R-166 filter[b], $\lambda_{max} = 2660$, $\Delta\lambda_{(1/2)} = 30$	BOHCL(L)	A-Ac(p, s)	Σ	0.3
			A-H(p)	Σ	0.15
			No-Ac(p, s)	Σ	2.0

[a] Without monochromator.
[b] Corion Instrument Corporation, Waltham, Mass. U.S.A.
Abbreviations: See p. 187.

the use of an interference filter with the R-166 photomultiplier offers little or no advantage.

The detection limit published for platinum when a 150-W xenon arc lamp source is used is of no analytical importance [265]. No data have been reported for electrodeless discharge lamp sources, although lamps with satisfactory spectral characteristics have been prepared [5].

5.26.2. INTERFERENCES

Platinum is known to suffer from strong suppression by most cations and anions [302, 327]. A preliminary study made in our laboratory has shown that the extent of these interferences is practically identical in both atomic fluorescence and atomic absorption [277]. To remove these interferences, it seems desirable to use a nitrous oxide-acetylene flame for the determination of platinum, even though the atomic fluorescence signal (as well as the atomic absorption signal) is much lower in this flame than in air-supported flames [259, 302, 327].

Serious spectral interferences from iron have been reported by Larkins [258]. The 2719.04 Å line of platinum overlaps the 2719.02 Å line of iron and gives resonance fluorescence of iron at 2719.02 Å and also direct-line fluorescence at 2750.14 Å. In addition the 2524.30 Å line of platinum overlaps the 2524.29 Å line of iron and also gives weak fluorescence. The importance of this interference in practical analysis may be judged from the fact that with the non-dispersive system with the R-166 photomultiplier the fluorescence signal from 1 μg/ml of iron is approximately equal to that from 1 μg/ml of platinum.

5.27. RHODIUM

Little is known about the atomic fluorescence spectrum of rhodium, but by analogy with the atomic absorption and emission spectra of all platinum metals, the fluorescence spectrum is probably complex and mechanisms are difficult to define. Manning and Heneage [265] were able to detect 10 and 3.0 μg/ml of rhodium, using a premixed air-hydrogen flame and a 150-W xenon arc lamp and shielded Perkin-Elmer hollow-cathode lamp, respectively, at the 3692.4 Å line. Larkins [258] could detect rhodium down to 3.0 μg/ml at 3434.9 Å, using an A.S.L. hollow-cathode lamp source and a separated air-acetylene flame with a conventional instrument, and down to 0.3 μg/ml with the same flame and excitation source by using a non-dispersive fluorescence system with an RCA 931 A photomultiplier and a Schott UG 1 glass absorption filter.

5.28. RUTHENIUM

Everything stated for rhodium is also valid for this element. Apart from several detection limits, no data on the atomic fluorescence characteristics of ruthenium are available. Ruthenium fluorescence has been measured at 3728.0 Å (in fact at the 3726.9, 3728.0 and 3730.4 Å triplet corresponding to the $a^5F - z^5F_0$ transitions) to give detection limits of 100 μg/ml when excited with a 150-W xenon arc lamp in an air-hydrogen flame [265] and 5.0 μg/ml when excited in a separated air-acetylene flame with a conventional hollow-cathode lamp [258]. In the latter flame, a limit of detection of 80 μg/ml was obtained with a non-dispersive system. These discouraging detection limits should be improved by at least a factor of 20 when using a hollow-cathode lamp source in a pulsed-operation mode.

5.29. SELENIUM

5.29.1. GENERAL CHARACTERISTICS AND FLUORESCENCE SPECTRUM

The most extensive analytical study of this element was performed by Dagnall, Thompson, and West [105, 106]. Details of the preparation of electrodeless discharge lamp sources were described in the first article [105], and a description of the fluorescence transitions and the relative fluorescence intensities of various selenium lines was given in the latter article [106]. Most other authors have also used electrodeless discharge lamps as excitation sources [58, 95, 114, 190, 268, 483]. Multielement lamps have also proved satisfactory [97, 270].

Spectral characteristics of the observed selenium fluorescence lines are listed in Table 5.41. Atomic fluorescence was observed at all three lines emanating from the $^3S_1$ state, that for the line at 2039.9 Å being the most intense. The fluorescence for the transitions from the $^3S_1$ excited state to the $^3P_0$ and $^3P_1$ states is probably predominantly due to direct-line fluorescence. Some atoms could thermally populate the metastable $^3P_1$ and $^3P_0$ states,

TABLE 5.41. Atomic fluorescence spectrum of selenium [106]

Wavelength (Å)	Energy levels (eV)	Transition	Relative source emission intensity[a, b]	Relative fluorescence intensity[a, b, c]
1960.9	0—6.323	$^3P_2 - {}^3S_1^0$	8	67
2039.9	0.247—6.323	$^3P_1 - {}^3S_1^0$	42	100
2062.8	0.314—6.323	$^3P_0 - {}^3S_1^0$	100	52
2074.8	0—5.974	$^3P_2 - {}^5S_2^0$	21	2

[a] Excited with an EDL.
[b] All intensities uncorrected for detector spectral response.
[c] In an A-P flame.

however, and radiative excitation from these states cannot be ruled out. Since the resonance line absorbs 5 times more strongly than lines emanating from the metastable states, and the energy gap between the ground-state and the first metastable state is 0.25 eV, direct-line fluorescence probably makes the major contribution, particularly in the air-propane flame used [106]. However, this fact was not confirmed experimentally. A very weak fluorescence signal was obtained at the intercombination resonance line at 2074.8 Å.

5.29.2. EXPERIMENTAL CONDITIONS AND DETECTION LIMITS

Optimum detection limits have been obtained with electrodeless discharge lamp sources, as expected, and as can be seen from Table 5.42. Continuous excitation sources are practically useless because of the extremely low energy output around 2000 Å [65, 190]. Since the selenium fluorescence lines lie in the far ultraviolet region, close proximity of the excitation source, flame, and monochromator entrance slit should be achieved to avoid considerable losses of both exciting and emitted radiation by atmospheric absorption. The use of lenses in entrance optics is also undesirable. Detectors should have a good spectral response in the region of 2000 Å. The best limit of detection of 0.036 μg/ml [190] was obtained with a system which adequately fulfils the above requirements. For the reasons above, non-dispersive systems and solar-blind photomultipliers can be used to advantage. Approximately one order of magnitude gain in sensitivity was obtained by replacing a dispersive system with a non-dispersive one [258].

TABLE 5.42. Papers contributing to the atomic fluorescence flame spectrometry of selenium

Ref.	Spectrometer Detector Band-width (Å)	Excitation source	Flame (burner)	Wave-length (Å)	Detection limit (μg/ml) S : N = 2
[58, 483]	L.I. RCA 1P-28 32	EDL	Ar-H, e.a.(t) Zeiss	1960.9	0.4[a]
[267, 268]	L.I. RCA 1P-28 32	EDL	O-H(t) Beckman 4020	1960.9	0.8[a]
[105, 106]	Unicam SP 900 A EMI 9601 B *ca.* 60	EDL	A-P(p)	1960.9	0.15
[114]	L.I. EMI 6256 B 17	EDL	A-H(p)	1960.9 2039.9	0.16[a] 0.20[a]
[95]	S.A. A 1740 HTV R-106 33	EDL	A-Ac(p)	1960.9	0.20
[270, 433]	S.A. A 1740 HTV R-106 3.3	EDL[b]	A-P(p)	2039.9	10
[65]	L.I. RCA 1P-28	150 W Xe-arc Eimac	Ar-H, e.a.(t) Zeiss	1960.9	1000
[143]	Technicon AFS-6 filter, $\lambda_{max} = 2175$ $\Delta\lambda_{(1/2)} = 225$	HCL(p)	A-Ac(p)	2039.9	1.5
[398, 399]	S.A. A 3000 HTV R-213 60	EDL (mod.)	A-Ac(p)	1960.9 2039.9	0.10[c] 0.15
[97]	S.A. A 1740 HTV R-106	EDL[d,e]	O-H(t) Beckman	1960.9	0.8[d] 0.6[e]
[190]	Beckman 440 RCA 1P-28	150 W Xe-arc	Ar-H, e.a.(p)	1960.9	134
		450 W Xe-arc	Ar-H, e.a.(p)	1960.9	3.9
		EDL	Ar-H, e.a.(p)	1960.9	0.60
		EDL	A-H(p)[f]	1960.9	0.036
[258]	Techtron AA-4 HTV R-166 9.9	HCL	A-Ac(p, s)	1960.9	50
	Techtron AA-4[g] HTV R-166 N.S.	HCL	A-Ac(p, s)	Σ	6.0

[a] S : N = 1.
[b] Multielement Bi—Hg—Se—Te source (iodides).
[c] Linearity range up to 50 μg/ml.
[d] Dual-element Te—Se source.
[e] Dual-element As—Se source.
[f] Hot mode, special optics.
[g] Without monochromator.

Abbreviations: See p. 187.

5.29.3. INTERFERENCES

No effect was found from 100-fold molar ratios of Ag, Al, Co(II), Cu(II), Fe(II), Hg(II), Mg, Na, NH_4^+, Pb, Sb(III), Zn, and Te in the air-propane flame. All solutions were acidified with 2*M* nitric acid to prevent any precipitation [106]. Thompson and Wildy [399] found a depressive effect (by about 40%) from 100-fold ratios of sodium and calcium in the air-acetylene flame.

5.29.4. EXTRACTION

Cresser and West [95] extracted selenium from acid solutions into MIBK. Maximum extraction was obtained from 8.5M hydrochloric acid. A linear calibration curve was obtained for aqueous solutions between 0.08 and 100 μg/ml following extraction into an equal volume of MIBK.

5.30. SILICON

5.30.1. GENERAL CHARACTERISTICS AND FLUORESCENCE SPECTRUM

The only atomic fluorescence results published for silicon are those of Dagnall *et al.* [117] and Kirkbright *et al.* [230]. Silicon fluorescence was excited with an electronically modulated silicon electrodeless discharge lamp which was shown to be a more intense excitation source than a silicon boosted-output hollow-cathode lamp (A.S.L.). The most satisfactory discharge lamps were prepared from elemental silicon and iodine in the ratio 7 : 1. The lamps were most intense and stable in a 1/4-wave cavity operated at 100 W. The atomic fluorescence was generated in a fuel-rich argon-shielded nitrous oxide-acetylene flame.

The only silicon atomic fluorescence lines obtained were those between 2500 and 2530 Å shown in Table 5.43. All fluorescence transitions correspond to resonance fluorescence. When a wide spectral band-pass is employed to obtain the greatest sensitivity, the monochromator transmits fluorescence emission at all six lines. Under the optimum conditions of fuel flow-rate, height of observation and spectral slit-width (0.33 Å), and with a modified Techtron AA4 atomic absorption spectrophotometer, the detection limits obtained at 2516.1 Å were 2.0 μg/ml for the conventional flame and 0.7 and

TABLE 5.43. Atomic fluorescence spectrum of silicon [117]

Wavelength (Å)	Energy levels (eV)	Transition	Relative source emission intensity[a]	Relative fluorescence intensity[b]
2506.9	0.010—4.953	$^3P_1 - {}^3P_2^0$	89	45
2514.3[c]	0—4.929	$^3P_0 - {}^3P_1^0$	69	100
2516.1[c]	0.028—4.953	$^3P_2 - {}^3P_2^0$	100	
2519.2	0.010—4.929	$^3P_1 - {}^3P_1^0$	64	27
2524.1	0.010—4.919	$^3P_1 - {}^3P_0^0$	70	36
2528.5	0.028—4.929	$^3P_2 - {}^3P_1^0$	73	48

[a] Excited with an EDL.
[b] In a No-Ac flame.
[c] Lines unresolved in fluorescence spectrum.

0.55 μg/ml in the nitrogen- and argon-shielded flames, respectively. A linear calibration graph was obtained over the range 2 – 100 μg/ml. Over the range 50 – 100 μg/ml, slight curvature of the calibration graphs towards the concentration axis was observed. Repetitive introduction (10 times) of a 50 μg/ml silicon solution resulted in a relative standard deviation of 1.8%.

No interference was observed from Al, Be, Cd, Cu, Hg^{2+}, K, Mg, Mn^{2+}, Mo^{6+}, Na, Ni, Pb, Te^{4+}, Zn, Cl^-, NO_3^-, F^-, and PO_4^{3-} (at 100-fold ratio by weight) on the atomic fluorescence intensity produced by 50 μg/ml of silicon.

The fact that all silicon fluorescence lines lie within a very narrow wavelength region in which the nitrous oxide-acetylene flame exhibits relatively low emission indicates that non-dispersive atomic fluorescence measurements should be particularly useful, especially if a narrow-band interference filter with peak transmission at 2500 Å were used in addition to the solar-blind photomultiplier.

5.31. SILVER

5.31.1. GENERAL CHARACTERISTICS AND FLUORESCENCE SPECTRUM

The atomic fluorescence of silver may be considered to be well understood. Many authors have reported at least limits of detection and optimum conditions for various experimental arrangements. A more complete investigation of the atomic fluorescence of silver was carried out by West and Williams [428] and later in our laboratory [244, 245, 393, 394]. Results from these studies and from other publications [97, 366, 443, 483] indicate that comparable limits of detection can be obtained in practically all commonly used analytical flames (turbulent or premixed) and both with boosted-output hollow-cathode and electrodeless discharge lamp sources. Despite very low detection limits attainable with the latter sources, their analytical use is limited by the short lifetime caused by a rapid plating-out of the metal on the lamp walls. Aldous *et al.* [5] described in detail the preparation and spectral characteristics of silver electrodeless discharge lamps but gave no analytical results.

The atomic fluorescence spectrum of silver consists only of the two resonance lines at 3280.7 and 3382.9 Å. It is interesting to note that the ratios of signals (corrected for detector response) for the two lines emitted by the source and from the air-acetylene flame are approximately the same (~ 1.7) (see Table 5.44). Our results show a higher quantum yield for the

TABLE 5.44. Atomic fluorescence spectrum of silver [245]

Wavelength (Å)	Energy levels (eV)	Transition	Relative source emission intensity[a, b]	Relative fluorescence intensity[a, b]	
				A-Ac	Ar-O-H
3280.7	0—3.778	$^2S_{1/2} - {}^2P^0_{3/2}$	100	100	100
3382.9	0—3.664	$^2S_{1/2} - {}^2P^0_{1/2}$	59	56	79

[a] BOHCL(S-W) source.
[b] All intensities corrected for detector spectral response.

3382.9 Å line in an argon-oxygen-hydrogen flame owing to smaller quenching of this line and/or a stepwise-line fluorescence contribution to the 3382.9 Å line.

5.31.2. EXPERIMENTAL CONDITIONS AND DETECTION LIMITS

Table 5.45 summarizes the basic instrumental data of systems used to investigate the atomic fluorescence determination of silver, and the detection limits attained. Line excitation sources allow the determination to be made

TABLE 5.45. Papers contributing to the atomic fluorescence flame spectrometry of silver

Ref.	Spectrometer Detector Band-width (Å)	Excitation source	Flame (burner)	Wave-length (Å)	Detection limit (μg/ml) S : N = 2
[103]	Unicam SP 900A EMI 9601B *ca.* 12.0	150 W Xe-arc	A-H(p)	3280.7	0.15[a]
				3382.9	0.10[a]
[411, 412]	L.I. RCA 1P-28 8.0	150 W Xe-arc	O-H(t) Beckman	3280.7	0.08[a]
[152]	L.I. RCA 1P-28	HHCL	A-H(t) Beckman	3280.7	0.001
[264, 265]	Perkin—Elmer 303 EMI 9592B 20	150 W Xe-arc	A-H(p)	3280.7	0.2
		HCL (Intensitron)	A-H(p)	3280.7	0.003[b]
[58, 267, 268]	L.I. RCA 1P-28 8.0	EDL	Ar-H, e.a.(t) O-H(t) Beckman 4020	3280.7	0.01[a]
[483]	L.I. RCA 1P-28 8.0	EDL	A-H(t) Zeiss	3280.7	0.000 1[a]
[142, 162, 164]	L.I. EMI 6255B[c] 2.8	450 W Xe-arc	H-e.a.(t)	3280.7	0.001[a]
			O-H(t) HETCO	3280.7	0.003[a]
[114]	L.I. EMI 6256B 17	EDL	A-H(p)	3280.7	0.02[a]
[428]	Techtron AA-4 HTV R-213 9.9	BOHCL(S-W)	A-P(p)	3280.7	0.001 7
			A-Ac(p)	3280.7	0.003 7
[349]	Jarrell—Ash (Versat.) RCA 1P-28 6.4	HHCL	A-H(t) HETCO	3280.7	0.03
[366]	L.I. RCA 1P-28 8.0	EDL	Ar-H, e.a.(t) HETCO	3280.7	0.000 5[a]
[65]	L.I. RCA 1P-28	150 W Xe-arc Eimac	A-H(t) Zeiss	3280.7	0.05
[97]	S.A. A 1740 HTV R-213	EDL[d, e, f]	O-H(t) Beckman	3280.7	0.002[d] 0.000 08[e] 0.004[f]

TABLE 5.45 (continued)

Ref.	Spectrometer Detector Band-width (Å)	Excitation source	Flame (burner)	Wave-length (Å)	Detection limit (μg/ml) S : N = 2
[291, 292, 393, 294]	L.I. RCA 1P-28 filter, $\lambda_{max} = 3375$ $\Delta\lambda_{(1/2)} = 70$	HCL(p)	A-H(p)	3382.9	0.002
			A-Ac(p)	3382.9	0.04[g]
[143, 145, 146, 147]	Technicon AFS-6 filter, $\lambda_{max} = 3287$ $\Delta\lambda_{(1/2)} = 42$	HCL(p)	A-Ac(p)	3280.7	0.07[h] 0.002[i]
[190]	Beckman 440 RCA 1P-28	150 W Xe-arc	A-H(p) Autolam	3280.7	0.06
		450 W Xe-arc	A-H(p) Autolam	3280.7	0.012
[253]	L.I. RCA 1P-28 4.0	150 W Xe-arc	A-H(t) Beckman 4020	3280.7	0.01[j]
[31]	Perkin—Elmer 403 HTV R-136 20	HCL(p) Intensitron	A-Ac(p, s)	3280.7	0.02
[258, 443]	Techtron AA-4[k] RCA 931A UG 1 Schott filter $\lambda_{max} = 3600$; $\Delta\lambda_{(1/2)} = 600$	BOHCL(L)	A-Ac(p)	Σ	0.001
			A-Ac(p, s)	Σ	0.000 15
[244, 245, 393, 394]	Techtron AA-4 HTV R-106 9.9	BOHCL(S-W)	A-H(p)	3280.7	0.000 3
				3382.9	0.000 5
			Ar-O-H(p, s)	3280.7	0.000 3
				3382.9	0.000 5
			A-Ac(p, s)	3280.7	0.000 8
				3382.9	0.001 5
				3280.7	0.000 13[l]

[a] S : N = 1.
[b] Reference [265].
[c] Photomultiplier cooled at —15 °C.
[d] Dual-element Ag—Zn source.
[e] Dual-element Ag—Cd source.
[f] Dual-element Ag—Hg source.
[g] Reference [294].

Abbreviations: See p. 187.

[h] Reference [143].
[i] Linear range 0—9 μg/ml.
[j] Photon counting detection.
[k] Without monochromator.
[l] With the "sampling cup technique"; sample volume of 100 μl, i.e. absolute detection limit 1.3×10^{-11} g.

with detection limits down to 0.1 μg/ml. The detection limit obtained with a 450-W xenon arc continuum source under favourable conditions is only about one order of magnitude inferior [142, 162]. Premixed fuel-lean hydrogen flames yield slightly better signal-to-noise ratios than acetylene flames [244, 245]. A surprisingly low detection limit was obtained with a simple non-monochromator system and an air-acetylene flame. With this system, the significant improvement in the signal-to-noise ratio on shielding the flame is obvious [258]. The high volatility of silver allows the sampling-cup technique (cf. Section 3.3.3) to be used to advantage [393].

The useful range of determination of silver by atomic fluorescence spectrometry varies slightly with the flame used. Kolihová [244, 245] reported upper limits of linearity of 2, 15 and 20 μg/ml for the premixed air-acetylene, argon-oxygen-hydrogen and air-hydrogen flame, respectively, at 3280.7 Å.

5.31.3. INTERFERENCES

No interference was found from 1000-fold ratios of a range of foreign ions in the air-propane flame, except for a slight depression of the silver signal in the presence of aluminium, which was attributed to solute volatilization interference [428]. When a continuous excitation source and wide monochromator band-width were used, the copper line at 3274.0 Å and a group of nickel lines around 3400 Å were found to interfere in the determination of silver at 3280.7 Å and 3382.4 Å, respectively, as expected [96].

5.31.4. EXTRACTION

Following extraction of silver as its di(n-butylamine) salicylate into MIBK, as little as 0.04 ng/ml was detected in the air-propane flame [428].

5.32. SODIUM

The atomic fluorescence characteristics of sodium (as well as of the other alkali metals) are so far of theoretical interest only. Christensen and Rollefson [86] excited sodium atomic vapour enclosed in a quartz container; stepwise-line fluorescence was observed at the sodium D-doublet, i.e. excitation at 3302.3 and 3303.0 Å ($^2S_{1/2} - 4p^2P_0$ transitions) followed by collisional deactivation to the $3p^2P_0$ states and fluorescence emission at 5890.0 and

5895.9 Å ($3p^2P_0 - {}^2S_{1/2}$ transitions) and this contributed to the resonance fluorescence at the D-doublet.

In analytical flames, weak resonance fluorescence at 3032.3 and 3303.0 Å and strong resonance fluorescence at 5890.0 and 5895.9 Å occur. Alkemade [10, 11], Hooymayers and Alkemade [197, 198], and Jenkins [211, 217] investigated sodium fluorescence with regard to quenching processes in flames. Alkemade [11] concluded that in flames diluted by a noble gas, a contribution from stepwise-line fluorescence at 5895.9 Å (after excitation at 5890.0 Å) can be expected, owing to the efficient collisional mixing and low quenching of the sodium D-doublet levels.

From the analytical point of view, there is no urgent practical need to develop a sensitive fluorescence technique for the determination of sodium because of its very satisfactory atomic absorption and atomic emission characteristics. Serious problems exist regarding the preparation of sufficiently intense and stable excitation sources. The sodium (and other alkali metals) Osram and Philips vapour discharge lamps are not usable for analytical atomic fluorescence in spite of their great intensities because they are highly self-reversed when operated under any practical conditions. Electrodeless discharge lamps (radiofrequency or microwave-excited) did not perform well, despite the considerable attention given to their fabrication [6, 34, 77, 194]. However, results published by Aldous *et al.* [6], who used a special, two-ply glass (Type X 93, Glass Tubing and Components Ltd., Chesterfield, England) which is not attacked by the alkali metal, and the results of Woodward [476], seem to be very hopeful.

A detection limit of analytical significance was reported by Hell and Ricchio [190]. These authors could detect 0.0075 μg/ml of sodium by atomic fluorescence at 5890.0 Å in an air-hydrogen flame with a 450-W xenon arc lamp source. Laser-excited atomic fluorescence of sodium has been reported by Kuhl and Marowsky [254].

5.33. STRONTIUM

Surprisingly little work on atomic fluorescence has been done for this element, mainly because of the lack of sufficiently intense excitation sources and its satisfactory atomic emission characteristics. The transition of importance in atomic fluorescence spectrometry for strontium (as well as for other alkaline-earth metals) is the ${}^1P_1^0 - {}^1S_0$ resonance transition (corresponding to 4607.3 Å for strontium).

Until recently, the results obtained were rather discouraging. However, it should be remembered that all results were obtained with use of low-temperature flames. Demers [142] reported a detection limit worse than 10 μg/ml of strontium with a 450-W xenon arc source and the hydrogen/entrained air flame. Zacha *et al.* [483] described the preparation of an electrodeless discharge lamp (strontium iodide-strontium) and obtained a detection limit of 0.03 μg/ml (signal-to-noise = 2) with this source and a turbulent hydrogen/argon/entrained air flame. The same detection limit was reported by Fraser and Winefordner [176] for laser-excited atomic fluorescence flame spectrometry in the premixed air-hydrogen flame. An excellent detection limit (0.8 ng/ml) was achieved with a Technicon AFS-6 atomic fluorescence spectrometer [145, 146] in a premixed air-acetylene flame. At the time of writing no interference studies in the flame atomic fluorescence determination of strontium have been published.

5.34. TELLURIUM

5.34.1. GENERAL CHARACTERISTICS AND FLUORESCENCE SPECTRUM

The atomic fluorescence characteristics of tellurium are very similar to those of selenium (cf. Section 5.29). Several authors have mentioned tellurium fluorescence, the most important data being reported by Dagnall *et al.* [105, 106], Kolihová [245], and Hell and Ricchio [190]. Electrodeless discharge lamps, usually prepared from the iodide [105, 106, 114, 270] or from the pure metal [58, 483], have been most often used as excitation sources.

TABLE 5.46. Atomic fluorescence spectrum of tellurium [245]

Wavelength (Å)	Energy levels (eV)	Transition	Relative source emission intensity[a, b]	Relative fluorescence intensity[a, b, c]
2142.7	0—5.783	$^3P_2 - {}^3S_1^0$	29	100
2259.0	0—5.486	$^3P_2 - {}^5S_2^0$	100	1.9
2383.3	0.584—5.783	$^3P_0 - {}^3S_1^0$	64	22.3
2385.8	0.589—5.783	$^3P_1 - {}^3S_1^0$	72	40.8

[a] Excited with a BOHCL(S-W).
[b] All intensities corrected for detector spectral response.
[c] In an A-H flame.

The spectral characteristics of lines exhibiting atomic fluorescence are listed in Table 5.46. The most intense fluorescence occurred at 2142.7 Å, from resonance fluorescence. The intercombination resonance line at 2259.0 Å gave only a weak fluorescence signal. Very weak fluorescence was also observed at 2081.0 Å. Since the separation of the ^3P triplet is larger than in the case of selenium, the fluorescence for transitions from the $^3S_1$ excited state to the $^3P_0$ and $^3P_1$ metastable states (2383.3 and 2385.8 Å lines, respectively) is probably due to virtually pure direct-line fluorescence. The fact that tellurium atoms absorb 10 times more strongly at 2142.7 Å than at 2383.3 and 2385.8 Å supports this [106]. The latter fluorescence lines are usually not resolved.

5.34.2. EXPERIMENTAL CONDITIONS AND DETECTION LIMITS

Detection limits attained with various instruments are summarized in Table 5.47. The lowest limit of detection (0.0055 μg/ml) was obtained with an electrodeless discharge lamp source positioned lengthwise and horizontally close to a premixed air-hydrogen flame burning on a slot burner (Autolam, in the hot-mode) [190]. Results from our laboratory [245] indicated that the boosted-output hollow-cathode lamp of Sullivan-Walsh design is also a sufficiently intense excitation source and that a nitrogen-sheathed premixed air-acetylene flame should be the flame atomizer of choice with respect to sensitivity and freedom from possible interferences. A surprisingly low detection limit was obtained with a 450-W xenon arc lamp source [190], and this could also be of practical use.

Calibration curves (log-log plots) for the air-acetylene flame were linear from the detection limit up to concentrations of 30, 50 and 800 μg/ml for the 2142.7 Å line, 2383.3 and 2385.8 doublet, and 2259.0 Å line, respectively. The last-mentioned line can be used with advantage to avoid dilution problems.

5.34.3. INTERFERENCES

A 100-fold molar ratio of Ag, Al, Co(II), Cu(II), Fe(II), Hg(II), Mg, NH_4^+, Sb(III), Se, or Zn had no effect on the atomic fluorescence signal of tellurium in a premixed air-propane flame [106]. The same ratios of sodium and calcium were reported to depress the signal by about 20% in the air-acetylene flame [399].

TABLE 5.47. Papers contributing to the atomic fluorescence flame spectrometry of tellurium

Ref.	Spectrometer Detector Band-width (Å)	Excitation source	Flame (burner)	Wave-length (Å)	Detection limit (μg/ml) S : N = 2
[58, 483]	L.I. RCA 1P-28 16	EDL	A-H(t) Beckman 4020	2142.7	0.5[a]
[267, 268]	L.I. RCA 1P-28 16	EDL	O-H(t) Beckman 4020	2142.7	0.5[a]
[105, 106]	Unicam SP 900 A EMI 9601 B *ca.* 60	EDL	A-P(p)	2142.7	0.05
[114]	L.I. EMI 6256 B 17	EDL	A-H(p)	2142.7	0.06[a]
[270, 433]	S.A. A 1740 HTV R-106 3.3	EDL[b]	A-P(p)	2259.0	1.5
[65]	L.I. RCA 1P-28	150 W Xe-arc Eimac	A-H(t) Zeiss	2142.7	50
[398, 399]	S.A. A 3000 HTV R-213 30	EDL (mod.)	A-Ac(p)	2142.7	0.05[c]
[97]	S.A. A 1740 HTV R-106	EDL[d]	O-H(t) Beckman	2142.7	1.5
[190]	Beckman 440 RCA 1P-28	150 W Xe-arc	Ar-H, e.a.(p)	2142.7	4.2
		450 W Xe-arc	Ar-H, e.a.(p)	2142.7	0.79
		EDL	Ar-H, e.a.(p)	2142.7	0.48
		EDL	A-H(p)[e]	2142.7	0.005 5
[245]	Techtron AA-4 HTV R-106 9.9	BOHCL(S-W)	A-Ac(p, s)	2142.7	0.05
				2385.8	0.5[f]
				2383.3	1.0[f]
				2259.0	4.0
				2159.8	20
				2081.0	70
			A-H(p)	2142.7	0.12
[258]	Techtron AA-4 HTV R-106 9.9	HCL	A-Ac(p, s)	2142.7	20
	Techtron AA-4[g] HTV R-166 N.S.	HCL	A-Ac(p, s)	Σ	3.0

[a] S : N = 1.
[b] Multielement Bi—Hg—Se—Te source (iodides).
[c] Linearity range up to 100 μg/ml.
[d] Dual-element Te—Se source.
[e] Hot mode, special optics.
[f] Spectral band-width 1.7 Å.
[g] Without monochromator.

Abbreviations: See p. 187.

5.35. THALLIUM

5.35.1. GENERAL CHARACTERISTICS AND FLUORESCENCE SPECTRUM

Intense fluorescence signals have been obtained from thallium in both premixed and turbulent hydrogen and acetylene flames. Bratzel *et al.* [62] examined the fluorescence height profiles in several premixed and unpremixed hydrogen flames burning on a Zeiss direct-injection burner, and suggested an optimum height of measurement of 7 cm for both types of air-hydrogen flame, and for the premixed nitrous oxide-hydrogen flame, while in the unpremixed nitrous oxide-hydrogen flame the signal reached a maximum at 9 cm.

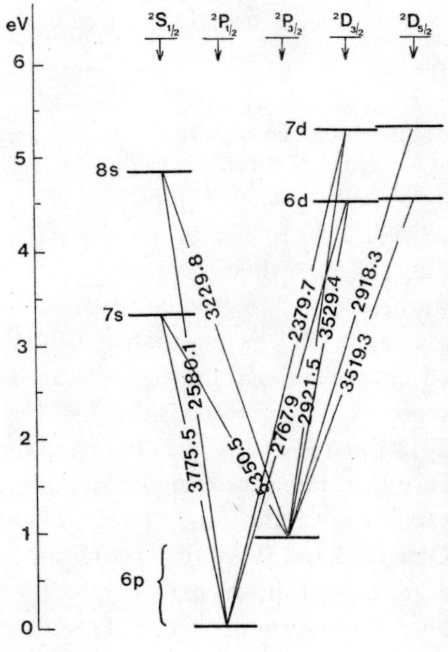

FIG. 5.4. Simplified atomic term diagram for thallium.

Electrodeless discharge lamps prepared from the metal [483] or chloride [73] as well as metal vapour discharge lamps [45, 186, 258, 266, 312] and a xenon arc continuum [65, 103, 142, 164, 190] have proved to be sufficiently intense excitation sources. Browner *et al.* [71] found a thallous chloride electrodeless discharge lamp approximately 5 times more intense than a Wotan thallium vapour discharge lamp.

Although the 3775.7 Å line has been almost exclusively used for atomic fluorescence measurements, the atomic fluorescence spectrum of thallium is quite complex (see Fig. 5.4). Mansfield *et al.* [266] observed fluorescence at seven lines, which they listed in the order of decreasing relative intensities: 3775.7, 5350.5, 3519.2, 3529.4, 2767.9, 2580.1, and 2379.7 Å. Browner *et al.* [73] reported relative fluorescence intensities of the three most intense lines excited with an electrodeless discharge lamp (Table 5.48). The lines at

TABLE 5.48. Atomic fluorescence spectrum of thallium [73]

Wavelength (Å)	Energy levels (eV)	Transition	Relative source emission intensity[a, b]	Relative fluorescence intensity[a, b, c]
2767.9	0—4.478	$^2P^0_{1/2} - 6d^2D_{3/2}$	13	18
3775.7	0—3.283	$^2P^0_{1/2} - 7s^2S_{1/2}$	100	100
5350.5	0.966—3.283	$^2P^0_{3/2} - 7s^2S_{1/2}$	80	36

[a] Excited with an EDL.
[b] Intensities uncorrected for detector spectral response.
[c] In an A-H flame.

3775.7, 2580.1, 2767.9, and 2379.7 Å are due predominantly to resonance fluorescence, while those at 5350.5, 3229.8, 2921.5, and 3529.4 Å are due to direct-line fluorescence. Because the $^2P_{3/2}$ metastable level lies too high above the $^2P_{1/2}$ ground-state (0.966 eV) to be significantly thermally populated in conventional low-temperature flames [313], only direct-line fluorescence could be responsible for the fluorescence emission. The lines at 2918.3 and 3519.3 Å must be due to thermally-assisted stepwise-line fluorescence (e.g. radiative excitation at 2767.9 Å to the $^2D_{3/2}$ level, then thermal excitation to the $^2D_{5/2}$ level, followed by emission at 3519.3 Å, etc.) Omenetto and Rossi [316] confirmed experimentally that in a high-temperature flame (nitrous oxide-acetylene), the $^2P_{3/2}$ metastable level is appreciably thermally populated, and the fluorescence signal at 5350.5 Å is a mixture of direct-line and resonance fluorescence. They were able to measure the thermally-assisted anti-Stokes fluorescence at 3775.7 Å (15.5% of the total fluorescence signal at this wavelength). However, the potential practical use of this type of fluorescence is considerably smaller than in the case of gallium and indium, because of the lower contribution of anti-Stokes fluorescence, and the unfavourable signal-to-noise ratio in the nitrous oxide-acetylene flame. Sufficiently sensitive and the most selective fluorescence signals can be obtained by irradiating the flame at 3775.7 Å via a filter which does not pass radiation of longer wavelength, and measuring the

TABLE 5.49. Basic papers contributing to the atomic fluorescence flame spectrometry of thallium

Ref.	Spectrometer Detector Band-width, Å	Excitation source	Flame (burner)	Wave-length (Å)	Detection limit (μg/ml) S : N = 2
[446]	L.I. RCA 1P-28	EDL	O-H(t) Beckman	3775.7	1.0[a]
[266]	L.I. RCA 1P-28 3.2	MVDL	O-H(t) Beckman	3775.7	0.04[a]
[411, 412]	L.I.; RCA 1P-28 5.0	150 W Xe-arc	O-H(t) Beckman 4020	3775.7	0.55[b]
[103]	Unicam SP 900A EMI 9601 B *ca.* 25	150 W Xe-arc	A-P(p) A-H(p)	3775.7	1.0[b] 0.5[b]
[186]	L.I. RCA 1P-28	MVDL	A-H(t) Beckman 4020	3775.7	0.5
[152, 152]	L.I. RCA 1P-28	HHCL	A-H(t) Beckman 4020	3775.7	5.0
[265]	Perkin—Elmer 303 EMI 9592 B 20	150 W Xe-arc	A-H(p)	3775.7	0.2
		HCL(Intensitron)	A-H(p)	3775.7	0.1
[312]	Jarrell—Ash (Versat.) RCA 1P-28 6.4	MVDL (Hg 90 W) Philips	A-H(t) HETCO	3775.7	0.3
[58, 267, 268]	L.I. RCA 1P-28 8.0	EDL	A-H(t) Beckman, Zeiss	3775.7	0.1[b]
[483]	L.I. RCA 1P-28 8.0	EDL	Ar-H e.a.(t)	3775.7	0.008[b]
[142, 164]	L.I. EMI 6255 B[c] 2.7	450 W Xe-arc	H-e.a.(t) HETCO	3775.7	0.07[b]

TABLE 5.49 (continued)

Ref.	Spectrometer Detector Band-width, Å	Excitation source	Flame (burner)	Wave-length (Å)	Detection limit (μg/ml) S : N = 2
[114]	L.I. EMI 6256 B 17	EDL	A-H(p)	3775.7	0.20[b]
[349]	Jarrell—Ash (Versat.) RCA 1P-28 6.4	MVDL	A-H(t) HETCO	3775.7	0.1
		HHCL	A-H(t) HETCO	3775.7	0.1
[73]	Perkin—Elmer 290 RCA 1P-28 20	EDL	A-H(p)	3775.7	0.12[b]
[190]	Beckman 440 RCA 1P-28	450 W Xe-arc	A-H(p) Autolam	3775.7	0.073
		150 W Xe-arc	A-H(p) Autolam	3775.7	0.13
		EDL	A-H(p)[d] Autolam	3775.7	0.027
[45]	Jarrell—Ash (Versat.) RCA 1P-28 6.4	MVDL Philips	A-Ac(t) HETCO	3775.7	0.005
[258]	Techtron AA-4 HTV R-106 1.0	MVDL	A-Ac(p, s)	3775.7 2767.8	0.1 0.5
	Techtron AA-4[e] HTV R-166 N.S.	MVDL	A-Ac(p, s)	Σ	0.3

[a] S : N = 0.75.
[b] S : N = 1.
[c] Photomultiplier cooled at —15 °C.
[d] Hot mode.
[e] Without monochromator.

Abbreviations: See p. 187.

direct-line fluorescence at 5350.5 Å. Under these conditions there is negligible source scatter and low flame background.

5.35.2. EXPERIMENTAL CONDITIONS AND DETECTION LIMITS

Detection limits attained for thallium with various instruments and sources are listed in Table 5.49. The results obtained under similar conditions agree quite well. It should be taken into account that the lowest detection limit reported for the use of a metal vapour discharge lamp source was obtained with entrance optics that were 5–10 times more efficient than those used in other systems [45]. Browner *et al.* [73] reported the fluorescence signal in the premixed air-hydrogen flame to be practically independent of the fuel flow and of the flame height up to 25 mm above the top of the burner head, and the calibration curve for the 3775.7 Å line to be linear up to 250 μg/ml. A relatively low limit of detection was obtained with a non-dispersive system and solar-blind photomultiplier [258].

5.35.3. INTERFERENCES

There are no data on "chemical" interferences associated with the determination of thallium by atomic fluorescence. However, serious spectral interference from iron lines at 3734.9 and 3737.1 Å was reported to affect the fluorescence signal at 3775.7 Å excited with a d.c. 500-W xenon arc source [96] when wide band-pass was used. Nickel did not interfere at 3775.6, 3783.5, and 3807.1 Å. This was attributed to the very low atomic population available in the excited state 0.423 eV above the ground-state, from which these lines originate.

5.36. TIN

5.36.1. GENERAL CHARACTERISTICS AND FLUORESCENCE SPECTRUM

Several papers have dealt briefly with the atomic fluorescence spectrometry of tin [65, 72, 114, 244, 245, 258, 265, 398]. Of these, only the study by Browner *et al.* [72] gives a more complete picture of the atomic fluorescence of tin in flames.

Generally, little is known about the atomization characteristics of tin in flames. Hydrogen flames (as in atomic absorption) yield higher analytical signals than acetylene flames. Kolihová [245] reported decreasing fluorescence signals in the following order of premixed flames: argon-oxygen-hydrogen > argon-hydrogen > air-hydrogen > air-acetylene. The fluorescence signal in the air-hydrogen and argon-hydrogen flames was found to be considerably dependent on the height of measurement in the flame but not significantly on the hydrogen flow. In an argon-oxygen-hydrogen flame, the oxygen flow-rate was quite critical. Richer flames exhibited higher fluorescence signals but were very prone to serious interference [72]. To overcome this, leaner flames should be used as a compromise between sensitivity and tendency to solute vaporization interferences. Slevin *et al.* [362] found greater atomization efficiency and more uniform distribution of tin atoms in a sheathed Alkemade burner supplied with air-acetylene mixtures than in the standard Méker-type burner. Bratzel *et al.* [61] studied the influence of the quenching properties of several nebulizing gases on the fluorescence signal and signal-to-noise ratio of tin in premixed laminar and unpremixed

TABLE 5.50. Atomic fluorescence spectrum of tin [72, 245]

Wavelength (Å)	Energy levels (eV)	Transition	Relative source emission intensity		Relative fluorescence intensity[a]	
			A[b]	B[c]	A[b]	B[c]
2246.1	0—5.518	$^3P_0 - {}^3D_1^0$	20	1.00	14	3.70
2354.8	0.210—5.473	$^3P_1 - {}^3D_2^0$	44	32.1	16	3.61
2429.5	0.425—5.526	$^3P_2 - {}^3D_3^0$	52	54.3	18	7.42
2546.6	0—4.867	$^3P_0 - {}^1P_1^0$	13	46.0	4	5.92
2706.5	0.210—4.789	$^3P_1 - {}^3P_2^0$	51	380	20	81.5
2840.0	0.425—4.789	$^3P_2 - {}^3P_2^0$	97	500	44	215
2863.3	0—4.329	$^3P_0 - {}^3P_1^0$	51	330	55	237
3009.1	0.210—4.329	$^3P_1 - {}^3P_1^0$	38	420	42	200
3034.1	0.210—4.295	$^3P_1 - {}^3P_0^0$	63	620	80	519
3175.1	0.425—4.329	$^3P_2 - {}^3P_1^0$	87	780	100	482
3262.3	1.068—4.867	$^1D_2 - {}^1P_1^0$	100	570	13	8.14
3330.6	1.068—4.789	$^1D_2 - {}^3P_2^0$	11	725	5	37.0
3801.0	1.068—4.329	$^1D_2 - {}^3P_1^0$	20	718	19	66.7

[a] In an Ar-O-H flame.

[b] Excited with a BOHCL(S-W); all intensities corrected for detector spectral response [245].

[c] Excited with an EDL; intensities uncorrected for detector spectral response [72].

turbulent air-hydrogen and argon-hydrogen-entrained air flames produced by using direct-injection nebulizer-burners. Best fluorescence intensities resulted when using argon rather than oxygen or nitrogen as the nebulizing gas.

Details of tin lines exhibiting atomic fluorescence are given in Table 5.50 together with the relative intensities of these lines when excited both with an electrodeless discharge lamp and a boosted-output hollow-cathode lamp source. It can be seen that the fluorescence transitions involved can be very complex and it is difficult to elucidate the mechanisms involved.

The lines at 2246.1, 2546.6, and 2863.3 Å are due to resonance fluorescence. No fluorescence signal was observed at the lowest resonance wavelength (2073.1 Å). Other lines originating from the same upper excited state as these resonance lines are due to direct-line fluorescence (3262.3, 3009.1, 3175.1, 3801.0 Å lines). It should be noted, however, that lines having the $^3P_1$ state as the lower level of the transition contain a large contribution from resonance fluorescence (due to the thermal population of the low-lying $^3P_1$ state). Absorption at 2706.5 Å gives rise to direct-line fluorescence at 2840.0 and 3330.6 Å. In this case, possible contributions from stepwise-line fluorescence after absorption at 2546.6 Å and collisional deactivation from the $^1P_1$ to the $^3P_2$ state cannot be excluded. The line of greatest analytical importance (at 3034.1 Å) is a mixture of stepwise-line fluorescence and resonance fluorescence. Filtering studies made by Browner *et al.* [72] for the group of fluorescence lines optically connected to the 2863.3 Å resonance line showed the existence of an appreciable population of tin atoms in the $^3P_1$ state and confirmed the existence of resonance fluorescence at 3034.1 and 3009.1 Å. The relative populations caused by thermal excitation at the estimated temperatures of the argon-oxygen-hydrogen flame (1440 K) and the argon-hydrogen flame ($\leq$ 1140 K) were calculated to be 0.7 : 1 and $<$ 0.6 : 1. The unexpectedly high ratio of the intensities at 3034.1 and 2863.3 Å in the argon-hydrogen flame implies that the $^3P_1$ state is populated by a quasiluminescence process, probably involving chemical reduction of the tin salts.

5.36.2. EXPERIMENTAL CONDITIONS AND DETECTION LIMITS

Detection limits attained for tin by atomic fluorescence are listed in Table 5.51. Of these, only values obtained with electrodeless discharge lamp sources seem to be promising. It has been stressed [72] that in order to obtain the really intense and long-lived source necessary for fluorescence work, the ratio of tin to iodine is critical and must lie between the limits of

TABLE 5.51. Papers contributing to the atomic fluorescence flame spectrometry of tin

Ref.	Spectrometer Detector Band-width (Å)	Excitation source	Flame (burner)	Wave-length (Å)	Detection limit (μg/ml) S : N = 2
[265]	Perkin—Elmer 303 EMI 9592 B 20	150 W Xe-arc	A-H(p)	2863.3	>1000
		HCL(Intensitron)	A-H(p)	2863.3	100
[114]	L.I. EMI 6256 B 17	EDL	A-H(p)	3034.1	0.60[a]
[398]	S.A. A 3000 HTV R-213 12.0	EDL(mod.)	A-Ac(p)	3034.1	0.5
[72]	Perkin—Elmer 290 RCA 1P-28 20.0	EDL(mod.)	Ar-O-H(p, s)	3034.1	0.10[a]
[65]	L.I. RCA 1P-28	150 W Xe-arc Eimac	A-H(t) Zeiss	3034.1	5.0
[258]	Techtron AA-4[b] HTV R-166 N.S.	HCL	A-Ac(p, s)	Σ	3.0
[244, 245, 393]	Techtron AA-4 HTV R-106 9.9	BOHCL(S-W)	Ar-H(p)	3034.1	1.0
				2863.3	2.0
				3009.1	2.0
				3175.1	2.0
				3263.3	5.0
			Ar-O-H(p, s)	3034.1	2.0
				2863.3	3.5
				3009.1	4.0
				3175.1	5.0
				3263.3	15
			A-H(p)	3034.1	10
				2863.3	35
				3009.1	15
				3175.1	15
				3263.3	50
			A-Ac(p, s)	3034.1	30
				2863.3	60
				3009.1	60
				3175.1	50
				3263.3	150

[a] S : N = 1.

[b] Without monochromator.

Abbreviations: See p. 187.

2.5 : 1 and 3.5 : 1 by weight. A boosted-output hollow-cathode lamp of Sullivan-Walsh design used in our laboratory had a very weak light output.

Although the highest fluorescence signals were obtained with an argon-oxygen-hydrogen flame, a better signal-to-noise ratio resulted from using an argon-hydrogen diffusion flame, as can be seen in Table 5.51 [245]. In all the flames investigated, the 3034.1 Å line gave the best detection limit. Browner *et al.* [72] reported the fluorescence emission at 3034.1 Å in a nitrogen-shielded argon-oxygen-hydrogen flame to give the best limit of detection (0.10 μg/ml) under almost entirely interference-free conditions.

The upper linear limit of calibration curves at 3034.1 Å was 250 μg/ml in the argon-oxygen-hydrogen flame [72] and 500 μg/ml in the argon-hydrogen diffusion flame [245]. The 3263.3 Å line (direct-line fluorescence) exhibited greatest linearity in all the flames investigated [245].

5.36.3. INTERFERENCES

Under favourable flame conditions (cf. [72]), 100-fold ratio of Al, Ba, Cd, Co, Cs, Cu, Fe(III), Hg, K, Li, Mn, Na, Ni, NH_4^+, Pb, Rb, Ti(IV), Tl, Zn, Zr, Br^-, Cl^-, HPO_4^-, I^-, MnO_4^-, NO_3^-, PO_4^{3-}, and SO_4^{2-} did not interfere in an argon-oxygen-hydrogen flame. Ca, Mg, Mo(IV), Sr, and U(VI) caused a signal depression of 12, 6, 5, 5, and 10%, respectively.

5.37. TITANIUM

Very little information exists on the determination of this element by atomic fluorescence spectrometry. This is primarily because the strongest absorbing lines of titanium lie in the 3200 – 4000 Å region where there is very strong NH and CN emission from the nitrous oxide-acetylene flame necessary to atomize this element. In addition, titanium has a very complex atomic spectrum with a multiplicity of lines with very similar energies and oscillator strengths and the photo-excited atom has therefore many transitions available for the dissipation of its acquired energy. A very intense excitation source is needed to get a sufficiently strong fluorescence signal at a given fluorescence wavelength. Dagnall *et al.* [112, 118] have developed a titanium electrodeless discharge lamp with very good spectral characteristics but were not able to detect titanium fluorescence in a nitrogen- or argon-shielded nitrous oxide-acetylene flame with that source [118]. Larkins and Willis [259] excited titanium fluorescence by means of a boosted-output hollow-cathode lamp of Löwe design in a nitrogen-shielded nitrous oxide-acetylene flame, and obtained fluorescence signals at fourteen wavelengths. Table 5.52 shows the relative intensities of these fluorescence lines.

TABLE 5.52. Atomic-fluorescence spectrum of titanium [259]

Wavelength (Å)	Energy levels (eV)	Transition	Relative source emission intensity[a, b]	Relative fluorescence intensity[a, b, c]
2646.6	0.048—4.731	$a^3F_4 - u^3D_3^0$	12	8
2942.0	0—4.211	$a^3F_2 - v^3F_2^0$	14	6
2948.3	0.021—4.225	$a^3F_3 - v^3F_3^0$	15	10
2956.8	0.021—4.213	$a^3F_3 - v^3F_2^0$	20	8
3186.5	0—3.890	$a^3F_2 - w^3G_3^0$	34	17
3192.0	0.021—3.904	$a^3F_3 - w^3G_4^0$	43	33
3199.9	0.048—3.921	$a^3F_4 - w^3G_5^0$	49	40
3341.9	0—3.709	$a^3F_2 - x^3G_3^0$	55	39
3354.6	0.021—3.715	$a^3F_3 - x^3G_4^0$	52	43
3371.5	0.048—3.724	$a^3F_4 - x^3G_5^0$	95	84
3635.5	0—3.409	$a^3F_2 - y^3G_3^0$	86	93
3642.7	0.021—3.424	$a^3F_3 - y^3G_4^0$	93	86
3653.5	0.048—3.433	$a^3F_4 - y^3G_5^0$	100	100
3998.6	0.048—3.148	$a^3F_4 - y^3F_4^0$	80	82

[a] Excited with a BOHCL(L).
[b] Intensities uncorrected for detector spectral response.
[c] In a No-Ac flame.

The most intense fluorescence occurred at 3653.5 Å, but the best detection limit (4.0 μg/ml) was obtained at 3199.9 Å. With a non-dispersive atomic fluorescence system, the limits of detection were also poor, because of the poor spectral response of the HTV R-166 photomultiplier in the spectral region at 3200 Å. These detection limits could be improved somewhat by use of an appropriate narrow-band interference filter [259].

For the reasons discussed above, the use of atomic fluorescence measurements with conventional excitation sources for the determination of titanium does not appear very promising. More optimistic results were reported very recently by Fraser and Winefordner [276] who could detect as little as 0.1 μg/ml of titanium (at 3998.6 Å) by laser-excited atomic fluorescence flame spectrometry.

5.38. VANADIUM

Very few atomic fluorescence experiments have been done with vanadium. Most of the problems discussed in Section 5.37 which make the determination of titanium by atomic fluorescence spectrometry very troublesome, also affect the atomic fluorescence spectrometry of vanadium. The first work dealing with vanadium fluorescence was done by Dagnall *et al.* [118]. This paper described the atomic fluorescence characteristics of this element, using a commercially available flame spectrophotometer in conjunction with an electronically modulated electrodeless discharge lamp source and a nitrogen- or argon-shielded nitrous oxide-acetylene flame. The maximum fluorescence emission occurred in the red interconal zone of a fuel-rich flame immediately above the primary reaction zone, at 3184.0 Å (an unresolved triplet 3183.4, 3183.9, and 3185.4 Å), resulting in the surprisingly good detection limits of 1.2 and 0.07 μg/ml (signal: noise = 1) for the unshielded and argon-shielded flame, respectively. Weak fluorescence was also detected at 3066.4 Å. No fluorescence signals were detected at other lines emitted by the excitation source. The range of linearity of the calibration curve was 1–250 μg/ml, and the relative standard deviation was found to be 1.75% at the 25 μg/ml level. Demers *et al.* [145, 146] reported a detection limit of 0.25 μg/ml (at 3184.0 Å), using a pulsed hollow-cathode lamp source and a Technicon AFS-6 atomic fluorescence spectrometer.

A thorough interference study with a wide range of cations and anions has shown that only three ions (at 100-fold ratio) affect the fluorescence signal produced by 25 μg/ml of vanadium. Aluminium and tungsten raised the fluorescence signal by about 18 and 14%, respectively, while cobalt reduced the signal by about 11% [118].

5.39. ZINC

5.39.1. GENERAL CHARACTERISTICS AND FLUORESCENCE SPECTRUM

Zinc, like cadmium, is an element particularly suited to the atomic fluorescence technique, primarily because of the availability of intense excitation sources, the favourable position of its main resonance wavelength, and its good atomization characteristics in all types of conventional flames. Since 1964 when the atomic fluorescence of zinc was first mentioned [445, 446], about sixty papers have been published dealing with the zinc fluorescence. The data collected from these papers, especially from several comprehensive general studies [98, 103, 203, 245, 272] in which optimum conditions have been established for the measurement of zinc fluorescence in various flame and non-flame atomizers allow one to state that the atomic fluorescence of zinc has been very well examined and that atomic fluorescence spectrometry should always be the technique of choice when extremely low amounts of zinc are to be determined.

The atomic fluorescence spectrum of zinc consists only of the 2138.6 Å resonance line ($^1S_0 - {}^1P_1$ transition). In contrast to cadmium, no fluorescence signal has been observed at the 3075.9 Å intercombination line due to both resonance fluorescence and stepwise-line fluorescence [245]. The latter result indicates a much shorter life-time (of the excited triplet state) than in the case of cadmium [330].

5.39.2. EXPERIMENTAL CONDITIONS AND DETECTION LIMITS

Spectacular limits of detection have been obtained by many workers with a variety of sources and flame atomizers, as can be seen from Table 5.53. Line sources, such as zinc vapour discharge lamps [103, 192, 266, 272, 415, etc.), electrodeless discharge lamps [58, 97, 98, 114, 483, etc.) and boosted-output hollow-cathode lamps [244, 245, 258] have mostly yielded detection limits of about 0.1 ng/ml (or lower), but with continuous sources, such as the 450-W xenon arc lamp, the lowest detectable concentration has generally been about 0.01 μg/ml [142, 190]. Zinc vapour discharge lamps are usually less self-reversed than cadmium ones, so that higher operating currents (800–900 mA) are recommended. The lowest detection limit of 1 pg/ml reported by Kuang-Pang Li [253] with this source, a premixed nitrous oxide-acetylene flame burning on a direct-injection burner and photon-counting detection is (together with the same value reported by Zacha *et al.* [483] for cadmium) the lowest limit of detection attained at present by flame atomic fluorescence. However, electrodeless discharge lamps (prepared from the metal or zinc chloride) both of the single [91, 98, 268, 483] and the dual-element type [97] seem to be the most promising excitation sources for zinc.

Since the main zinc resonance wavelength lies in the region of maximum spectral response of the solar-blind photomultiplier and in a region of low background of all common analytical flames, the use of non-dispersive systems [258, 415, 420, 421] and systems with interference filters [143, 144, 420, 421] is very advantageous.

Many authors have examined the effect of flame composition and burner-

TABLE 5.53. Basic papers contributing to the atomic fluorescence flame spectrometry of zinc at 2138.6 Å

Ref.	Spectrometer Detector Band-width (Å)	Excitation source	Flame (burner)	Detection limit (μg/ml) S : N = 2
[375, 445, 446]	L.I. RCA 1P-28	MVDL (Osram)	O-Ac(t) O-H(t) Beckman	0.04[a]
[266]	L.I. RCA 1P-28 32	MVDL (Osram) (Philips)	O-H(t) Bekmann	0.000 1[a]
[103]	Unicam SP 900 A EMI 9601 B *ca.* 60	MVDL (Osram)	A-P(p)	0.000 6[b]
		150 W Xe-arc	A-P(p) A-H(p)	0.6[b]
[142]	L.I. EMI 6255 B[c] 6.0	450 W Xe-arc	H-e.a.(t) Hetco	0.01[b]
			O-H(t)	0.1[b]
			A-H(p)[d] Tri-flame	0.2[b,e]
[185, 186]	L.I. RCA 1P-28	MVDL (Philips)	A-H(t) Hetco	0.005
[58, 483]	L.I. RCA 1P-28 16	EDL	A-H(t) Zeiss	0.000 04[b]
[58, 267, 268]	L.I. RCA 1P-28 16	EDL	A-H(t) Beckman 4020	0.000 1[b]
[411, 412]	L.I. RCA 1P-28 45	150 W Xe-arc	O-H(t) Beckman 4020	0.6[b]
[203]	Techtron AA-3 RCA 1P-28 10	MVDL (Philips)	O-H(t) Beckman 4020	0.009[a]
[312]	Jarrell—Ash (Max. Versat.) RCA 1P-28 6.4	MVDL (Philips)	A-H(t) Hetco	0.000 5
[114]	L.I. EMI 6256 B 17	EDL	A-H(p)	0.000 2
[192, 226, 227]	Unicam SP 900A EMI 9601 B *ca.* 60	MVDL (Osram)	A-Ac(p, s)	0.000 2
[265]	Perkin—Elmer 303 EMI 9592 B 20	HCL (Intensitron)	A-H(p)	0.05
		150 W Xe-arc	A-H(p)	20

TABLE 5.53 (continued)

Ref.	Spectrometer Detector Band-width (Å)	Excitation source	Flame (burner)	Detection limit (μg/ml) S : N = 2
[206]	Unicam SP 900A EMI 9601 B	R.F. plasma	A-Ac(p)	0.05
			No-H(p)	0.8
[398, 399]	S.A. A 3000 HTV R-213 30	EDL (mod.)	A-Ac(p)	0.000 5
[65]	L.I. RCA 1P-28	150 W Xe-arc (Eimac)	A-H(t) Zeiss	7.0
[415]	L.I. HTV R-166 N.S.	MVDL (Philips)	A-H(t) Beckman 4020	0.000 2
[190]	Beckman 440 RCA 1P-28	450 W Xe-arc	A-H(p)	0.015
		150 W Xe-arc	A-H(p)	0.080
[98]	S.A. A 1740 HTV R-213	EDL	A-H(p)	0.001
			A-Ac(p)	0.002
			A-H(t) Beckman	0.000 5
			O-H(t) Beckman	0.000 1
[420, 421]	L.I. RCA 1P-28 32	MVDL (Philips)	A-TG(p)	0.007
	L.I. RCA 1P-28 filter[g]	MVDL (Philips)	A-TG(p)	0.000 01
	L.I. HTV R-166 N.S.	MVDL (Philips)	A-TG(p)	0.000 02
[270, 433]	S.A. A 1740 HTV R-213 1.3	EDL[h]	A-P(p)	0.002
[96, 433]	S.A. A 1740 HTV R-213 60	500 W Xe-arc	A-Ac(p)	10
[97]	S.A. A 1740 HTV R-213	EDL[i,j]	O-H(t) Beckman	0.000 1[i] 0.001[j]
[165]	L.I. HTV R-166 N.S.	EDL	Ar-H-e.a.(t) HETCO	0.000 45

TABLE 5.53 (continued)

Ref.	Spectrometer Detector Band-width (Å)	Excitation source	Flame (burner)	Detection limit (μg/ml) S : N = 2
[253]	L.I. RCA 1P-28 4	MVDL (Osram)	A-H(t) Beckman 4020	0.000 01[k]
			No-Ac(p, d.i.)	0.000 001[k]
	L.I. RCA 1P-28 12	150 W Xe-arc	A-H(t) Beckman	~0.1[k]
[32]	Perkin—Elmer 403 HTV R-136 20	EDL(mod.)	A-Ac(p)	0.005
	L.I. HTV R-166 N.S.	EDL(mod.)	A-Ac(p)	0.002
[143, 144, 145, 146, 147]	Technicon AFS-6 filter, $\lambda_{max} = 2175$ $\Delta\lambda_{(1/2)} = 225$	HCL(p)	A-Ac(p)	0.07[l] 0.001[m]
[272]	L.I. RCA 931 A 37	MVDL (Philips)	A-Ac(p)	0.003 2[a]
			A-Ac(p, s)	0.000 79[a]
			O-Ac(t) Zeiss	0.004 6[a]
[258]	Techtron AA-4 HTV R-106 10	BOHCL(S-W)	A-Ac(p, s)	0.003
			A-Ac(p)	0.005
	Techtron AA-4[n] HTV R-166 N.S.	BOHCL(S-W)	A-Ac(p, s)	0.000 3[o]
		HCL	A-Ac(p)	0.003
[244, 245, 393]	Techtron AA-4 HTV R-106 10	BOHCL(S-W)	A-Ac(p, s)	0.02
			A-Ac(p, s)	0.001 3
			A-H(p)	0.003
			O-Ar-H(p)	0.000 6
			A-Ac(p, s)	0.000 05[p]

[a] S : N = 0.75.
[b] S : N = 1.
[c] Photomultiplier cooled at —15 °C.
[d] With "thermal concentrator" producing dry aerosol.
[e] Cf. Ref. [142].
[f] Linearity range 0.001—2.0 μg/ml.
[g] Baird-Atomic prominent ultraviolet line filter A.1.
[h] Dual element Cd—Zn source.
[i] Dual element Ag—Zn source.
[j] Dual element Zn—Hg source
[k] Photon-counting technique
[l] Cf. Ref. [143]
[m] Linear range 0—6 μg/ml
[n] Without monochromator
[o] Upper linear limit 0.3 μg/ml
[p] With the sampling-cup technique (see Chapter 3.3.3.); sample volume of 100 μl (i.e. absolute detection limit 5×10^{-12}g)

Abbreviations: See p. 187

nebulizer type on the sensitivity of zinc determination. Cresser and West [93] reported a slightly higher sensitivity in turbulent hydrogen-supported flames than in laminar premixed hydrogen-supported flames, but noted the earlier incidence of curvature in the calibration curves obtained with the former flames, which arises because of the existence of appreciable populations of ground-state atoms outside this type of flame at the heights used for measurement. Martin *et al.* [272] compared premixed, separated premixed, and turbulent acetylene flames, and obtained the lowest limit of detection with the separated flame. The turbulent flame with oxy-acetylene was found to be superior to the premixed flame, with respect to strength of fluorescence signal, but had a large noisy scattering component which adversely affected the detection limit.

Kolihová [245] examined the use of various premixed flames with respect to detection limits and shapes of calibration curves, and obtained the lowest limit of detection in an oxy-hydrogen flame diluted with argon. The linear range of the calibration curves extended to about 2 μg/ml for all flames investigated, which corresponds well with some conclusions made in Section 2.2.2.

To avoid any effects of diverse ions on the atomic fluorescence signal of zinc (cf. Section 5.39.3), the separated air-acetylene flame should be used.

5.39.3. INTERFERENCES

Dagnall *et al.* [103] found no effect from 100-fold molar ratios of 41 common cations and 18 anions on 0.6 μg/ml of zinc in a premixed air-propane flame. Hsu [203] also found no cationic interferences (100-fold ratios) in a turbulent oxy-hydrogen flame, but observed interference from some anions, which could be effectively overcome by the addition of EDTA.

A detailed study of cationic interferences with the zinc fluorescence signal in various premixed and turbulent flames was undertaken by Cresser and West [98]. Their results indicated that the laminar, premixed air-acetylene flame was more suitable for the determination of zinc in the presence of variable amounts of diverse ions, and that this flame also appeared to be much less susceptible to interference caused by variation in ground-state atomic distribution than the turbulent hydrogen flames.

All of the cationic interferences encountered in the turbulent flames were enhancements, attributable to scatter of the source radiation, which is very significant at the zinc main resonance wavelength. However, 20 000-fold amounts of aluminium gave enhancements of 50% and 100% in turbulent oxy-hydrogen and air-hydrogen flames, respectively, which could not be

TABLE 5.54 Summary of atomic fluorescence detection limits in non-flame atomizers

Ref.	Element	Wave-length (Å)	Source	Atomizer	Detection limit pg	Notes
274]	Antimony	2311.5	HCL[a]	GFA	200	
[16]	Arsenic	2311.5	HCL[b]	CRA	500	
[64]	Beryllium	2348.6	EDL	MFLA	4000	
[16]	Beryllium	2348.6	HCL[b]	CRA	30	
[19]	Bismuth	3067.7	EDL	CFA	10[c]	
[64]	Bismuth	3067.7	EDL	MFLA	40[c]	
8, 324, 325]	Bismuth	3067.7		GEA		solid samples
4, 275]	Cadmium	2288.0	HCL[a]	GFA	0.25	
3, 64]	Cadmium	2288.0	EDL	MFLA	0.02	[f]
[2]	Cadmium	2288.0	HCL	CFA	0.001[d]	[e]
[3]	Cadmium	2288.0	EDL	CFA	0.015	argon atmosphere, [f]
[3]	Cadmium	3261.1	EDL	CFA	5.0	argon atmosphere, [f]
[3]	Cadmium	2288.0	EDL	CFA	0.03	nitrogen atmosphere
[3]	Cadmium	3261.1	EDL	CFA	0.03	nitrogen atmosphere
9, 40, , 324]	Cadmium	2288.0	EDL	GEA	0.26	solid samples
188]	Cadmium	2288.0	EDL	GEA	10	solid samples
[16]	Cadmium	2288.0	HCL[b]	CRA	0.03	
7, 293]	Cadmium	2288.0	BOHCL(S-W)	MFLA	0.04	
[49]	Cadmium	2288.0	EDL	Pt-F	0.0002[d]	
[16]	Calcium	4226.7	HCL[b]	CRA	0.1	
[2]	Copper	3247.5	BOHCL(S-W)	CFA	0.004[d]	[e]
274]	Copper	3247.5	HCL[a]	GFA	450	
[16]	Copper	3247.5	HCL[b]	CRA	5	
188]	Copper	3247.5		GEA	100	solid samples
[64]	Copper	3247.5	EDL	MFLA	200[c]	
[49]	Copper	3247.5	EDL	Pt-F	0.02[d]	
[19]	Gallium	4172.1	EDL	CFA	50[c]	
[64]	Gallium	4172.1	EDL	MFLA	20000[c]	
[1]	Gold	2428.0	BOHCL(ASL)	CFA	5	[f]
[1]	Gold	2676.0	BOHCL(ASL)	CFA	4	[f]
188]	Gold			GEA		solid samples
[16]	Gold	2428.0	BOHCL(S-W)	CRA	10	
393]	Gold	2428.0	BOHCL(S-W)	MFLA	8	
[16]	Iron	2483.3	HCL[b]	CRA	2	
274]	Iron	2483.3	HCL[a]	GFA	3000	
[49]	Iron	2483.3	EDL	Pt-F	4.0[d]	
274]	Lead	2833.1	HCL[a]	GFA	35	
[64]	Lead	4057.8	EDL	MFLA	4000	
[2]	Lead	2833.1	HCL	CFA	0.01[d]	[e]
[19]	Lead	4057.8	BOHCL(S-W)	CFA	10	

TABLE 5.54 (continued)

Ref.	Element	Wave-length Å	Source	Atomizer	Detection limit pg	Notes
[16, 261]	Lead	4057.8	BOHCL(S-W)	CRA	3	f
[245, 393]	Lead	4057.8	BOHCL(S-W)	MFLA	10	
[274]	Magnesium	2852.1	HCL[a]	GFA	3.5	
[64]	Magnesium	2852.1	EDL	MFLA	4000	
[16]	Magnesium	2852.1	HCL[b]	CRA	0.03	
[430]	Magnesium	2852.1	BOHCL(S-W)	CFA	0.0001	
[19]	Magnesium	2852.1	BOHCL(S-W)	CFA	1	
[39, 40]	Manganese	2794.8		GEA	150	solid samples
[188]	Manganese	2794.8		GEA	1000	solid samples
[63, 64]	Mercury	2536.5	EDL	MFLA	20[c]	
[324]	Mercury	2536.6		GEA	14	solid samples, a
[324]	Mercury	2536.5		GEA	910	solid samples, ai
[49]	Mercury	2536.5	Hg(pen)	Pt-F	0.008[d]	
[300]	Mercury	2536.5	Hg(pen)	Quartz cell	600	
[301]	Mercury	2536.5	Hg(pen)	Quartz cell	3000	
[361]	Mercury	2536.5	Hg(pen)	MFLA	100	
[401]	Mercury	2536.5	Hg-g. lamp	Quartz cell	2000	
[274]	Silver	3280.7	HCL[a]	GFA	1.5	
[64]	Silver	3280.7	EDL	MFLA	2.0[c]	
[430]	Silver	3280.7	BOHCL(S-W)	CFA	30	
[19]	Silver	3280.7	BOHCL(S-W)	CFA	1[c]	
[324]	Silver	3280.7	EDL	GEA	1.2	solid samples, a
[324]	Silver	3280.7		GEA	27	solid samples, ai
[188]	Silver	3280.7		GEA	10	solid samples
[16]	Silver	3280.7	HCL[b]	CRA	0.5	
[245, 393]	Silver	3280.7	BOHCL(S-W)	MFLA	0.07	
[274]	Thallium	2767.9	HCL[a]	GFA	2000	
[19]	Thallium	3775.7	EDL	CFA	50[c]	
[64]	Thallium	3775.7	EDL	MFLA	2000[c]	
[324]	Thallium	3775.7		GEA	0.9	
[274, 275]	Zinc	2138.6	HCL[b]	GFA	0.04	
[2]	Zinc	2138.6	HCL	CFA	0.001[d]	e
[64]	Zinc	2138.6	EDL	MFLA	20[c]	
[188]	Zinc	2138.6		GEA	10	solid samples
[16]	Zinc	2138.6	HCL_1	CRA	0.1	
[19]	Zinc	2138.6	EDL	CFA	0.02[c]	
[245, 247, 293]	Zinc	2138.6	BOHCL(S-W)	MFLA	0.08	
[49]	Zinc	2138.6	EDL	Pt-F	0.0004[d]	

[a] Hollow-cathods lamp run at high current.
[b] Normal or high-intensity, not specified.
[c] Signal: noise = 1.
[d] Expressed in μg/ml.
[e] Solvent extraction methods described.
[f] Interference studies included.

Abbreviations: See p. 187.

attributed to scatter alone. The effect was found to be due to appreciable changes in the fluorescence emission-height profile. In a premixed air-acetylene flame, the same concentration of aluminium gave a 6% decrease in signal. The effect of hydrochloric acid indicated that zinc compound formation plays an important part in the interference mechanism in turbulent flames, by affecting the ground-state atom distribution.

Kolihová and Sychra [245, 393] investigated the effect of the main inorganic components of some biological materials on the atomic fluorescence signal of zinc determined by the sampling-cup technique. They found no interference from 3000-fold ratios of Na, K, Ca, Mg, and phosphate.

5.39.4. ORGANIC SOLVENTS AND EXTRACTION

Hsu [203] studied the effect of 29 organic solvents on the fluorescence signal of zinc in the turbulent oxy-hydrogen flame. In most cases, the fluorescence signal was found to be enhanced. At high percentages of alcohol, the flame composition had little effect on the fluorescence intensity. Dagnall *et al.* [103] extracted the zinc chelate with diethyldithiocarbamate into equal volumes of MIBK, and obtained a detection limit 5 times lower than that obtained with an aqueous solution in a premixed air-propane flame.

5.40. NON-FLAME CELLS

Non-flame atomizers (carbon rod, carbon filament, L'vov and Massmann furnaces, etc.) have been applied in atomic fluorescence determination, and the available literature is summarized in Table 5.54.

Chapter 6

APPLICATIONS

Relatively few applications of atomic fluorescence spectrometry to the analysis of "real" samples have so far been reported. This is not surprising, because in the early stages of development of every new analytical technique, more attention is given to fundamental research. In addition, there is strong competition with other flame analytical methods, particularly with atomic absorption. It is logical to assume that the widely used atomic absorption technique will be preferred in solving conventional problems for which suitable atomic absorption procedures have been developed. Another limiting factor is that simple commercial instrumentation is still not generally available for atomic fluorescence spectrometry.

However, the potential range of application of atomic fluorescence spectrometry is broader than that of atomic absorption. Very often higher sensitivity is available for a particular element by atomic fluorescence than by atomic absorption, and in some instances a direct analysis of materials for traces of an element may be possible without the use of preconcentration procedures which almost always result in contamination problems, blank problems, etc. Other areas of application of atomic fluorescence spectrometry arise from its special suitability for use with non-flame atomizers [459, 463, 466] and for multielement analysis [143, 144, 291, 292, 293, 466].

A wide expansion of the practical applications of atomic fluorescence spectrometry is, therefore, to be expected during the next few years, particularly in those instances where the use of this technique will enable a particular analysis to be made more sensitively, precisely, rapidly, and simply.

6.1. ANALYSIS OF BIOLOGICAL MATERIALS

6.1.1. DETERMINATION OF METALS IN BLOOD SERUM, BLOOD, AND URINE

The first application of atomic fluorescence flame spectrometry to the analysis of biological material was published by Klaus [238], who described relatively simple procedures for the determination of zinc and cadmium in human blood serum and human and rat urine by using a modified Zeiss PMQ II spectrophotometer and zinc and cadmium metal vapour-discharge lamp sources. Zinc was determined without prior concentration and there was good agreement between the results obtained by direct analysis and those obtained after oxidation of the sample. The results obtained for cadmium after wet ashing of the sample are questionable, because the normal physiological level of this element is reported to be of about one order of magnitude lower than the levels reported [87].

A method developed in our laboratory for the direct determination of zinc and copper in blood serum at normal physiological levels [245, 246] clearly demonstrates the advantages of the application of atomic fluorescence over that of atomic absorption to this analysis. To avoid transport interferences, blood serum samples must be diluted at least 25-fold. This means that both elements are measured over a concentration range of about 0.03 – 0.06 μg/ml. These concentration values lie too close to the atomic absorption detection limits attainable for these elements (particularly for copper) with the commercial instrument used (Varian-Techtron Model AA-4 atomic absorption spectrometer) and the atomic absorption determination therefore exhibits very low precision. The atomic fluorescence technique with a modified instrument and boosted-output hollow-cathode lamps of the Sullivan-Walsh type as excitation sources yields detection limits of about one order of magnitude lower (~ 1 ng/ml) for both elements, and thus allows serum zinc and copper to be determined directly with much higher sensitivity and precision (relative standard deviation less than ±5%). No matrix interferences or interferences due to scatter are observed at this dilution. The results listed in Table 6.1 indicate that the determination can be simply and rapidly performed by using calibration curves prepared from aqueous standard solutions without the addition of concomitants.

The method described was tested on hundreds of clinical samples with very good results. Since the same dilution of the serum is usually also used for the direct determination of sodium and potassium, all four elements can be simultaneously determined from one sample solution.

TABLE 6.1. Results of the direct determination of zinc and copper in blood serum (dilution 1 : 25)[a]

Serum No.	Cu, μg/ml[b]			Zn, μg/ml		
	A	B	C	A	B	C
1	2.20	2.26	2.2	1.18	1.15	1.1
2	2.15	2.00	2.1	1.35	1.28	1.3
3	2.25	2.35	2.4	0.78	0.73	0.75
4	1.95	2.05	1.9	0.85	0.85	0.90

[a] Air-hydrogen flame; Cu 3247.5 Å, Zn 2138.6 Å.
[b] Blood serum of pregnant women with an elevated copper level.
A Calibration curve technique.
B Standard addition technique.
C Mean value determined by atomic absorption technique after the mineralization of 5 ml of serum.

The analysis of blood serum for zinc has also been reported by Kuang-Pang Li [253], who measured the atomic fluorescence excited by an Osram zinc vapour-discharge lamp in a premixed nitrous oxide-acetylene flame burning in a direct-injection Zeiss burner, and monitored the atomic fluorescence signals by photon-counting. The high sensitivity of this system allowed the determination to be made with blood serum diluted 1 : 100. Because of the dilution, it was possible to perform the analysis with less than 50 μl of blood serum. The recovery was very satisfactory.

The calcium content of the blood serum was determined by atomic fluorescence spectrometry using a 450-W xenon arc source and a turbulent hydrogen entrained air flame. No advantages were found for atomic fluorescence over atomic absorption spectrometry with regard to sensitivity, precision, and accuracy. Vickers and Merrick [414] described a very rapid and sensitive method for screening urine samples for mercury, with a detection limit of 2 ng/ml. Mercury was extracted as the dithizonate into chloroform or MIBK and determined by aspiration of the extract into a turbulent oxygen-hydrogen flame, with a mercury pen lamp as the excitation source. Matoušek and Sychra [277, 278] have shown that the high sensitivity of the determination of gold by atomic fluorescence in a hydrogen-oxygen-argon flame can be applied to the determination of gold in the urine of subjects, being treated with gold-containing substances.

Because of the extremely low content of most metals present in biological samples, and because of the very small sample volumes which are frequently available for analysis, more efficient atomization techniques than the conventional flame atomizers are often required to improve the sensitivity of the trace element analysis of biological materials.

A rapid, simple, and sufficiently precise micromethod for the determination of zinc and cadmium in blood serum by using the "sampling-cup technique" (cf. Section 3.3.3) was developed in our laboratory [245, 393]. To overcome the effect of the organic matrix, the procedure described by Delves [139] was used for sample oxidation. In this procedure the samples are ashed directly in the cup with 30% hydrogen peroxide.

For the determination of zinc and cadmium, 10 µl of 50–100-fold diluted blood serum and 5 µl of undiluted serum, respectively, are pipetted into the cup and dried by placing the cup close to the nitrogen-shielded air-acetylene flame. After drying, the cup is cooled, 20 µl of hydrogen peroxide are added and the cup is again placed near to the flame and heated carefully until a dry residue is obtained. This generally requires not more than 2 min. The sample is thus ready for analysis. The cup is then introduced quickly into the flame and the transient fluorescence signal measured at the appropriate fluorescence wavelength (Zn 2138.6, Cd, 2288.0 Å). The calibration curve is prepared either by putting aqueous reference solutions through the same procedure, or by using the addition technique. Care must be taken to use high-purity hydrogen peroxide and demineralized water for oxidizing and diluting the samples, otherwise high blank values result. The relative standard deviation does not exceed 8 and 20%, for the zinc and cadmium determinations, respectively. Table 6.2 shows the results of the analysis of two serum

TABLE 6.2. Determination of zinc and cadmium in human blood serum by atomic fluorescence spectrometry using the "sampling-cup technique" [393]

Serum No.	Sampling-cup technique (µg/ml)				Usual AFS in flame (µg/ml)	
	Calibration curve		Addition technique			
	Zn	Cd	Zn	Cd	Zn	Cd
1	1.3	0.003	1.5	0.003	1.4	0.003[a]
2	2.0	0.003	2.1	0.004	2.2	0.003[a]

[a] Semiquantitative results.

samples for zinc and cadmium by the "sampling-cup technique" in comparison with those obtained by the usual flame atomic fluorescence technique. The results obtained both by means of the calibration curve and the addition techniques are in good agreement with those obtained by nebulizing diluted serum samples into a nitrogen-shielded air-acetylene flame [246].

The main advantages of the method described are the small sample volume required (less than 1 μl of serum for the determination of zinc), the relatively very good precision of the cadmium determination (cadmium is very difficult to determine directly and precisely by the usual flame atomic fluorescence method with a conventional experimental arrangement because of the extremely low cadmium content of normal serum), and the general freedom from interferences. By evaporation and atomization of the partially ashed samples from the cup into a sufficiently hot environment, i.e. the flame, chemical interferences due to incomplete dissociation, as well as spectral interferences due to light-scattering, are avoided.

Preliminary results from our laboratory have shown that the "sampling-cup technique" can also be applied to the determination of elevated lead levels in whole blood according to the method suggested by Delves [139]. An additional advantage of the "fluorescence mode" in this instance is the possibility of using direct-line fluorescence.

The carbon rod atomizer (cf. Section 3.3.4) has been shown to be suitable for the direct determination of lead in biological fluids (about 2 μl of sample), [16, 261] when used in conjunction with a boosted-output hollow-cathode lamp of the Sullivan-Walsh type. The following steps were taken to overcome the effects of organic matter and high salt content, which would cause substantial scatter of the exciting radiation and chemical interferences.

(*a*) Dilution of the sample with water, by a factor of 2.5.

(*b*) Preheating the sample to remove most of the organic material at a temperature which is just below that at which lead will vaporize. This step provides dry ashing and also removes part of the matrix by fractional vaporization. It should be stressed that the preheating parameters are critical and must be chosen very carefully to avoid losses of lead.

(*c*) Measurement of direct-line fluorescence at 4057.8 Å excited by the 2833.1 Å line, while the 4057.8 Å line is filtered from the source radiation with a UG-5 filter (6 mm filter transmits 1.2% of incident radiation at 4057.8 Å and 76% at 2833.1 Å).

The procedure described was applied to the determination of lead in whole blood and urine samples. Blood samples do not present any basic difficulty because the lead level in blood is of the order of 0.2 μg/ml and could be higher in some patients. The determination was carried out by both the standard addition technique (addition of lead nitrate solution and no other preparation), and constructing the calibration curve with aqueous lead nitrate reference solutions. The good agreement of the results obtained for both peak signal and integrated signal measurements, and an independent check of the results by ammonium pyrroldinedithiocarbamate (APDC)-MIBK extraction followed by atomic absorption flame determination indicated that the determination of lead in blood serum can be performed simply by using aqueous lead reference solutions. The detection limit for Pb was found to be 5 ng/ml in whole blood, and the relative standard deviation was approximately 5%.

The determination of lead in urine is more difficult than it is in blood, mainly because the normal urine lead level is more than one order of magnitude lower while the salt content is higher. Even when the sample was preheated and the 4057.8 Å line was selectively attenuated from the source radiation, the signal due to scattering the exciting radiation was observed to be of the same order as the actual lead signal. A simple correction for scatter may be made by comparing the higher lead fluorescence signal intensity obtained with argon as carrier gas with that obtained with nitrogen (cf. Section 4.2.7). The ratio of the signals is 13 : 1 for the 4057.8 Å line. When the lead was extracted from the urine with the APDC-MIBK system, the same background signal was observed from the remaining urine matrix for both nitrogen and argon as carrier gas. For urine samples containing lead, the signal measured for the argon system consists of lead fluorescence and scattered radiation while that measured for the nitrogen system represents the same scatter signal together with a fluorescence signal which can be neglected to a first approximation. Subtraction of the two signals will therefore give the true lead signal.

Urine samples were collected directly in polyethylene bottles containing acetic acid (about 1% of their volume) and diluted 2.5-fold. The calibration with aqueous lead reference solutions was found to be adequate, as in the case of blood analysis. The detection limit in undiluted urine was about 5 ng/ml.

Bratzel *et al.* [64] have stated that a metal filament loop atomizer seems to be especially useful for trace metal analysis in small biological samples. When we studied practical applications of this type of atomizer to the determination of zinc and cadmium in blood serum and lead in whole blood, serious matrix interferences were observed which were almost independent of the loop temperature even when the samples were preheated [245, 393]. These interferences are analogous to those observed by Amos and co-workers [16] and are due to both incomplete dissociation owing to the evaporation of the sample into the cooler environment around the loop, and to different vaporization rates caused by a large quantity of extraneous ions and organic compounds. For the applications considered, the determination via a calibration curve technique is not possible, because it is very difficult to match precisely the standard solutions to the biological samples. Use of the addition technique is very time-consuming, particularly if account is taken of the fact that additions cannot be applied directly to the loop because of the strong dependence of the fluorescence signals on the sampling volume and the salt concentration. Attempts to overcome this problem by combining an argon-hydrogen diffusion flame with the loop [16] were also unsuccessful, because of the lower fluorescence signals due to greater quenching caused by entrained atmospheric nitrogen and the very short lifetime of the metallic loops.

6.1.2. DETERMINATION OF MERCURY IN WHEAT FLOUR

Mercury has been determined by a flameless (dispersive or non-dispersive) atomic fluorescence system using a combustion furnace technique for the generation of mercury vapour and a silver amalgamator to collect the mercury before fluorescence measurement [300]. Mercury vapour in the carrier gas stream was brought into close contact with a silver wire at ambient temperatures and was effectively removed from the gas stream by amalgam formation. The mercury was subsequently released for the final measurement step by heating (see Fig. 6.1). Excitation was realized in a fluorescence cell by

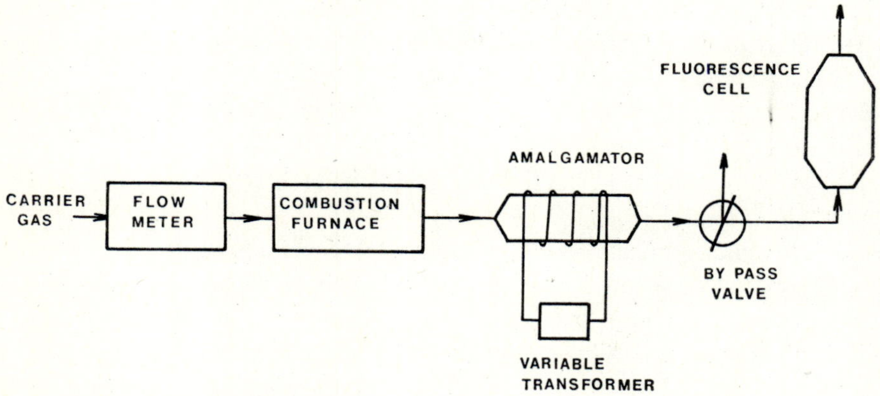

FIG. 6.1. Schematic diagram of a gas train for mercury determination by the furnace-amalgamator technique [300].
Reprinted from Ref. [300] by permission of The American Chemical Society

means of a mercury pen lamp. The system was capable of quantitative determination of mercury in samples containing as little as 0.6 ng of mercury. Mercury vapour standards were used to prepare the calibration curve. Air in a polyethylene container was saturated with mercury vapour at room temperature. Known volumes of air were drawn from the container by a syringe and injected, through a rubber septum, into an argon stream flowing into the amalgamator. Addition of oxygen to the argon stream during the combustion, and inclusion of a lead acetate filter between the furnace and the amalgamator were essential for the successful application

TABLE 6.3. Analysis of standard wheat samples [300][a]
Reprinted from Ref. [300] by permission of The American Chemical Society

Sample	Weight, mg	Hg found, ng	ppm Hg
1	18.3	95	5.2
2	23.5	130	5.5
3	30.0	135	4.5

[a] International Atomic Energy Agency, Code 66/10.

of this technique to the analysis of wheat samples. Results for replicate samples are shown in Table 6.3. Data accompanying the wheat flour sample report the mercury content to be 4.59 ± 0.32 ppm. This value is a pooled mean of neutron-activation analysis results from 15 laboratories.

6.2. INDUSTRIAL APPLICATIONS

6.2.1. APPLICATIONS TO METALLURGY

Arsenic, bismuth, lead, selenium, and tellurium in nickel-base alloys and lead in steels have been determined at levels down to 1 ppm in the sample [121]. Atomic fluorescence was excited with electrodeless discharge lamps. The determination of small amounts of lead in steels required a preconcentration procedure which involved acid dissolution and subsequent extraction into MIBK from an iodide-hydrochloric acid medium; the lead content of the organic phase was determined by direct aspiration into a premixed air-propane flame. Matrix elements are mostly left in the aqueous phase. Typical results are shown in Table 6.4.

A method for the assay of silicon in low alloy steels has been described by Kirkbright *et al.* [230]. The steels were dissolved in a mixture of hydrofluoric and nitric acids and aspirated directly into a nitrogen-shielded

TABLE 6.4. Determination of lead in British standard steel samples by atomic fluorescence spectrometry

Steel sample No.	Lead found, %[a]	Average certificate value, %	Range of the certificate value, %
BCS 326	0.0148	0.014	0.012—0.017
BCS 327	0.0110	0.010	0.009—0.012
BCS 328	0.0146	0.015	0.012—0.018
BCS 329	0.0480	0.050	0.042—0.053

[a] Average of 3 determinations.

nitrous oxide-acetylene flame. Atomic fluorescence was excited by a modulated silicon electrodeless discharge lamp source. The method reported was simple, rapid, and more sensitive than the corresponding atomic absorption method, the detection limit being 0.7 μg/ml at 2516.1 Å, compared to 4.0 μg/ml for atomic absorption at the same wavelength. The results obtained were in good agreement with the certificate values and with those obtained by atomic absorption spectrometry with similar apparatus, as can be seen in Table 6.5. No interference in the atomic fluorescence determination of

TABLE 6.5. Determination of silicon in low alloy steels [230]

B.C.S. Sample No.	Certificate value, %	AAS %	AFS[a] %	AFS spread, %
251/1	0.41	0.41	0.42	0.40—0.43
252/1	0.28	0.27	0.27	0.26—0.28
253/1	0.65	0.64	0.65	0.58—0.70
255/1	0.15	0.14	0.15	0.14—0.15
256/1	0.23	0.22	0.21	0.19—0.23
258/1	0.96	0.98	0.96	0.92—0.97

[a] Average values obtained from 15 determinations on each steel.

silicon was encountered from large amounts of iron. This observation is contrary to that reported for the determination of silicon by atomic absorption [256, 329]. It is possible that the absence of interference from iron was caused by the presence of relatively large amounts of hydrofluoric acid in the solutions nebulized, which leads to the formation of relatively stable iron fluorides in the flame. It is necessary, however, to use the same concentration of hydrofluoric acid in both the reference and sample solutions, because the silicon fluorescence signal is slightly dependent on the hydrofluoric acid concentration.

Cobalt and nickel can be determined in a variety of mild, alloy, high-speed and stainless steels according to the method described by Norris and West [306]. The procedure is extremely simple, consisting of sample dissolution in a mixture of concentrated hydrochloric and nitric acids, dilution, and direct aspiration into an argon-shielded air-acetylene flame. The fluorescence of cobalt and nickel at 2407.3 and 2320.0 Å, respectively, is excited by a dual-element electrodeless discharge lamp. The cobalt and nickel concentrations are read sequentially from calibration curves prepared from pure cobalt and nickel reference solutions containing approximately the same concentration of acids. The method is rapid, precise, and accurate. Iron and over 20 common trace and minor metal constituents of steels do not interfere. Some data concerning the precision and accuracy for one of the standard steels (B.C.S. 277 Mild steel) are given in Table 6.6.

TABLE 6.6. Ranges and relative standard deviations for the atomic fluorescence determination of cobalt and nickel in steel [B.C.S. 277], [306]
Reprinted from Ref. [306] by permission of Elsevier Publishing Company

	Cobalt	Nickel
Certificate value	0.120	0.230
Mean value of atomic fluorescence analysis	0.127	0.233
Range in certificate values	0.119—0.137	0.227—0.242
Range by AFS	0.120—0.133	0.225—0.237
Relative standard deviation (%) by AFS	3.5	2.0

The application which best demonstrates the freedom from spectral interference due to scattered source-radiation is the determination of bismuth in aluminium alloys [193]. Bismuth is determined in a separated air-acetylene flame by measurement of the direct-line fluorescence at 3024.6 Å excited by radiation from an iodine electrodeless discharge lamp source at 2061.7 Å (cf. Section 5.6). Aluminium must be added to the bismuth reference solutions; this corrects for the small decrease in the bismuth fluorescence observed in the presence of large amounts of aluminium and caused by change in the nebulization efficiency. The results obtained for the determination of bismuth in Alcan 28 S aluminium alloy samples by both atomic fluorescence spectrometry and solution spectrophotometry with thiourea are listed in Table 6.7 and are in good agreement.

The rapid determination of copper, iron, magnesium, manganese, nickel, and zinc in aluminium alloys with a prototype six-channel atomic fluorescence spectrometer (Technicon AFS-6) has been reported by Dagnall *et al.* [127]. Simultaneous determination of all six metals at a single dilution was possible with this technique. To overcome the chemical interference with magnesium from the aluminium matrix in a nitrogen-sheathed air-acetylene

TABLE 6.7. Determination of bismuth in Alcan 28 S aluminium alloy [193]

Reprinted from Ref. [193] by permission of Pergamon Press Ltd.

Sample	Bismuth found, %	
	Spectrophotometric method with thiourea	Atomic fluorescence method
R 8771/1	0.45, 0.46	0.46, 0.48
R 8771/3	0.46, 0.48	0.49, 0.49
R 8772/2	0.50, 0.52	0.52, 0.52
R 8789/1	0.52	0.52, 0.52
R 8790/3	0.45	0.47, 0.46
R 8821/3	0.57	0.56, 0.53
NH 760 A	0.60	0.60, 0.64
NH 760	0.60	0.60, 0.57

flame, strontium was used as a releasing agent. At a strontium concentration of 500 μg/ml complete release of magnesium occurred, and the thermal emission from the strontium was insufficient to produce enough shot-noise on the photomultiplier tube to have any serious adverse effect on the standard deviation of the fluorescence signals. The results compared favourably with the published values for the British chemical standards.

Copper from several different sources has been investigated for the presence of zinc impurity, by spraying fairly concentrated copper solutions (as nitrate) directly into a premixed air-acetylene flame, and determining the zinc content by atomic resonance fluorescence at the 2138.6 Å line excited with a Philips zinc vapour-discharge lamp [421]. The minimum concentration of zinc in copper that could be determined in this way was approximately 10^{-5}%, the limit being set by scatter of primary radiation, from within the flame, which produced a signal that prevented the amplifier from being used at maximum gain. The fluorescence technique was shown to be more sensitive than atomic absorption spectrometry (Table 6.8).

TABLE 6.8. Zinc content of various samples of copper determined by atomic absorption and atomic fluorescence flame spectrometry [421]
Reprinted from Ref. [421] by permission of Pergamon Press Ltd.

Copper sample	Zinc content %	
	AFS	AAS
A (electrolytic)	0 (assumed)	0 (assumed)
B (vacuum melted)	0.000 03	none found
C (electrolytic)	0.000 06	none found
D (vacuum melted)	0.001 1	0.001 1

A solution containing 10 mg/ml of copper produced a scatter-signal equivalent to the fluorescence signal from 0.01 μg/ml of zinc. The incomplete decomposition of compounds entering the flame was suggested to be the major source of the scatter. This agrees with observations made in our laboratory. However, our experience indicates that the presence and intensity of the scatter-signal under given conditions is strongly affected by the fineness of the aerosol produced by a nebulizer. For example, no scatter-signal was observed under optimized flame conditions (shielded air-acetylene flame) and aspiration rate (5 ml/min) with a Techtron premixed atomization system when spraying copper nitrate solutions containing 20 mg/ml of copper.

Massmann [274] has shown the effect of matrix elements on the direct determination of traces of silver in zinc and lead by non-flame atomic fluorescence and the necessity of integration of the fluorescence signal to overcome this effect.

Steels and iron-base alloys can be analysed for minor metallic constituents by a simple non-flame and non-dispersive system consisting of a sputtering chamber and a solar-blind photomultiplier [443] (cf. Fig. 3.39). The sputtered atomic vapour can be simultaneously excited by three different excitation lamps modulated at different frequencies and the output signals due to each isolated by synchronous demodulation [419].

6.2.2. APPLICATIONS TO PETROLEUM PRODUCTS AND FUELS

Several workers have demonstrated that atomic fluorescence offers greater accuracy and simplicity of operation, and better reliability and simpler maintainance in spectrometric oil analysis than the other two flame spectrometric methods [130, 366, 389].

Sychra and Matoušek [387, 389] determined nickel in gas oils and petroleum distillation residues by direct nebulization of samples diluted with xylene into a separated air-acetylene flame. The detection limits in the oil solutions were about 0.01 μg/ml, and the difference between duplicate readings did not generally exceed 0.01 μg/ml. In the petroleum distillation residues, nickel was determined with a relative standard deviation of 5–7%. The main advantage of the fluorescence method compared to atomic absorption is the fact that no correction for non-specific molecular absorption and light-scattering is needed; this is nearly always necessary in atomic absorption. No measurable signal caused by scattering of the excitation radiation was observed, when the optimal combination of sample dilution and flame parameters was chosen. The fluorescence method is, therefore, simpler, more precise, and time-saving. Some results are shown in Table 6.9; all values listed are the average of 2 or 3 determinations by the addition technique.

TABLE 6.9. Results for nickel in gas oils (GO) and petroleum distillation residues (PDR)
Reprinted from Ref. [389] by permission of Elsevier Publishing Company

Sample	Sample solution, μg/ml		Approximate dilution	Total, μg/ml	
	AFS	AAS		AFS	AAS[a]
GO 1	0.04	0.03	10	0.40	0.37
GO 2	0.08	0.08	4	0.34	0.37
GO 3	0.24	0.22	4	0.99	0.94
GO 4	0.30	0.32	9	2.58	2.78
PDR 1	1.15	1.19	10	11.4	11.3
PDR 2	0.80	0.76	40	33.0	28.9
PDR 3	0.82	0.85	400	320	330
PDR 4	0.92	0.90	400	350	320

[a] After mineralization.

Satisfactory methods of analysis of jet-engine lubricating oils for silver, copper, iron, magnesium, nickel, and lead were provided by the use of the HETCO direct-injection burner with a hydrogen/argon/entrained air or hydrogen-air flame [366]. Electrodeless discharge lamps were employed as spectral excitation sources. No pretreatment of the lubricationg oil was necessary. Breakdown of the oil viscosity was not noted for the used oils examined; however, adaptation of this method to the analysis of other types of lubricating oils is not recommended.

TABLE 6.10. Detection limits and conditions for determination of metals in untreated jet-engine lubricating oils, with a direct-injection burner

Metal, wavelength, Å	Detection limit μg/ml[a]	Height in flame, cm	Gas flow, 1/min	
			H_2	Ar
Ag 3281	0.001 (0.0005)	9.5	13.0	6.7
Cu 3248	0.01 (0.005)	9.5	13.0	6.7
Fe 2483	1.0 (0.3)	8.0	13.0	6.7
Mg 2852	0.1 (0.01)	8.5	14.0	6.7
Ni 2320	1.0 (0.04)	4.0	11.0	6.7
Pb 4058	2.0 (0.2)	8.5	14.5	9.0

[a] Detection limits in parenthesis refer to aqueous solutions.

In Table 6.10, a comparison is given of the detection limits obtained from aqueous solution with those obtained from oils. The higher detection limits found for the oils can be attributed to the following factors: viscosity effects giving rise to larger aerosol particles and lower solvent evaporation and uptake rates; higher flame background noise and slightly increased fluorescence quenching due to the presence of large polyatomic species, e.g. CO, CO_2, CH.

The results listed in Table 6.11 for copper, iron, magnesium, and silver indicate a good agreement with existing results obtained by spectrographic and atomic absorption determinations and apparently no obvious interference effects arising from such sources as mismatching of standards, viscosity breakdown of oils, etc.

The authors of this book would like to point out, however, that their results with direct nebulization of any type of undiluted or insufficiently diluted oil sample into the hydrogen-supported turbulent flames indicate the presence of a serious amount of scatter at any height in the flame. For this reason, dilution of oil samples with xylene or MIBK (at least 1 : 5) and the use of a premixed air-acetylene flame is invariably employed in their laboratory.

TABLE 6.11. Determination of wear metals in untreated correlation samples with a direct-injection burner [366]

Sample	Concentration, μg/ml											
	Ag			Cu			Fe			Mg		
	AF	AE	AA	AF	AE	AA	AF	AE	AA	AF	AE	AA
67—2	0.71	0.9	0.7	7.9	8.5	5.6	23	30.1	28.3	4.7	2.4	3.0
67—3	0.66	0.8	0.5	15.0	16.2	10.6	20	15.8	19.1	2.1	0.8	1.1
67—6	0.44	0.3	0.7	8.4	9.6	6.0	16	16.3	17.0	2.1	2.2	2.1
67—9	0.46	0.1	0.4	6.2	9.3	4.1	34	27.5	28.7	10.8	11.4	10.1
67—10	0.32	0.4	0.4	9.5	8.3	4.7	26	27.5	25.6	4.3	3.6	3.8
67—11	1.14	0.9	1.1	3.8	3.0	3.0	13	9.3	9.5	1.6	2.1	2.2
67—12	0.46	0.5	0.5	3.2	4.7	2.7	33	27.5	22.7	2.6	1.7	2.5
68—2	0.87	0.6	0.9	17.3	15.5	15.9	24	15.0	21.5	2.6	0.8	2.8
68—3	0.59	0.6	0.6	4.1	4.6	2.8	43	22.6	29.0	3.2	1.0	2.0
68—4	0.40	0.4	0.4	2.0	2.9	1.5	75	80.7	58.1	3.7	3.1	2.5

AF = atomic fluorescence; AE = atomic emission; AA = atomic absorption.

The combination of flame atomic fluorescence with a 150-W xenon arc continuum source and flame atomic emission was found to provide a satisfactory method for determination of iron, copper, silver, and magnesium in jet-engine lubricating oils in the ppm range [288]. Dilution of the oils and oil standards to four times the original volume with carbon tetrachloride or MIBK was found to be optimum for sample preparation. For oil samples, detection limits of 2.0, 0.9, 0.4 and 0.3 μg/ml at the 2483, 3248, 3383 and 2852 Å lines, respectively, were achieved in a hydrogen/argon/entrained air flame burning on a HETCO direct-injection burner. The simple d.c. electrometer read-out system used was found to be unreliable for determination of lead, chromium, and nickel, owing to serious spectral interferences.

Traces of copper, iron and lead in kerosene have been determined by atomic fluorescence excited with a 450-W xenon arc continuum source by using a burner of a new design [93]. This burner gives flat premixed flames with hydrogen or with a liquid hydrocarbon as fuel; the flame is non-luminous with kerosene or benzene as the hydrocarbon fuel and has low background emission. The detection limits for copper, iron and lead in kerosene were found to be 0.004, 0.04 and 0.06 μg/ml, respectively.

An ideal application of atomic fluorescence spectrometry to monitoring wear metals in used lubricating oils seems to be the use of the Technicon AFS-6 multichannel instrument (cf. Section 3.5) [143, 144] because it combines the major advantage of spark emission spectroscopy, namely, multi-element analysis, with the well-known advantages of flame methods.

Two groups of elements were tested simultaneously in oils after 10-fold dilution of the oils and reference solutions with MIBK [144]: nickel, iron, chromium, silver, copper, and tin were determined in an air-acetylene flame, aluminium, magnesium, and iron in an argon-separated nitrous oxide-acetylene flame. Some data on detection limits, linearity of calibration curves, and precision are shown in Table 6.12. The detection limits of all

TABLE 6.12. Detection limits, linearity, and precision of the simultaneous determination of wear metals in oils

Metal, wavelength, Å	Detection limit, µg/ml[c]	Upper linearity limit, µg/ml	Relative standard deviation, %[d]		
			10 µg/ml	1 µg/ml	0.1 µg/ml
Ni 2320[a]	0.015	20	0.35	1.03	6.68
Fe 2483[a]	0.02	25	0.69	1.20	8.64
Cr 3579[a]	0.02	50	1.26	1.76	9.47
Ag 3281[a]	0.006	9	0.23	0.74	3.76
Cu 3248[a]	0.004	9	0.17	0.42	2.50
Sn 2246[a]	0.2	50	1.69	9.48	—
Mg 2852[b]	0.01	4	0.35	0.96	2.02
Al 3962[b]	0.2	50	1.25	7.40	—
Fe 2483[b]	0.1	30	1.81	3.4	—

[a] Determined simultaneously in an air-acetylene flame.
[b] Determined simultaneously in an argon-separated nitrous oxide-acetylene flame.
[c] $S:N = 2$.
[d] 12—15 successive readings, sample aspiration period 16 sec.

the metals easily meet the sensitivity requirements of an oil analysis programme. Moreover, the linearity of the calibration curves is maintained up to relatively high concentrations and double dilution was not at any time needed in the oil analyses. The precisions for all the metals are also excellent, even at low concentrations.

The results for 12 samples are listed in Table 6.13. Good agreement between atomic absorption and atomic fluorescence results was achieved. At higher sample concentrations, the results for iron in the nitrous oxide-acetylene flame are generally slightly higher than those obtained in the air-acetylene flame, both by atomic absorption and atomic fluorescence. Since the higher results are probably closer to the true values (larger iron particles are more efficiently volatilized in the hotter flame), the nitrous oxide-acetylene flame

TABLE 6.13. Wear metals in oils by simultaneous multi-element atomic fluorescence analysis and single-element atomic absorption analysis [144]

Sample	Concentration, μg/ml[a]													
	Ag		Cu		Cr		Fe			Al		Mg	
	AFS	AAS	AFS	AAS	AFS	AAS	AFS	AAS	AFS[c]	AFS	AAS	AFS	AAS
952,801	0.3	0.3	N.D.	N.D.	N.D.	N.D.	3.6	3.6	3.4	N.D.	N.D.	0.67	0.71
952,802	0.3	0.2	0.1	N.D.	N.D.	N.D.	4.9	5.1	5.2	N.D.	N.D.	1.22	1.18
952,803	0.4	0.3	0.1	N.D.	N.D.	N.D.	6.3	6.2	6.1	N.D.	N.D.	0.84	0.85
952,804	0.1	N.D.	N.D.	N.D.	N.D.	N.D.	N.D.	0.1	N.D.	N.D.	N.D.	0.68	0.80
70–5A	0.5	0.3	4.9	5.8	3.0	3.0	11.1	12.2	10.9	N.D.	N.D.	7.20	7.40
70–5B	0.7	0.4	13.8	14.3	6.9	7.2	44.7	42.3	46.9	10.9	10.4	1.73	1.68
70–9A	0.5	0.3	16.6	17.8	3.1	2.1	21.0	20.7	22.6	4.8	3.9	5.81	6.01
70–9B	2.0	2.2	25.1	25.0	7.7	7.2	72.1	74.8	79.0	12.6	12.9	4.03	2.0
70–12A	0.7	0.5	3.1	2.9	3.2	3.5	6.8	7.0	6.4	4.7	3.6	2.95	2.92
70–12B	1.1	1.2	16.8	19.8	5.2	5.8	39.2	48.0	51.1	11.1	9.9	1.87	1.80
71–2A	0.4	0.3	11.6	12.0	1.2	1.0	13.8	14.5	15.0	N.D.	N.D.	1.31	1.20
71–2B	1.5	2.1	33.2	32.5	6.0	6.3	70.6	75.8	83.8	21.5	22.5	2.63	2.47

[a] Evaluation by calibration curve (Conostan D-12 organometallic oil standards).
[b] Ag, Cu, Cr and Fe determined simultaneously in an air-acetylene flame.
[c] Al, Mg and Fe determined simultaneously in an argon-separated nitrous oxide-acetylene flame (750 ppm of potassium, as an organometallic, added to ensure a reproducible ionization of sodium in the flame).
N.D. Not detectable.

should be the preferred flame for determining iron in oils. The analysis is free from chemical interference, as checked from recovery tests.

The use of non-flame atomization techniques for the determination of wear metals in jet-engine oils should also be of great analytical utility, as shown by Fraser and Winefordner [177].

6.2.3. APPLICATIONS TO CHEMICAL INDUSTRY

Klaus [237] determined zinc at around the $10^{-4}\%$ level in various metal compounds by means of a modified PMQ II spectrophotometer and Osram zinc vapour-discharge lamp source. The limit of detection was 0.02 μg/ml, and the relative standard deviation 5%. Optimization of experimental parameters, interferences and various evaluation techniques were also discussed.

A method was developed for the determination of mercury in a low-temperature propane-butane-air flame, which made use of the reduction of mercury to the metal with stannous chloride [417]. The method was applied to the determination of mercury in water-soluble salts of a number of metals, which contained not less than $10^{-4}\%$ of mercury. The relative standard deviation was not more than 5%.

Pleva [328] compared the precision and accuracy of atomic fluorescence and atomic absorption techniques for the determination of trace metallic constituents in Margruder standard fertilizer samples. Atomic fluorescence was excited in a turbulent hydrogen/entrained air flame with a 450-W xenon arc continuum source. No advantage of atomic fluorescence over atomic absorption was found except for the possibility of using a single source for all elements tested. Both techniques are equally precise and accurate, as can be seen from the results shown in Table 6.14.

TABLE 6.14. Results for fertilizer analyses by atomic fluorescence and atomic absorption spectrometry [328]

Element	Stated value, ppm	AFS value, [% deviation]	AAS value, [% deviation]
Cu	5.46	5.4 [1.1]	5.5 [0.7]
Fe	41.04	37 [9.9]	40 [2.5]
Zn	23.29	23 [1.2]	23 [1.2]
Mn	17.18	17 [1.0]	17 [1.0]
Mg	138	140 [1.4]	140 [1.4]
Ca	1225	1200 [2.0]	1200 [2.0]

TABLE 6.15. Non-flame atomic fluorescence determination of copper in silicon carbide [188]

Sample No.	Concentration, %	
	AFS	Spectrographic analysis
94	1.7×10^{-4}	1×10^{-4}
96	2.5×10^{-4}	5×10^{-4}
102	2.0×10^{-3}	5×10^{-4}
155	1×10^{-6}	traces
159	1×10^{-5}	traces
A	8.0×10^{-5}	1×10^{-4}
B	1.0×10^{-4}	3—9×10^{-5}

Guzeev and Nedler [188] determined traces of copper in silicon carbide by using non-flame thermal-impulse atomization of solid samples (cf. Section 3.3.4). Some results are listed in Table 6.15 and compared with those obtained spectrographically.

6.3. GEOCHEMICAL APPLICATIONS

6.3.1. APPLICATIONS TO ROCK, SEDIMENT, AND SOIL ANALYSES

Muscat and Vickers determined mercury in the nanogram range in rocks and natural sediments by a flameless atomic fluorescence system which used either a simple reduction-aeration method for generating mercury vapour [301] or the more sensitive combustion furnace or reduction-aeration-silver amalgamator technique (cf. Section 6.1.2) [300]. A schematic diagram of the apparatus for the simple reduction-aeration method is shown in Fig. 6.2. For this procedure mercury must be present in solution as the Hg(II) ion. Samples requiring digestion were treated by the $KMnO_4-K_2S_2O_8$ oxidation method in $H_2SO_4-HNO_3$ (1 : 4) medium. Excess of oxidant in standards and digested samples was removed by addition of hydroxylamine hydrochloride solution. When measurements were to be made, 50 ml of the solution were placed in the reduction-aeration flask, the system was closed,

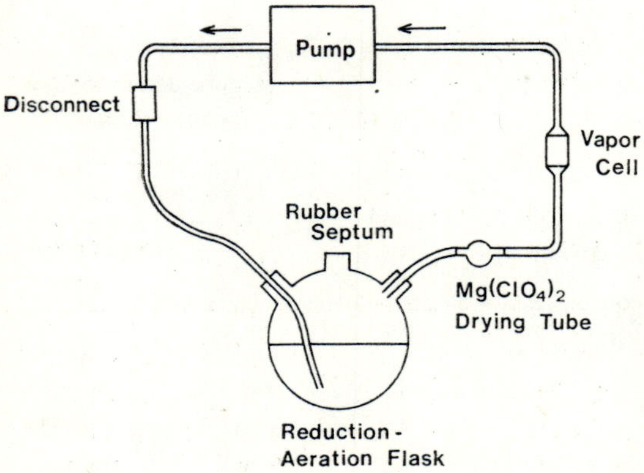

FIG. 6.2. Schematic diagram of a gas train for the reduction-aeration method for mercury determination [301].

Reprinted from Ref. [301] by permission of Elsevier Publishing Company

3 ml of the stannous chloride solution were added to the solution by syringe through the rubber septum, and the pump was started immediately. Mercury vapour in the fluorescence cell was excited with a mercury pen lamp. A steady signal was obtained within 1 min. For the combination of the reduction-aeration method with the silver amalgamator, the resulting mercury vapour was swept into the amalgamator by bubbling the selected carrier gas through the solution. Reference solutions of mercuric chloride were used for preparing calibration curves.

Initial attempts to apply the combustion furnace-amalgamator technique to natural sediment samples proved unsuccessful, and the difficulty was ascribed to the chloride content of the samples. The interference was eliminated by placing a second amalgamator operated at 400 °C in the gas train nearer the furnace, which removed the interfering species from the gas stream without hindering the progress of mercury vapour into the "analytical amalgamator". Some results of analyses of rocks and sediments are shown in Tables 6.16 and 6.17.

Dagnall *et al.* [124] determined exchangeable calcium, copper, magnesium, manganese and zinc in ammonium acetate, acetic acid, and EDTA soil-extracts by using a Technicon AFS-6 multichannel atomic fluorescence

TABLE 6.16. Comparison of results for mercury in standard rock samples by different generation techniques

Reprinted from Ref. [300] by permission of The American Chemical Society

Sample[a]	Hg found, ppm				
	Simple reduction-aeration method	Reduction-aeration method + amalgamator	Furnace technique + amalgamator	AAS	NAA[b]
W — 1	0.18[c]	0.18	0.18	0.18	0.17
CSP — 1	—	—	0.015	0.015	0.021

[a] United States Geological Survey.
[b] Neutron-activation analysis.
[c] Mean of six determinations.

TABLE 6.17. Comparison of results of determinations of mercury in natural sediment samples [300]

Reprinted from Ref. [300] by permission of The American Chemical Society

Method	Sample weight g	Hg found ng	ppm Hg
Reduction-aeration + amalgamator	1.123	338	0.34
AAS	0.808	220	0.27
Reduction-aeration + amalgamator	0.620	186	0.30
AFS	0.731	190	0.26
Combustion furnace + amalgamator	0.329	92	0.28
AFS	0.268	67	0.25

spectrometer. Determination of all the elements in a single extract at two dilutions was possible by this technique. The method was shown to permit rapid, sensitive and sufficiently accurate analysis for these samples.

6.3.2. APPLICATIONS TO HYDROGEOLOGY AND WATER-QUALITY CONTROL

Demers [142] determined calcium, magnesium, cobalt, copper, iron, manganese, nickel and zinc in lake-water samples either directly (calcium and magnesium) or by using various preconcentration techniques, such as distillation, extraction or a thermal concentration device which introduced the sample into the flame in the form of a dry concentrated aerosol. Atomic fluorescence was excited with a 450-W xenon arc lamp in the hydrogen/entrained air flame. To avoid chemical interferences due to high concentrations of other ions in the sample, which affect the determination of calcium and magnesium, the addition of strontium chloride (making the solution 0.1% with respect to strontium chloride hexahydrate) was necessary.

The high sensitivity of the determination of gold in the hydrogen-oxygen-argon flame, and freedom from interferences in this flame permitted the direct determination of gold in mine-water containing 0.045 μg/ml of gold [277, 278]. Following the extraction procedure described by Butler *et al.* [78] and aspiration of the extracts into the flame specified, gold can be determined in mine-waters down to a level of 10^{-5} μg/ml.

Mercury was determined in a reference water sample (Federal Water Quality Administration, Analytical Quality Control Laboratory, Cincinnati, Ohio, U.S.A.) by the non-flame reduction-aeration-silver amalgamator technique described in Sections 6.1.2 and 6.3.1. A measured value of 4.2 ± 0.4 ng/ml is in a very good agreement with a "true" value of 4.2 ng/ml.

Marshall and Smith [271] determined trace metallic impurities (iron, copper, nickel and zinc) in high-purity boiler feed-water over the range $10^{-1}-10^{-2}$ μg/ml, using a specially designed non-dispersive atomic fluorescence instrument and electrodeless discharge lamp sources.

A Technicon AFS-6 multichannel atomic fluorescence spectrometer has been shown to be suitable for the simultaneous determination of metals in a water-quality or water-pollution analysis programme [143, 144]. For some trace metals, a concentration step is necessary.

TABLE 6.18. Detection limits, linearity, and precision for the simultaneous determination of some metals in water

Metal, wavelength, Å	Detection limit μg/ml[a, b]	Upper linearity limit, μg/ml	Relative standard deviation, %		
			10 μg/ml	1 μg/ml	0.1 μg/ml
Mg 2852	0.002	2	0.18	0.40	2.33
Ca 4227	0.001	6	0.48	0.61	1.58
Ag 3281	0.01	9	0.54	0.86	4.82
Ni 2320	0.09	20	0.96	3.74	—
Mn 2795	0.008	15	0.68	0.75	3.56

[a] S : N = 2.

[b] Note that more recent detection limits for the improved instrument are much better for some elements [145, 146].

Precision and sensitivity obtained for one group of metals investigated is shown in Table 6.18.

Manganese and silver can be determined simultaneously in the original water sample, calcium and magnesium after a 50-fold dilution and addition of a releasing agent (0.1% strontium) and nickel usually after a preconcentration step. A comparison of the results for manganese, calcium and magnesium obtained by atomic absorption analysis and by simultaneous atomic fluorescence analysis is given in Table 6.19.

TABLE 6.19. The determination of some metals in water by atomic absorption analysis and by simultaneous atomic fluorescence analysis

Sample	Concentration, μg/ml					
	Mn		Ca		Mg	
	AFS	AAS	AFS	AAS	AFS	AS
1	0.07	0.06	122	120	46.0	45.4
2	N.D.	0.01	124	121	29.9	29.6
3	0.07	0.08	48.6	48.8	12.8	12.8
4	0.03	0.02	149	145	31.5	31.1
5	N.D.	N.D.	59.5	60.0	13.3	13.4
6	0.09	0.08	102	100	22.7	23.0
7	0.04	0.03	45.1	44.7	13.4	13.4
8	0.03	0.03	46.7	46.7	13.5	13.4
9	0.04	0.02	185	182	63.4	62.8
10	N.D.	N.D.	45.1	45.3	13.9	13.6

N.D. Not detectable.

6.4. OTHER APPLICATIONS

6.4.1. DIRECT ISOTOPIC ANALYSIS BY ATOMIC FLUORESCENCE SPECTROMETRY

Atomic fluorescence and atomic absorption spectrometry offer a means of isotopic analysis, utilizing the hyperfine structural lines to distinguish between the various stable isotopes of a given element. These two techniques are similar in that selective fluorescence (or absorption) by different isotopes of the same element can occur if the primary source contains only one of the isotopes and the hyperfine emission or absorption lines do not overlap to any great extent. Therefore, the applicability of these techniques is mainly determined by the shape and width of the hyperfine line profiles, especially the Doppler width in the flame atomization cell. Isotopic analysis by atomic absorption spectrometry has been reported for uranium [181, 348, 350], lithium [182, 263, 435, 485], and lead [70], i.e. for the elements with a relatively large isotopic line separation. Isotopic analysis by atomic fluorescence spectrometry seems to offer some advantages over atomic absorption because of its simpler instrumental requirements, e.g. a low-dispersion monochromator (due to the fact that the source is not viewed directly). Electrodeless discharge sources are readily prepared from very small quantities of the isotopically pure elements.

Veillon and Park [413] carried out a satisfactory direct isotopic analysis of mercury, using isotopically-pure electrodeless discharge lamp sources (prepared from the metal) operated at low temperatures and low power levels. To minimize line broadening, the arrangement for cooling the source with water was used as illustrated in Fig. 6.3, with the coolant flow adjusted

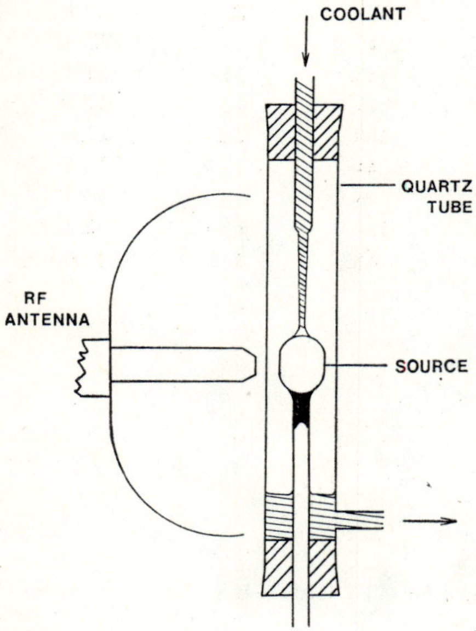

FIG. 6.3. Source-lamp cooling arrangement [413].
Reprinted from Ref. [413] by permission of The American Chemical Society

so that only a thin stream of water flowed over the lamp surface. This was necessary in order to minimize absorption of the microwave energy by the coolant. To decrease Doppler broadening of the hyperfine absorption lines, atomic fluorescence of mercury was generated in a low-temperature hydrogen-argon-oxygen flame.

The hyperfine system of mercury results from contributions by its isotopes and is somewhat complex, as shown in Table 6.20. The displacements given in this table represent the distance in wavelength units of each

TABLE 6.20. Hyperfine structure of mercury spectrum
Reprinted from Ref. [413] by permission of The American Chemical Society

Component and displacement, Å	Contributing isotopes
—0.025	199, 201
—0.010	198, 201
0.0	200
+0.012	202
+0.022	199, 201, 204

component from the line centre of the ^{200}Hg component. Considering the component spacing, the probable component absorption line-widths in the flame, and the contribution to some components by more than one isotope, it may be expected that complete separation of the hyperfine absorption components cannot be obtained.

Veillon and Park performed the analysis as follows: they measured fluorescence signals for all source-isotope/solution-isotope combinations, corrected them for the difference in source intensities, and thus obtained fluorescence signals which represent the extent of overlap of the hyperfine absorption lines of each isotope. Comparing these results with those for the

TABLE 6.21. Overlap observed with various source-isotope combinations [413]
Reprinted from Ref. [413] by permission of The American Chemical Society

Solution-isotope used	Source-isotope used					
	198	199	200	201	202	204
198	100	61	68	52	40	29
199	59	100	33	83	33	42
200	66	35	100	44	56	40
201	52	80	46	100	60	75
202	37	38	56	62	100	70
204	29	42	38	78	68	100

isotopic source used with a solution of the same isotope they calculated a per cent overlap for each isotope combination. These results are shown in Table 6.21. Measurement of the fluorescence of real mercury samples with each isotopic source then yielded six fluorescence intensity values, each of which is the sum of the six isotope concentrations, when multiplied by the appropriate factor from Table 6.21. This system of six equations with six unknowns was solved by computer. Some results of analyses performed on natural and synthetic samples are shown in Table 6.22. The main sources of errors in these isotopic analyses are due primarily to inaccuracies in the overlap measurements and the measurement of relative source intensities.

TABLE 6.22. Analysis of natural mercury and a synthetic mixture [413]
Reprinted from Ref. [413] by permission of The American Chemical Society

Isotope	Natural Hg sample		Synthetic Hg sample	
	abundance, %	found, %	concentration, μg/ml	found, μg/ml
198	10.02	9.70	50	51.5
199	16.92	18.4	50	52.0
200	23.10	23.8	50	47.5
201	13.22	12.6	50	51.0
202	29.72	30.0	50	48.0
204	6.84	6.71	50	53.0

6.4.2. NON-ANALYTICAL APPLICATIONS

The phenomenon of atomic fluorescence has been widely applied to the measurement of some constants of atomic spectra, such as damping constants of atoms in flames [402, 403], lifetime of excited atoms [305, 324], effective atomic cross-sections of second-order collisions [202, 296], specific quenching cross-sections of excited atoms for various gases [10, 197, 198, 199, 201, 211, 214, 215, 216, 217], diffusion studies [69], etc. Omenetto *et al.* [318, 525] have suggested a simple technique for flame temperature measurements, utilizing thermally-assisted excitation processes.

Chapter 7

COMPARISON OF ATOMIC FLUORESCENCE SPECTROMETRY WITH ATOMIC ABSORPTION AND FLAME EMISSION SPECTROMETRY

For the practising analyst, it is important to know how atomic fluorescence spectrometry compares in performance with the two other methods of atomic spectrometric analysis, and for which elements better results may be attained with either atomic absorption spectrometry or flame emission spectrometry. In this chapter we shall try to summarize briefly the answers to these questions.

At first glance it is evident that the acceptance of atomic fluorescence spectrometry in chemical analysis has been much slower than that of atomic absorption spectrometry. At the time of writing this book, i.e. eight years after the publication of the original paper by Winefordner and Vickers [444] the number of analytical applications of atomic fluorescence spectrometry is rather limited, as enumerated in the preceding chapter. In comparison, eight years after publication of the original paper of Walsh on the use of atomic absorption spectra in 1955, the number of analytical applications of atomic absorption spectrometry included more than 100 papers and a number of commercial instruments were already sold.

However, this difference should not be overemphasized. To a great extent atomic fluorescence spectrometry and atomic absorption spectrometry offer solutions to the same type of analytical problem and the development and application of atomic fluorescence spectrometry should be considered against the background of existing atomic absorption spectrometry as a broadly exploited analytical method satisfying many of the existing analytical requirements which could otherwise be solved by atomic fluorescence spectrometry. Thus the pressure for a simple, expedient, and broadly applicable analytical method is much lower than that existing in the late fifties, at the time of the development of atomic absorption spectrometry.

The merits of atomic fluorescence spectrometry in comparison with the other methods have been considered by different authors on the basis of theoretically calculated signal strengths and/or limits of detection [11, 459,

466]. The paper by Winefordner and others [466] is the most extensive and covers all three methods.

Assuming identical conditions (i.e. resonance line measured, entrance optics, monochromator, detector system) the ratio of signals from a dilute atomic vapour measured by atomic fluorescence and atomic absorption spectrometry is given by [466]:

$$\frac{S_{\text{AAS}}}{S_{\text{AFS}}} = \frac{1}{Y\Omega/4\pi}$$

where Y is the fluorescence quantum yield for the particular transition and $\Omega/4\pi$, the fractional solid angle of excitation radiation collected by the entrance optics and falling on the atomic vapour. For commonly employed monochromators $\Omega/4\pi$ is of the order of 0.01. When working with flames the fluorescence yield may be around 0.1 so that the signal measured by atomic absorption spectrometry is approximately a thousandfold stronger than the fluorescence signal. When a large aperture system is used to select the wavelength of fluorescence emission, $\Omega/4\pi$ may be increased considerably [143, 145, 258, 291]. Nevertheless, the signals observed in atomic fluorescence are invariably smaller than in atomic absorption.

For flame temperatures between 2000 and 3000 K and radiances of currently available primary light-sources (a 450-W high-pressure xenon lamp as a continuum source and boosted-output hollow-cathode lamp or electrodeless discharge lamp as line source), Winefordner *et al.* [466] concluded that "if signals alone are considered, atomic absorption or atomic emission spectrometry should be used for the measurement of analyte atoms in essentially all flames".

The situation changes if noise is taken into account and the signal-to-noise ratio is considered. The total noise in emission and fluorescence flame spectrometry may be expressed as

$$N_{\text{tot}} = \sqrt{N_{\text{shot}}^2 + N_{\text{flame}}^2}$$

where N_{shot} is the shot-noise of the photodetector output as a result of the quantum nature of light and the particulate nature of electron emission from the photocathode and N_{flame} is the flame-flicker noise, which is the result of random fluctuations in the background intensity emitted by the flame. In comparison the total noise in atomic absorption is greater because the square of the light-source noise must be added in the equation above, and simultaneously the shot-noise value is greater than for emission or fluorescence because a greater light flux is usually falling upon the photodetector surface.

Therefore, considering signal-to-noise ratios for the three methods under identical conditions, i.e. instrumental system and flame (which in practice is rarely the case) it can be concluded that limits of detection should be lower for atomic fluorescence spectrometry than for atomic absorption spectrometry at all wavelengths with low-temperature flames and comparable with high-temperature flames. In the latter, atomic emission detection limits should be superior to those of both atomic fluorescence and atomic absorption spectrometry at wavelengths longer than about 3250 Å [459].

These results are of limited practical validity as may be seen from Table 7.1, where experimentally found values for limits of detection for all three methods by different workers are listed. Exceptions to this rule are rather more frequent than compliances with it because the quality of light-sources for different elements proves to be of decisive importance.

The situation with heated graphite atomizers is similar. With most commercial apparatus the thermal background emission of the hot atomizer body cannot usually be efficiently screened off unless a rather cold region of the atomizing device is viewed and in such regions strong interferences usually occur. However, a big advantage which the use of non-flame atomizers brings to atomic fluorescence spectrometry is the application of an inert gas atmosphere and the corresponding diminution of quenching.

In Table 7.2 experimental limits of detection in atomic fluorescence spectrometry and atomic absorption spectrometry with non-flame atomizers are compared.

On the basis of experimental limits of detection in air-acetylene flames (which for practical analysis find the broadest use) and with existing primary light sources, several workers have expressed the view that atomic fluorescence spectrometry has little to offer over atomic absorption spectrometry [32, 493]. This statement is, however, somewhat oversimplified. There are several examples where the use of available commercial atomic absorption/flame emission spectrometers in the "fluorescence mode" (even without specialized primary sources) yields satisfactory gain in sensitivity over atomic absorption for a particular element. This practice may obviate the need for concentration techniques (which would be necessary for atomic absorption measurements) [106, 117, 119, 245, 246, 276, 278, 428] and thus justifies the changeover of the instrument to an atomic fluorescence spectrometer.

The attainable limits of detection are not the only merit of an analytical method. Depending on the particular analytical problem, other properties may be rather more important. One of the advantages of atomic fluorescence spectrometry over atomic absorption spectrometry is its ready application to simultaneous multielement analysis. Atomic fluorescence spectrometry

TABLE 7.1. Comparison of experimental limits of detection in flame spectrometric methods

Element	Wavelength (Å) and detection limit [μg/ml][a]						
	λ	AFS(L)[b]	AFS(C)[c]	λ	AAS(L)[d,e]	λ	AES[f,g]
Ag	3281	0.000 1	0.001	3281	0.003 (0.036)[h]	3281	0.02
Al	3962	0.03	4.0	3093	0.04 (0.76)	3962	0.005
As	1973	0.07	2.0	1937	0.25 (0.78)	2350	50
Au	2428	0.005	3.5	2428	0.01 (0.13)	2676	4.0
Ba	5536	0.2	N.M.[j]	5536	0.01 (0.20)	5536	0.002
Be	2349	0.008	N.M.[j]	2349	0.002 (0.017)	2349	0.1[k]
Bi	3068	0.004 7	0.17	2231	0.06 (0.22)	2231	2.0[l]
Ca	4227	0.000 3	0.02	4227	0.0005(0.021)	4227	0.000 1
Cd	2288	0.000 001[i]	0.015	2288	0.0006(0.011)	3261	2.0
Co	2407	0.0015	0.021	2407	0.007 (0.066)	3454	0.05
Cr	3579	0.000 3	0.3	3579	0.005 (0.055)	4255	0.005
Cu	3248	0.000 3	0.015	3248	0.003 (0.040)	3274	0.01
Fe	2483	0.000 6	0.035	2483	0.005 (0.062)	3720	0.05
Ga	4172	0.007 2	0.14	2944	0.10 (1.1)	4172	0.01
Ge	2652	0.1	1.8	2652	0.15 (1.5)	2652	0.5
Hg	2537	0.07	100	2537	0.20 (2.2)	2537	40
In	4102	0.015	2.0	3039	0.05 (0.38)	4511	0.005
Ir	2544	4.0	100	2089	0.36 (0.77)	3800	100
Mg	2852	0.000 1	0.0024	2852	0.0003(0.003)	2852	0.005
Mn	2795	0.000 5	0.0025	2795	0.003 (0.024)	4031	0.005
Mo	3133	0.2	N.M.[j]	3133	0.04 (0.33)	3903	0.1
Na	5890	N.M.	0.0075	5890	0.0003(0.003)	5890	0.000 1
Ni	2320	0.001	1.0[j]	2320	0.008 (0.066)	3415	0.03
Pb	4058	0.01[i]	0.61[m]	2170	0.02 (0.11)	4058	0.2
Pd	3405	0.04	50	2448	0.02 (0.092)	3635	0.05
Pt	2659	0.5	500	2659	0.12 (1.2)	2659	2.0
Rh	3435	0.3	10[n]	3435	0.006 (0.15)	3692	0.9
Ru	3728	5.0	100	3499	0.16 (0.72)	3728	0.3
Sb	2312	0.04	100	2176	0.07 (0.29)	2598	20
Se	1961	0.15[i]	3.9	1961	0.48 (1.1)	2040	23[o]
Si	2516	0.55	N.M.[j]	2516	0.30 (1.6)	2516	5.0
Sn	3034	0.12[i]	5.0	2246	0.03 (0.46)	2840	0.3
Sr	4607	0.000	10[i]	4607	0.004 (0.044)	4607	0.000 2
Te	2143	0.05	0.79	2143	0.12 (0.26)	2386	200
Ti	3200	0.1	N.M.[j]	3643	0.10 (1.4)	3200	0.2
Tl	3776	0.005	0.073	2768	0.02 (0.28)	5351	0.02
V	3185	0.07[i]	N.M.[j]	3185	0.11 (0.88)	4379	0.01
Zn	2139	0.000 001	0.015	2139	0.002 (0.009)	2139	50

[a] Signal-to-noise ratio = 2 when not otherwise stated.
[b] Atomic fluorescence spectrometry, excited with a line source.
[c] Atomic fluorescence spectrometry, excited with a continuous source.
[d] Atomic absorption spectrometry with a line source.
[e] Values reported by manufacturers for Varian-Techtron AA-5 spectrometer.
[f] Atomic emission spectrometry.
[g] Results from references [441, 442] when not otherwise stated.
[h] Values in parenthesis represent characteristic concentration (μg/ml/1 %).
[i] Signal-to-noise ratio = 1.
[j] N.M. = not measured.
[k] See reference [84].
[l] See reference [85].
[m] At the 2833 Å line.
[n] At the 3692 Å line.
[o] See reference [443].

TABLE 7.2. Comparison of experimental limits of detection in non-flame atomic fluorescence and atomic absorption spectrometry [with primary line source]

Element	Wavelength (λ), (Å) and absolute detection limit (pg)[a]				
	λ	AFS	λ	AAS	
				CRA 61	HGA 70[c]
Ag	3281	0.07	3281	0.2	20
As	1973	500	1937	100	160
Au	2676	4.0	2428	10	60
Be	2349	30	2349	0.9	3.4
Bi	3068	10	2231	7.0	280
Ca	4227	0.1	4227	0.3	3.1
Cd	2288	0.02	2288	0.1	0.8
Cu	3248	5.0	3248	7.0	45
Fe	2483	2.0	2483	3.0	50
Ga	4172	50	4172	20	1 200
Hg	2537	20	2537	100	15 000
Mg	2852	0.03	2852	0.06	1.0
Pb	4058	3.0	2170	5.0	23
Sb	2312	200	2176	30	510
Tl	3776	0.9	2768	3.0	90
Zn	2139	0.04	2139	0.08	2.1

[a] Signal-to-noise ratio = 2 when not otherwise stated.
[b] Carbon rod atomizer, model 61, Varian-Techtron.
[c] Hot graphite furnace, model 70, Perkin-Elmer.

is more promising in this sphere than atomic absorption spectrometry and several instruments have been constructed (see Section 3.5). The virtual specificity of atomic fluorescence spectrometry allows even non-dispersive systems to be used [165, 258, 259, 425, 426] with a large gain in sensitivity for many elements over that obtainable by atomic absorption spectrometry. Another advantage of atomic fluorescence spectrometry over atomic absorption is the fact that continuous sources may be used in conjuction with medium resolution monochromators without loss in the fluorescence signal, while in atomic absorption, the spectral band-width of the monochromator must be less than about 0.2 Å to yield satisfactory limits of detection.

All these advantages must be taken into account when selecting the analytical method for a particular problem. In many cases atomic fluorescence spectrometry may offer the best solution from the point of view of analytical as well as economical parameters. Whatever the present situation it is generally acknowledged that atomic fluorescence spectrometry has greater potential developments in the future than either atomic absorption or flame emission spectrometry.

List of Frequently Used Symbols

a	classical damping constant
A_c	cross-sectional area of the exciting beam
A_F	surface area of fluorescence cell
A_{ul}	Einstein coefficient of spontaneous emission (for transition u to l)
c	speed of the light
C	concentration
e	base of natural logarithm
e	charge of electron
E	voltage on the photomultiplier anode resistance
E_0	incident radiance coming from continuum source
E_0'	incident radiance coming from line source
E_i	energy of electronic state i
E_j	mean amplitude of Johnson noise
f	absorption oscillator strength
f	frequency (in electronics)
f_p	pulse repetition frequency
f_s	reabsorption factor (accounting for the fluorescence radiance loss)
F	apparatus distribution function
g	transfer efficiency between dynodes
g_i	statistical weight of electronic state i
G	current gain of a photomultiplier
G_A	gain of electronic current
h	Planck's constant
i_0	steady-state current of a hollow-cathode lamp
i_{pk}	peak current of a hollow-cathode lamp
I_A	anode current
I_s	measured photodetector signal
k	Boltzmann constant
k_v	absorption coefficient of a single atomic line at relative frequency v
k_0	peak atomic absorption coefficient for Doppler broadening
m	mass of the electron

n	concentration of analyte atoms in all states within the flame cell
n_0	concentration of analyte atoms in ground state within the flame cell
n_i	concentration of atoms in electronic state i within the flame cell
Q_D	total radiant energy reaching the photocathode
Q_X	quantum efficiency
r_u	total pseudo first order rate constant for quenching from state u
R	universal gas constant
R	anode resistance of a photomultiplier (load resistor)
R_A	instrumental read-out
R_λ	radiant sensitivity
S/N	signal-to-noise ratio
t	total distance travelled by the beam of radiation through the homogenous absorbing medium
T	temperature in K
v	relative frequency
w	slit width
w_p	pulse width
X_A	deflection of a measuring instrument corresponding to signal plus background
X_B	deflection of a measuring instrument corresponding to background
X_i	ratio of atoms in state i to all atoms of the same kind
Y	fluorescence quantum yield
Y_{ul}	quantum yield for a fluorescence transition from upper state u to lower state l
α_c	absorption function of continuum radiation (def. by Eq. 2.8)
α_L	absorption function of line radiation (def. by Eq. 2.9)
β	efficiency of converting submicroscopic forms in flame cells into atoms
γ	photoanodic sensitivity
δ_v	Voigt profile function
Δ_f	frequency bandwidth
$\Delta\lambda_L$	Lorentzian half-width
$\Delta\lambda_D$	Doppler half-width
$\Delta\nu_D$	Doppler half-width
$\Delta\nu_L$	Lorentzian half-width
$\Delta\Phi_{A,C}$	radiant flux coming from continuum source absorbed in elementary cell
$\Delta\Phi_{Fi}$	observed fluorescence radiant flux
$\Delta\Phi_{A,L}$	radiant flux coming from line source absorbed in elementary cell
$\Delta\Phi_A$	radiant flux absorbed by the analyte atoms (within the elementary cell)

$\Delta\Phi_{\mathrm{Fi}\nu}$	observed spectral fluorescence radiant flux of elementary cell
$\Delta\Omega$	solid angle (of incident radiation)
ε	efficiency of introducing the sample analyte into the atomizer in submicroscopic form (molecules or atoms)
η	average secondary emission coefficient
$\varkappa$	factor characterizing the ratio of the optical densities for the excitation and the fluorescence processes
λ_0	wavelength at line centre
ν	frequency
ν_0	frequency at line centre
Φ_{F}	total observed fluorescence radiant flux (emitted into the solid angle Ω)
$\Phi_{\mathrm{F}\nu}$	total observed spectral fluorescence radiant flux
$\Phi_{\lambda 0}$	spectral radiant flux (continuously emitted by a hollow-cathode lamp)
$\Phi_{\lambda \mathrm{F}}$	spectral fluorescence radiant flux
$\Phi_{\lambda \mathrm{pk}}$	peak spectral radiant flux of the pulsed hollow-cathode lamp
σ_{c}^2	collisional broadening cross-section
σ_{Q}^2	quenching cross-section
σ_{s}	standard deviation of signal read-out
τ	transmission factor of the spectrometer
τ_{c}	time constant
τ_{e}	transmission factor of the entrance optics
Ψ	transmittance of a photomultiplier tube window
Ω'	solid angle collected by spectrometer
Ω_{e}''	solid angle of the entrance beam from the fluorescence cell

BIBLIOGRAPHY

1. J. AGGETT and T. S. WEST, *Anal. Chim. Acta*, **55,** 349 (1971). Atomic absorption and fluorescence spectroscopy with a carbon filament atom reservoir. IV. The determination of gold by atomic fluorescence and atomic absorption spectroscopy with an unenclosed atom reservoir.
2. J. AGGETT and T. S. WEST, *Anal. Chim. Acta*, **57,** 15 (1971). Atomic absorption and fluorescence spectroscopy with a carbon filament atom reservoir. V. Determination of elements in organic solvents.
3. J. F. ALDER and T. S. WEST, *Anal. Chim. Acta*, **51,** 365 (1970). Atomic absorption and fluorescence spectroscopy with a carbon filament atom reservoir. Determination of cadmium by atomic fluorescence spectroscopy with an unenclosed atom reservoir.
4. K. M. ALDOUS, R. M. DAGNALL, K. C. THOMPSON and T. S. WEST, *Anal. Chim. Acta*, **41,** 380 (1968). Some preliminary studies of the analytical potentialities of the atomic-hydrogen plasma torch.
5. K. M. ALDOUS, R. M. DAGNALL and T. S. WEST, *Analyst*, **94,** 347 (1969). Preparation and spectral characteristics of microwave-excited electrodeless discharge tubes for palladium, silver, platinum, and gold.
6. K. M. ALDOUS, R. M. DAGNALL and T. S. WEST, *Anal. Chim. Acta*, **44,** 457 (1969). The preparation of electrodeless discharge tubes for some alkali and alkaline earth elements.
7. K. M. ALDOUS, R. F. BROWNER, R. M. DAGNALL and T. S. WEST, *Anal. Chem.*, **42,** 939 (1970). Application of capillary burners in flame spectrometry.
8. K. M. ALDOUS, R. M. DAGNALL and T. S. WEST, *Lab. Pract.*, **19,** 587 (1970). Preparation and operation of microwave excited electrodeless discharge tubes for use in atomic spectroscopy.
9. K. M. ALDOUS and B. W. BAILEY, Paper presented to the 3rd International Conference on Atomic Absorption and Atomic Fluorescence Spectrometry, Paris, 1971. A comparison of photon counting and lock-in amplifier detection methods in atomic fluorescence spectrometry.
10. C. T. J. ALKEMADE, *Proceedings of the 10th Colloquium Spectroscopicum Internationale*, p. 143, Spartan Books, Washington, D.C., 1963. Excitation and related phenomena in flames.
11. C. T. J. ALKEMADE, *Atomic Absorption Spectroscopy*, ed. by R. M. Dagnall and G. K. Kirkbright, Butterworths, London, 1970. A theoretical discussion on some aspects of atomic fluorescence spectroscopy in flames.
12. C. TH. ALKEMADE and P. T. J. ZEEGERS, pp. 3–125 in *Spectrochemical Methods of Analysis*, ed. by J. D. Winefordner, Wiley, New York, 1971. Excitation and de-excitation processes in flames.

13. M. D. AMOS, P. L. BENNETT and P. E. THOMAS, Paper presented to the 15th Colloquium Spectroscopicum Internationale, Madrid, 1969. Some investigations of the parameters of atomic fluorescence spectroscopy.
14. M. D. AMOS and P. E. THOMAS, Paper presented to the Conference of Flame Spectroscopy, Ostrava, 1970, Czechoslovakia. Some studies in non-flame atomisation for atomic absorption and atomic fluorescence spectroscopy.
15. M. D. AMOS, P. A. BENNETT and K. G. BRODIE, *Resonance Lines*, **2,** 3 (1970). Investigations into the use of shielded flames in spectroscopy.
16. M. D. AMOS, P. A. BENNETT, K. G. BRODIE, P. W. Y. LUNG and J. P. MATOUŠEK, *Anal. Chem.* **43,** 211 (1971). Carbon rod atomizer in atomic absorption and atomic fluorescence spectrometry and its clinical application.
17. M. D. AMOS, K. G. BRODIE and J. P. MATOUŠEK, Paper presented to the Pittsburgh Conference on Analytical Chemistry and Applied Spectroscopy, 1971. Development with the carbon rod atomizer and its application to the petroleum industry.
18. M. D. AMOS, K. G. BRODIE and J. P. MATOUŠEK, Paper presented to the 3rd International Conference on Atomic Absorption and Atomic Fluorescence Spectrometry, Paris, 1971. Some applications of non-flame atomisation in analytical spectroscopy.
19. R. G. ANDERSON, I. S. MAINES and T. S. WEST, *Anal. Chim. Acta*, **51,** 355 (1970). Atomic absorption and fluorescence spectroscopy with a carbon-filament atom reservoir. II. Use of the atom reservoir in atomic fluorescence spectroscopy.
20. R. G. ANDERSON, I. S. MAINES and T. S. WEST, Paper presented to the International Atomic Absorption Spectroscopy Conference, Sheffield, 1969. The carbon filament atom reservoir in the atomic absorption and fluorescence spectroscopy of sub-nanogram amounts of metals.
21. R. J. ANDERSON and J. CLEARY, *J. Opt. Soc. Am.*, **60,** 531 (1970). Detection of weak light signals.
22. M. J. AL ANI, R. M. DAGNALL and T. S. WEST, *Analyst*, **92,** 597 (1967). Preparation of metal halide-mercury microwave-excited electrodeless discharge tubes as spectral line sources.
23. D. N. ARMENTROUT, *Anal. Chem.*, **38,** 1235 (1966). Determination of nickel by atomic fluorescence flame spectrometry.
24. R. J. ATKINSON, G. D. CHAPMAN and L. KRAUSE, *J. Opt. Soc. Am.*, **55,** 1269 (1965). Light sources for the excitation of atomic resonance fluorescence in potassium and rubidium.
25. R. M. BADGER, *Z. Physik*, **55,** 56 (1929). Flame fluorescence and quenching of fluorescence in gas mixtures at high pressure.
26. B. W. BAILEY, *Spectrosc. Letters*, **2,** 81 (1969). Selection of wavelengths for use in excitation and emission in atomic fluorescence spectroscopy.
27. B. W. BAILEY, *Anal. Chim. Acta*, **54,** 537 (1971). Atomic fluorescence spectroscopy of lead.
28. B. W. BAILEY and J. M. RANKIN, Paper presented to the 3rd International Conference on Atomic Absorption and Atomic Fluorescence Spectrometry, Paris, 1971. The use of organic liquid fuel flames in atomic absorption and atomic fluorescence spectrometry.
29. S. V. BARANOV, L. K. MASHTAROV and L. G. POFRALIDI, *Zh. Prikl. Spektrosk.*, **10,** 595 (1969). General characteristics of high-frequency excited electrodeless discharge lamps and power supplies.
30. S. V. BARANOV, N. P. IVANOV, L. G. POFRALIDI, V. V. KNYAZEV, B. M. TALALAEV

and E. N. Vasiliev, *Zh. Analit. Khim.*, **24,** 1649 (1969). Electrodeless tubes with high-frequency excitation of the spectrum as a radiation source in atomic absorption analysis.

31. R. M. Barnes and R. F. Jarrell, pp. 288—305 in *Analytical Emission Spectroscopy*, ed. by E. L. Grove, Vol. I, Part 1, M. Dekker, New York, 1971. Gratings and Grating Instruments.
32. W. B. Barnett and H. L. Kahn, *Anal. Chem.*, **44,** 935 (1972). Comparison of atomic fluorescence with atomic absorption as an analytical technique.
33. F. H. Barr and E. H. Eberhardt, *Appl. Optics*, **6,** 1575 (1967). On the elimination of the lumen from the responsivity calibration of photodetectors.
34. A. S. Bazhov and Y. A. Lazarev, *Zavodsk. Lab.*, **32,** 761 (1966). High-frequency electrodeless discharge lamps with metallic vapor.
35. A. S. Bazhov, V. K. Zakharov and P. A. Koka, *Zavodsk. Lab.*, **34,** 245 (1968). High-frequency electrodeless discharge at atmospheric pressure as a light source for emission spectral analysis.
36. A. S. Bazhov and A. V. Zherebenko, *Zh. Prikl. Spektrosk.*, **12,** 760 (1970). Spectral characteristics of high-frequency electrodeless discharge lamps with vapour of lead and its compounds.
37. A. S. Bazhov and A. V. Zherebenko, *Zh. Prikl. Spektrosk.*, **14,** 1131 (1971). High-frequency electrodeless lamps with vapours of metals and their salts.
38. K. H. Becker and W. Groth and W. Jud, *Z. Naturforsch.*, A **24,** 1953 (1969). Measurement of small atomic concentrations in the gas phase by resonance fluorescence in the vacuum ultraviolet.
39. Y. I. Belyaev, Paper presented to the 15th Colloquium Spectroscopicum Internationale, Madrid, 1969. On the increase in sensitivity of direct spectrochemical. trace analysis by atomic fluorescence and by direct current arc atomizer.
40. Y. I. Belyaev, A. V. Karyakin and A. M. Pchelintsev, *Zh. Analit. Khim.*, **25,** 852 (1970). Increasing the sensitivity of the atomic fluorescence determination of traces of elements by using an arc impulse atomizer for solid samples.
41. Y. I. Belyaev and A. M. Pchelintsev, *Zh. Analit. Khim.*, **25,** 2094 (1970). Estimation of accuracy and sensitivity of the atomic fluorescence determination of traces of elements under the conditions of the impulse thermal atomisation of solid samples.
42. Y. I. Belyaev, A. V. Karyakin, A. M. Pchelintsev and T. A. Koveschnikova, Paper presented to the 16th Colloquium Spectroscopicum Internationale, Heidelberg, 1971. The use of atomic absorption and atomic fluorescence under conditions of the impulse atomisation of solid sample for the determination of impurities in pure substances.
43. Y. I. Belyaev and A. M. Pchelintsev, *Zh. Analit. Khim.*, **25,** 2238 (1970). Determination of concentration by the reading of the atomic fluorescence set-up with impulse thermal atomisation of solid samples.
44. P. Bender, *Phys. Rev.*, **36,** 1543 (1930). Optical excitation of cadmium by cadmium hydride and zinc hydride bands.
45. P. Benetti, N. Omenetto and G. Rossi, *Appl. Spectry.* **25,** 57 (1971). A new optical system for flame spectroscopy with special reference to thermally assisted anti-Stokes fluorescence applications.
46. P. A. Bennett, *Resonance Lines*, **1,** 1 (1969). An evaluation of light sources for atomic fluorescence spectroscopy.
47. Yu. P. Betin, B. I. Verkhovskii, V. I. Kanunnikova, E. E. Maizil and G. I. Sata-

RINA, *Zh. Prikl. Spektrosk.*, **15,** 583 (1971). Atomic fluorescence analysis using high-frequency electrodeless lamp VSB-2 and the xenon arc lamp DKSSH-1000.

48. M. A. BIANCIFIORI and C. BORDONALI, Paper presented to the International Atomic Absorption Spectroscopy Conference, Sheffield, 1969. A direct analysis of solid materials by means of atomic absorption and fluorescence; atomisation by ionic bombardment.
49. M. S. BLACK, T. H. GLENN, M. P. BRATZEL and J. D. WINEFORDNER, *Anal. Chem.*, **43,** 1769 (1971). Atomic fluorescence spectrometry with continuous nebulization into a platinum furnace.
50. R. BLEEKRODE and W. VAN BENTHEM, *J. Chem. Phys.*, **51,** 2757 (1969). Resonance fluorescence of copper atoms in the gas phase.
51. C. BOECKNER, *Bur. Stds. J. Res.* **5,** 13 (1930).
52. A. L. BOERS, C. T. J. ALKEMADE and J. A. SMIT, *Physica*, **22,** 358 (1956). The yield of resonance fluorescence of sodium in a flame.
53. A. BOGROS, *Compt. Rend.*, **183,** 124 (1926).
54. A. R. BOILEAU and F. D. MILLER, *Appl. Optics*, **6,** 1179 (1967). Changes of spectral sensitivity of multiplier phototubes resulting from changes in temperature.
55. P. W. J. M. BOUMANS, Paper presented to the 16th Colloquium Spectroscopicum Internationale, Heidelberg, 1971. Studies of glow discharge for spectrochemical analysis.
56. H. BRANDENBERGER and H. BADER, *Helv. Chim. Acta*, **50,** 1409 (1967). Determinaton of nanogram amounts of mercury in solutions by a flameless atomic absorption method.
57. H. BRANDENBERGER and H. BADER, *Chimia*, **22,** 449 (1968). Sensitivity increase of atomic absorption analysis by flameless atomization.
58. M. P. BRATZEL JR. and J. D. WINEFORDNER, *Anal. Letters*, **1,** 43 (1967). Influence of type of turbulent flame on limits of detection in atomic fluorescence flame spectrometry.
59. M. P. BRATZEL JR., J. M. MANSFIELD and J. D. WINEFORDNER, *Anal. Chim. Acta*, **39,** 394 (1967). Influence of acids concentration on atomic fluorescence of cadmium.
60. M. P. BRATZEL JR., R. M. DAGNALL and J. D. WINEFORDNER, Paper presented to the International Atomic Absorption Spectroscopy Conference, Sheffield, 1969. Some improved turbulent flames and their use in atomic fluorescence spectrometry.
61. M. P. BRATZEL JR., R. M. DAGNALL and J. D. WINEFORDNER, *Anal. Chem.* **41,** 713 (1969). A comparative study of premixed and turbulent air-hydrogen flames in atomic fluorescence spectrometry.
62. M. P. BRATZEL JR., R. M. DAGNALL and J. D. WINEFORDNER, *Anal. Chem.*, **41,** 1527 (1969). Evaluation of premixed flames produced using a total consumption nebulizer burner in atomic fluorescence spectrometry.
63. M. P. BRATZEL JR., R. M. DAGNALL and J. D. WINEFORDNER, *Anal. Chim. Acta*, **48,** 197 (1969). A new, simple atom resevoir for atomic fluorescence spectrometry.
64. M. P. BRATZEL JR., R. M. DAGNALL and J. D. WINEFORDNER, *Appl. Spectry.*, **24,** 518 (1970). A hot wire loop atomizer for atomic fluorescence spectrometry.
65. M. P. BRATZEL JR., R. M. DAGNALL and J. D. WINEFORDNER, *Anal. Chim. Acta*, **52,** 157 (1970). Evaluation of 150-W Eimac xenon lamp for atomic fluorescence spectrometry.
66. F. BRECH, Paper presented to the 7th National Meeting of Society for Applied Spectroscopy, Chicago, 1968. A review of atomic fluorescence, flame emission, and atomic absorption spectroscopy.

67. F. Brech, Paper presented to the 9th National Meeting of Society for Applied Spectroscopy, New Orleans, 1970. The challenge and the future for atomic absorption, flame, emission and related techniques.

68. R. G. Brewer, *Rev. Sci. Instr.*, **32,** 1356 (1961). High intensity low noise rubidium light source.

69. L. Brewer and J. B. Tellinghuisen, J. Chem. Phys., 54, 5133 (1971). Detection of iodine atoms by an atomic fluorescence technique. Application to study of diffusion and wall recombination.

70. W. H. Brimhall, *Anal. Chem.*, **41,** 1349 (1969). Measurement of lead isotopes by differential atomic absorption.

71. R. F. Browner, R. M. Dagnall and T. S. West, *Anal. Chim. Acta*, **45,** 163 (1969). Electronically modulated microwave-excited electrodeless discharge tubes as sources in atomic absorption spectroscopy.

72. R. F. Browner, R. M. Dagnall and T. S. West, *Anal. Chim. Acta*, **46,** 207 (1969). The determination of tin by atomic fluorescence spectroscopy, with an electronically modulated electrodeless discharge tube as source.

73. R. F. Browner, R. M. Dagnall and T. S. West, *Talanta*, **16,** 75 (1969). Atomic fluorescence spectroscopy – VIII. Atomic fluorescence and atomic absorption of thallium and mercury with electrodeless discharge tubes as sources.

74. R. F. Browner, R. M. Dagnall and T. S. West, *Anal. Chim. Acta*, **50,** 375 (1970). The determination of lead by atomic fluorescence spectroscopy.

75. E. M. Bulewicz and P. J. Padley, *Proc. Roy. Soc. London*, **A, 323,** 377 (1971). Photometric investigations of the behaviour of chromium additives in premixed $H_2 + O_2 + N_2$ flames.

76. J. C. Burger, W. Gillies and G. K. Yamasaki, Chapter 12 in *Analytical Flame Spectrometry, Selected Topics*, ed. by R. Mavrodineanu, Philips Technical Library, McMillan, London, 1970. Hollow cathode discharge devices.

77. D. H. Burling, M. Czajkowski and L. Krause, *J. Opt. Soc. Am.*, **57,** 1162 (1967). Light sources for the excitation of atomic resonance fluorescence in cesium and sodium.

78. L. P. R. Butler, A. Strasheim, F. W. E. Strelow, P. E. Mathews and E. C. Feast, *Colloq. Spectros. Intern.*, 12th, Exeter, 1965. The determination of gold in mine solutions by atomic absorption spectroscopy.

79. L. R. P. Butler and J. A. Brink, Chapter 2 in *Flame Emission and Atomic Absorption Spectrometry*, Vol. 2, ed. by J. A. Deans and T. C. Rains, Dekker, New York, 1971. Light sources for atomic absorption and atomic fluorescence spectrometry.

80. M. W. P. Cann, *Appl. Optics*, **8,** 1645 (1969). Light sources in the 0.15–20 μ spectral range.

81. G. Cario and J. Franck, *Z. Phys.*, **11,** 161 (1922). Decomposition of hydrogen molecules with excited mercury atoms.

82. P. G. Cath and A. M. Peabody, *Anal. Chem.*, **43,**(11), 91A (1971). High speed current measurements.

83. J. J. Cetorelli and J. D. Winefordner, *Appl. Spectry.*, **21,** 421 (1967). Use of signal-to-noise ratio theory to evaluate parameters characteristics of multiplier phototubes.

84. J. Champy, *Method. Phys. Anal.*, **5,** 311 (1969). Atomic fluorescence theory. Applications.

85. D. H. Christopher, F. W. Noble and R. L. Bowman, Paper presented to the Pittsburgh Conference on Analytical Chemistry and Applied Spectroscopy, 1971

Electronic modulation and automatic intensity control of electrodeless discharge tubes.
86. C. J. Christensen and G. K. Rollefson, *Phys. Rev.*, **34,** 1157 (1929). The influence of the method of excitation on transition probabilities in sodium vapor.
87. G. D. Christian, *Anal. Chem.*, **41** (1), 24A (1969). Medicine, trace elements, and atomic absorption spectroscopy.
88. R. W. Claus, Beckman Instruments Inc., Fullerton, Calif., unpublished work.
89. P. Clechet, *Chim. Mod.*, **11,** 207 (1966). Atomic absorption spectrophotometry and atomic fluorescence spectroscopy.
90. D. O. Cooke, R. M. Dagnall, B. L. Sharp and T. S. West, *Spectrosc. Letters*, **4,** 91 (1971). Application of photon counting as a detection system in atomic fluorescence and emission spectroscopy.
91. D. O. Cooke, R. M. Dagnall and T. S. West, *Anal. Chim. Acta*, **54,** 381 (1971). Optimisation of some experimental parameters in the preparation and operation of microwave-excited electrodeless discharge lamps.
92. D. O. Cooke, R. M. Dagnall and T. S. West, *Anal. Chim. Acta*, **56,** 17 (1971). Evaluation of some three-quarter-wave microwave cavities for the operation of electrodeless discharge lamps.
93. D. H. Cotton and D. R. Jenkins, *Spectrochim. Acta*, **25B,** 283 (1970). The determination of very low concentrations of copper, iron, and lead in hydrocarbon fuels by atomic fluorescence spectrometry.
94. T. G. Cowley, V. A. Fassel and R. N. Kniseley, *Spectrochim. Acta*, **23B,** 771 (1968). Free atom formation processes in premixed fuel-rich and stoichiometric oxygen-acetylene flames employed in atomic emission and atomic absorption spectroscopy.
95. M. S. Cresser and T. S. West, *Spectrosc. Letters*, **2,** 9 (1969). Determination of selenium by atomic fluorescence in air-acetylene flame with an integrating atomic emission spectrophotometer using automatic background correction.
96. M. S. Cresser and T. S. West, *Spectrochim. Acta*, **25B,** 61 (1970). Some interference studies in atomic fluorescence spectroscopy with a continuum source.
97. M. S. Cresser and T. S. West, *Anal. Chim. Acta*, **51,** 530 (1970). Multi-element atomic fluorescence spectroscopy. Part III. A new type of dual-element microwave-excited electrodeless discharge tube.
98. M. S. Cresser and T. S. West, *Anal. Chim. Acta*, **50,** 517 (1970). Some observations on the determination of zinc by atomic fluorescence spectrometry under conditions favorable for analysis.
99. C.S.I.R.O., Division of Chemical Physics, Annual Report 1970–71, p. 2. Atomic Absorption Spectroscopy.
100. L. A. Currie, *Anal. Chem.*, **40,** 586 (1968). Limits for qualitative detection and quantitative determination.
101. M. Czerny and A. F. Turner, *Z. Phys.*, **61,** 792 (1930). On the astigmatism of grating spectrometers.
102. R. M. Dagnall, T. S. West and P. Young, *Talanta*, **13,** 803 (1966). Determination of cadmium by atomic fluorescence and atomic absorption spectrophotometry.
103. R. M. Dagnall, K. C. Thompson and T. S. West, *Anal. Chim. Acta*, **36,** 269 (1966). An investigation of some experimental parameters in atomic fluorescence spectrophotometry.
104. R. M. Dagnall, K. C. Thompson and T. S. West, Paper presented to the 1st International Conference on Atomic Absorption Spectroscopy, Prague, 1967. Atomic-

fluorescence and atomic absorption studies employing electrodeless discharge tubes as sources.

105. R. M. DAGNALL, K. C. THOMPSON and T. S. WEST, *Talanta*, **14,** 551 (1967). Studies in atomic fluorescence spectroscopy — III. Microwave-excited electrodeless discharge tubes as spectral sources for atomic fluorescence and atomic absorption spectroscopy.

106. R. M. DAGNALL, K. C. THOMPSON and T. S. WEST, *Talanta*, **14,** 557 (1967). Studies in atomic fluorescence spectroscopy — IV. The atomic fluorescence spectroscopic determination of selenium and tellurium.

107. R. M. DAGNALL, K. C. THOMPSON and T. S. WEST, *Talanta*, **14,** 1151 (1967). Studies in atomic fluorescence spectroscopy — V. The fluorescence characteristics and determination of antimony.

108. R. M. DAGNALL, K. C. THOMPSON and T. S. WEST, *Talanta*, **14,** 1467 (1967). Studies in atomic fluorescence — VI. The fluorescence characteristics and analytical determination of bismuth with an iodine electrodeless discharge tube as source.

109. R. M. DAGNALL, K. C. THOMPSON and T. S. WEST, *Analyst*, **92,** 506 (1967). Molecular emission spectroscopy in cool flames. Part I. The behaviour of sulphur containing compounds.

110. R. M. DAGNALL, K. C. THOMPSON and T. S. WEST, *Analyst*, **93,** 153 (1968). The nitrous oxide-hydrogen flame in spectroscopic analysis.

111. R. M. DAGNALL, K. C. THOMPSON and T. S. WEST, *Talanta*, **15,** 677 (1968). Studies in atomic fluorescence spectroscopy — VII. Fluorescence and analytical characteristics of arsenic, with a microwave excited electrodeless discharge tube as source.

112. R. M. DAGNALL, R. PŘIBIL, JR. and T. S. WEST, *Analyst*, **93,** 281 (1968). Preparation of microwave-excited electrodeless discharge tubes for titanium, vanadium, and zirconium for use as spectral-line sources.

113. R. M. DAGNALL, K. C. THOMPSON and T. S. WEST, *Anal. Chim. Acta*, **41,** 551 (1968). The thermal and atomic fluorescence emission of germanium in a nitrogen-oxygen-acetylene flame.

114. R. M. DAGNALL, M. R. G. TAYLOR and T. S. WEST, *Spectrosc. Letters*, **1,** 397 (1968). Some detection limits in atomic fluorescence spectroscopy with microwave excited electrodeless discharge tubes as excitation sources.

115. R. M. DAGNALL and T. S. WEST, *Appl. Optics*, **7,** 1287 (1968). Some applications of microwave excited electrodeless discharge tubes in atomic spectroscopy.

116. R. M. DAGNALL, (Cl. G. 01n) 12 Mar. 1970, Brit. Appl. 0.5 July (1968). Spectral analysis with electrodeless discharge tubes.

117. R. M. DAGNALL, G. F. KIRKBRIGHT, T. S. WEST and R. WOOD, *Anal. Chim. Acta*, **47,** 407 (1969). The determination of silicon by flame photometry and atomic fluorescence spectroscopy with a separated nitrous oxide-acetylene flame.

118. R. M. DAGNALL, G. F. KIRKBRIGHT, T. S. WEST and R. WOOD, *Anal. Chem.*, **42,** 1029 (1970). Atomic fluorescence spectrometry of aluminium, molybdenum, titanium, vanadium, and zicronium in inert gas separated nitrous oxide-acetylene flames.

119. R. M. DAGNALL, G. F. KIRKBRIGHT, T. S. WEST and R. WOOD, *Analyst*, **95,** 4 (1970). An investigation of the determination of germanium by flame photometry and atomic fluorescence spectroscopy by using a separated nitrous-oxide acetylene flame.

120. R. M. DAGNALL, Paper presented to the Conference on Flame Spectroscopy, Ostrava (Czechoslovakia), 1970. Photon counting detection systems in atomic fluorescence spectrometry.

121. R. M. DAGNALL, M. R. G. TAYLOR and T. S. WEST, *Lab. Pract.*, **20,** 209 (1971).

Method for the determination of low melting point elements in nickel-base alloys by atomic fluorescence spectroscopy.

122. R. M. Dagnall, Paper presented to the 3rd International Conference on Atomic Absorption and Atomic Fluorescence Spectrometry, Paris, 1971. The use of photon counting in flame and non-flame atomic fluorescence spectrometry.

123. R. M. Dagnall, M. Silvester and T. S. West, Paper presented to the 16th Colloquium Spectroscopicum Internationale, Heidelberg, 1971. Some recent developments in electronically modulated and pulsed electrodeless discharge lamps used in flame atomic fluorescence spectrometry.

124. R. M. Dagnall, G. F. Kirkbright, T. S. West and R. Wood, *Anal. Chem.*, **43,** 1765 (1971). Multichannel atomic fluorescence and flame photometric determination of calcium, copper, magnesium, manganese, potassium, and zinc in soil extracts.

125. R. M. Dagnall, B. L. Sharp and T. S. West, *Nature*, **234,** 59 (1971). Measuring of transient atomic fluorescence and atomic absorption signals using photon counting.

126. R. M. Dagnall, M. D. Silvester and T. S. West, *Talanta*, **18,** 1103 (1971). Electronic modulation of microwave-excited electrodeless discharge tubes for use in atomic fluorescence spectrometry.

127. R. M. Dagnall, G. K. Kirkbright, T. S. West and R. Wood, *Analyst*, **97,** 245 (1972). The simultaneous determination of six metals in aluminium alloys by atomic fluorescence spectrometry.

128. R. M. Dagnall and M. D. Silvester, *Anal. Chem.*, **44,** 204 (1972). Evaluation of an attenuator for improved operation of electrodeless discharge lamps.

129. D. K. Davies, *J. Appl. Phys.*, **38,** 4713 (1967). An interferometric study of a high-intensity hollow cathode source.

130. W. O. Davis, *Natl. Bur. Stds. Rept.* No. AD-670.568 (1968). A preliminary evaluation of the application of atomic fluorescence spectroscopy to spectrometric oil analysis.

131. J. B. Dawson and D. J. Ellis, *Spectrochim. Acta*, **23A,** 565 (1967). Pulsed current operation of hollow-cathode lamps to increase the intensity of resonance lines for atomic absorption spectroscopy.

132. D. B. Dawson, *Meth. Phys. d'Analyse* (*G.A.M.S.*). *Special No.*, p. 32 (1971). The application of photon counting technique to flame photometry.

133. J. B. Dawson and B. D. Tucker, Paper presented to the 16th Colloquium Spectroscopicum Internationale, Heidelberg, 1971. Relative detection limits in emission, absorption, and fluorescence, flame photometry.

134. L. de Galan and J. D. Winefordner, *J. Quant. Spectrosc. Radiat. Transfer*, **7,** 251 (1967). Measurement of the free atom fraction of 22 elements in an acetylene-air flame.

135. L. de Galan and J. D. Winefordner, *J. Quant. Spectrosc. Radiat. Transfer*, **7,** 703 (1967). Thermal equilibrium in turbulent diffusion flames.

136. L. de Galan and G. F. Samaey, *Spectrochim. Acta*, **25B,** 245 (1970). Measurement of degrees of atomisation in premixed laminar flames.

137. L. de Galan, *Spectrosc. Letters*, **3,** 123 (1970). On sensitivity and detection limit with reference to flame spectrometry.

138. L. de Galan and H. G. Wagenaar, *Meth. Phys. d'Analyse* (*G.A.M.S.*), Special No., p. 10, 1971. The profile of atomic spectral lines.

139. H. T. Delves, *Analyst*, **95,** 431 (1970). A micro-sampling method for the rapid determination of lead in blood by atomic absorption spectrometry.

140. D. R. DEMERS and D. W. ELLIS, *Anal. Chem.*, **40,** 860 (1968). Interferences affecting the atomic fluorescence of calcium and magnesium.

141. D. R. DEMERS, *Appl. Spectry.*, **22,** 797 (1968). Atomic fluorescence flame spectrometry: potential tool for trace analysis.

142. D. R. DEMERS, *Ph. D. Thesis*, University of New Hampshire, Durham, 1969. Studies in atomic fluorescence flame spectrometry.

143. D. R. DEMERS and D. G. MITCHELL, Paper presented to the Technicon International Congress, New York, 1970. An atomic fluorescence spectrometer for simultaneous multi-element analysis.

144. D. R. DEMERS, M. GARDELS and D. MITCHELL, Paper presented to the Pittsburgh Conference on Analytical Chemistry and Applied Spectroscopy, 1971. Simultaneous multi-element analysis of water quality and lubricating oils by atomic fluorescence spectroscopy.

145. D. R. DEMERS, M. GARDELS and D. MITCHELL, Paper presented to the Pittsburg Conference on Analytical Chemistry and Applied Spectroscopy, 1971. Simultaneous multielement analysis by atomic fluorescence spectroscopy.

146. D. R. DEMERS, R. HERRON and D. MITCHELL, Paper presented to the 9th National Meeting of Society for Applied Spectroscopy, New Orleans, 1970. A new instrumental approach to multielement analysis.

147. D. R. DEMERS, M. GARDELS and D. MITCHELL, Paper presented to the Pittsburgh Conference on Analytical Chemistry and Applied Spectroscopy, 1971. Optimalization of instrument design parameters in multielement atomic fluorescence spectroscopy.

148. M. B. DENTON and H. V. MALMSTADT, *Appl. Phys. Letters*, **18,** 485 (1971). Tunable organic dye laser as an excitation source for atomic flame fluorescence spectrometry.

149. G. DIJKSTRA, *Geol. Mijnbouw*, **47,** 379 (1969). Instrumental methods in the chemical analysis of minerals.

150. M. DILLARD and H. R. ROBL, *Phys. Rev.*, **184,** 312 (1969). Transient effects in resonance fluorescence and spontaneous emission by a system of identical atoms.

151. J. I. DINNIN and A. W. HELZ, *Anal. Chem.*, **39,** 1489 (1967). Demountable hot hollow cathode lamp as an excitation source in atomic fluorescence flame spectrometry.

152. J. I. DINNIN, *Anal. Chem.*, **39,** 1491 (1967). Atomic fluorescence flame spectrometric detection of palladium, titanium, zirconium, chromium, and aluminium using a hot hollow cathode lamp.

153. J. N. DODD, W. J. SANDLE and D. ZISSERMANN, *Proc. Phys. Soc.*, **92,** 497, (1967). Resonance fluorescence in cadmium. Modulation effects and lifetime measurements.

154. J. N. DODD, E. ENEMARK and A. GALLAGHER, *J. Chem. Phys.*, **50,** 4838, (1969). Quenching of cesium resonance radiation by helium.

155. L. DUNOYER, *Radium*, **10,** 400 (1913).

156. L. DUNOYER, *Compt. Rend.*, **178,** 1475 (1924).

157. L. EBDON, G. F. KIRKBRIGHT and T. S. WEST, *Anal. Chim. Acta*, **47,** 563 (1969). The determination of iron by atomic fluorescence spectroscopy with a modulated microwave-excited electrodeless discharge tube.

158. L. EBDON, G. F. KIRKBRIGHT and T. S. WEST, *Talanta*, **17,** 965 (1970). Atomic fluorescence characteristics and analytical determination of manganese in an air-acetylene flame.

159. E. H. EBERHARDT, *Appl. Optics*, **6,** 161 (1967). Multiplier phototubes as quantum counters.

160. E. H. EBERHARDT, *Appl. Optics*, **6,** 251 (1967). Threshold sensitivity and noise ratings of multiplier phototubes.

161. E. H. EBERHARDT, *Appl. Optics*, **6,** 359 (1967). Noise factor measurements in multiplier phototubes.

162. D. W. ELLIS and D. R. DEMERS, *Anal. Chem.* **38,** 1943 (1966). Use of the hydrogen-entrained-air flame in atomic fluorescence flame spectrometry.

163. D. W. ELLIS, Paper presented to the 153rd National Meeting of American Chemical Society, Miami Beach, Florida, 1967. Use of the hydrogen/entrained air flame in atomic fluorescence flame spectrometry.

164. D. W. ELLIS and D. R. DEMERS, Chap. 20 in *Trace Inorganics in Water*, ed. by R. A. Baker, Advances in Chemistry Series 73, American Chemical Society, Washington, D. C., 1968. Atomic fluorescence flame spectrometry.

165. R. C. ELSER and J. D. WINEFORDNER, *Appl. Spectry.*, **25,** 345 (1971). A simple monochromatorless atomic fluorescence flame spectrometer.

166. E.M.I. Photomultiplier Tubes, Brochure 30M (6–67), EMI Electronics, Ltd.

167. V. A. FASSEL and D. W. GOLIGHTLY, *Anal. Chem.*, **39,** 466 (1967). Detection limits of elements in the spectra of premixed, oxy-acetylene flames.

168. F. C. FEHSENFELD, K. M. EVENSON and H. P. BROIDA, *Rev. Sci. Instr.*, **36,** 294 (1965). Microwave discharge cavities operating at 2450 MHz.

169. F. FEIGL, *Mikrochemie*, **1,** 4 (1923).

170. F. J. FERNANDEZ, D. C. MANNING, J. D. KERBER and H. L. KAHN, Paper presented to the 3rd International Conference on Atomic Absorption and Atomic Fluorescence Spectrometry, Paris, 1971. The place of atomic fluorescence in flame spectroscopy.

171. J. A. FIORINO, R. N. KNISELEY and V. A. FASSEL, Paper presented to the Pittsburgh Conference on Analytical Chemistry and Applied Spectroscopy, 1971. Molecular absorption and light scattering in flames employed for atomic absorption spectroscopy.

172. A. A. FISCHER and G. C. HAYWARD, Paper presented to the Pittsburgh Conference on Analytical Chemistry and Applied Spectroscopy, 1970. Microwave-excited electrodeless discharge lamps as sources for atomic absorption spectroscopy.

173. B. FLEET, K. V. LIBERTY and T. S. WEST, *Anal. Chim. Acta*, **45,** 205 (1969). The atomic fluorescence spectroscopy of cobalt with a high-intensity hollow-cathode lamp and a microwave-excited electrodeless discharge tube as sources.

174. A. T. FORRESTER, R. A. GUDMUNDSEN and P. O. JOHNSON, *J. Opt. Soc. Am.*, **46,** 339 (1956). High efficiency microwave excitation of light sources.

175. M. L. FRANKLIN, G. HORLICK and H. V. MALMSTADT, *Anal. Chem.*, **41,** 2 (1969). Basic and practical considerations in utilizing photon counting for quantitative spectrochemical methods.

176. L. M. FRASER and J. D. WINEFORDNER, *Anal. Chem.*, **43,** 1693 (1971). Laser-excited atomic fluorescence flame spectrometry.

177. L. M. FRASER and J. D. WINEFORDNER, Paper presented to the 9th National Meeting of Society for Applied Spectroscopy, New Orleans, 1970. Graphite furnace for atomic fluorescence spectrometry.

178. C. FUCHTBAUER, *Phys. Z.*, **21,** 635 (1920). New means for emission of spectral lines due to light irradiation (fluorescence).

179. A. FULTON, K. C. THOMPSON and T. S. WEST, *Anal. Chim. Acta*, **51,** 373 (1970). Multielement atomic fluorescence spectroscopy. II. Dual-element arsenic-antimony electrodeless discharge tube spectral line source for atomic fluorescence and atomic absorption spectroscopy.

180. V. B. Gerard, *J. Sci. Instr.*, **39**, 217 (1962). Laboratory alkali metal vapour lamps for optical pumping experiments.

181. J. A. Goleb, *Anal. Chem.*, **35**, 1978 (1963). Uranium isotope investigations by atomic absorption.

182. J. A. Goleb and Y. Yokoyama, *Anal. Chim. Acta*, **30**, 213 (1964). The use of a discharge tube as an absorption source for the determination of lithium-6, lithium-7 isotopes by atomic absorption spectrophotometry.

183. J. A. Goleb, *Appl. Spectry.*, **25**, 94 (1971). The use of permanent mercury source for spectroscopic studies of varying amounts of mercury.

184. S. R. Goode and S. R. Crouch, Paper presented to the Pittsburgh Conference on Analytical Chemistry and Applied Spectroscopy, 1971. An automated atomic absorption-atomic fluorescence system for use with non-flame cells.

185. G. I. Goodfellow, *Anal. Chim. Acta*, **36**, 132 (1966). Some experimental observations on inter-element effects in atomic fluorescence spectrometry.

186. G. I. Goodfellow, *U. K. At. Energy Authority*, *Rept.* No. 0 87/66, AWRE. An assessment of atomic fluorescence spectrometry as a method of trace analysis.

187. C. L. Grant, *Am. Soc. Test. Mater.*, *Spec. Tech. Publ.* No. 443, 1969. Comparison of atomic absorption with other spectrochemical methods.

188. I. D. Guzeev and V. V. Nedler, *Zavodsk. Lab.*, **36**, 1458 (1970). The use of atomic fluorescence for analysis of high-purity substances.

189. A. T. Hattenburg, *Appl. Optics*, **6**, 95 (1967). Spectral radiance of a low current graphite arc.

190. A. Hell and S. Ricchio, Paper presented to the Conference on Analytical Chemistry and Applied Spectroscopy, 1970. Evaluation of atomic fluorescence excitation sources.

191. D. N. Hingle, G. R. Kirkbright and T. S. West, *Analyst*, **93**, 522 (1968). The determination of beryllium by thermal-emission and atomic fluorescence spectroscopy in a separated nitrous-oxide-acetylene flame.

192. R. S. Hobbs, G. F. Kirkbright, M. Sargent and T. S. West, *Talanta*, **15**, 997 (1968). Spectroscopy in separated flames — IV. Application of the nitrogen-separated air-acetylene flame in flame emission and atomic fluorescence spectroscopy.

193. R. S. Hobbs, G. F. Kirkbright and T. S. West, *Talanta*, **18**, 859 (1971). Spectroscopy in separated flames — VII. Determination of bismuth by atomic fluorescence spectroscopy in a separated air-acetylene flame with electronically modulated electrodeless discharge tube sources.

194. K. Hoffmann and R. Seiwert, *Exp. Technik Phys.*, **8**, 161 (1960). Light sources for the excitation of resonance fluorescence of alkali metals.

195. M. E. Hofton and D. P. Hubbard, Paper presented to the 3rd International Conference on Atomic Absorption and Atomic Fluorescence Spectrometry, Paris, 1971. Some observations on the determination of low levels of silver by flame spectroscopic methods with particular reference to steels.

196. J. R. Hollahan, *J. Chem. Educ.*, **43**, 401A, 297A (1966). Analytical application of electrodelessly discharged gases. Research with electrodelessly discharged gases.

197. H. P. Hooymayers and C. T. J. Alkemade, *J. Quant. Spectry. Radiative Transfer*, **6**. 501 (1966). Quenching of excited alkali atoms and related effects in flames. Part I, Theoretical analysis.

198. H. P. Hooymayers and C. T. J. Alkemade, *J. Quant. Spectry. Radiative Transfer*, **6**, 847 (1966). Quenching of excited alkali atoms and related effects in flames: Part II. Measurements and discussion.

199. H. P. HOOYMAYERS and G. NIENIHUS, *J. Quant. Spectry. Radiative Transfer*, **8**, 955 (1968). Quenching of excited rubidium (5^2 P) atoms in flames.

200. H. P. HOOYMAYERS, *Spectrochim. Acta*, **23B**, 567 (1968). Theory of analytical curves in atomic fluorescence flame spectrometry.

201. H. P. HOOYMAYERS and P. L. LIJNSE, *J. Quant. Spectry. Radiative Transfer*, **9**, 995 (1969). The relationship between the fluorescence yield and the under-population of doublet excited states.

202. E. S. HRYCYSHYN and L. KRAUSE, *Can. J. Phys.*, **47**, 215 (1969). Senzitized fluorescence in vapors of alkali metals-XI. Energy transfer in potassium-rubidium collisions.

203. CH. J. HSU, *Ph. D. Thesis*, Louisiana State University, Baton Rouge, 1968. Studies of the atomic fluorescence of zinc and beryllium and the spectroscopic studies of the chromotropic acid-nitrate reaction.

204. HTV Photosensitive Devices. Hamamatsu TV Co.

204. H. G. C. HUMAN, Paper presented to the 3rd International Conference on Atomic Absorption and Atomic Fluorescence Spectrometry, Paris, 1971. Development of a high intensity spectral line source.

206. CH. A. M. HUSSEIN and G. NICKLESS, Paper presented to the 2nd International Conference on Atomic Absorption Spectroscopy, Sheffield, 1969. An investigation into the R.F. plasma as an excitation source in atomic absorption and fluorescence spectrometry.

207. N. P. IVANOV and V. Z. KRASLSHCHIK, *Metody Analiza Khim. Reakt. i Preparatov, Gos. Kom. Sov. Min. SSR po Khim.*, **7**, 1 (1963). Basic properties and analytical use of a hollow cathode.

208. N. P. IVANOV, L. V. MINERVINA, S. V. BARANOV, L. G. POFRALIDI and I. I. OLIKOV, *Zh. Analit. Khim.*, **21**, 1129 (1966). Tubes without electrodes with high-frequency excitation of the spectrum of In, Ga, Bi, Sb, Tl, Pb, Mg, Ca, Cu, as a radiation source in atomic absorption analysis.

209. N. P. IVANOV and G. K. KOZYREVA, *Vestn. Mosk. Univ. Ser. II.*, **23**, 127 (1968). Some new applications of electrodeless lamps in atomic absorption spectroscopy.

210. E. JACOBSEN and G. R. HARRISON, Paper presented to the 34th Annual Meeting of the Optical Society of America, *J. Opt. Soc. Am.*, **39**, 1054 (1949). Ultrafrequency excitation of Hg 198 lamps for interferometer illumination.

211. D. R. JENKINS, *Proc. Roy. Soc.*, **A 293**, 493 (1966). The determination of cross sections for the quenching of resonance radiation of metal atoms. I. Experimental method and results for sodium.

212. D. R. JENKINS, *Spectrochim. Acta*, **23B**, 167 (1967). An analysis of the optimum conditions for the detection of metals in flames by atomic fluorescence.

213. D. R. JENKINS and T. M. SUGDEN in *Flame Emission and Atomic Absorption Spectrometry*, ed. by J. A. Dean and T. C. Rains, Vol. I. p. 151, Dekker, New York, 1969.

214. D. R. JENKINS, *Proc. Roy. Soc.*, **303A**, 453 (1968). The determination of cross sections for the quenching of resonance radiation of metal atoms. II. Results for potassium, rubidium and caesium.

215. D. R. JENKINS, *Proc. Roy. Soc.*, **303A**, 467 (1968). The determination of cross sections for quenching of resonance radiation of metal atoms. III. Results for thallium.

216. D. R. JENKINS, *Proc. Roy. Soc.*, **306A**, 413 (1968). The determination of cross-sections for the quenching of resonance radiation of metal atoms. IV. Results for lithium.

217. D. R. Jenkins, *Trans. Faraday Soc.*, **64,** 36 (1968). Relationships between measured and true quenching cross-sections for atoms with doublet excited states.

218. D. R. Jenkins, *Spectrochim. Acta*, **25B,** 47 (1970). The influence of flame composition on the intensity of atomic fluorescence.

219. J. D. Johnson, H. R. Gram and G. D. Propster, Paper presented to the 9th National Meeting of Society for Applied Spectroscopy, New Orleans, 1970. The design and application of a demountable hollow cathode source.

220. H. L. Kahn, in *Trace Inorganics in Water*, ed. by R. A. Baker, Advances in Chemistry Series No. 73, American Chemical Society, Washington D. C., 1968. Principles and practice of atomic absorption.

221. H. Kaiser, *Spectrochim. Acta*, **3,** 40 (1947). The calculation of a detection limit.

222. H. Kaiser, *Z. Anal. Chem.*, **216** (1966). A contribution to definition of limit of detection, guarantee of purity and associated problems.

223. D. A. Katskov, G. G. Lebedev, and B. V. L'vov, *Zh. Prikl. Spektrosk.*, **10,** 215 (1969). Spectral characteristics of impulse lamps with hollow cathode for atomic absorption measurements.

224. J. P. Keene, *Rev. Sci. Instr.*, **34,** 1220 (1963). Fatigue and saturation in photomultipliers.

225. M. A. Kelly, *Ger. Offen.* 2,045, 134 (Cl G 01n), 1971. Atomic fluorescence spectrometer.

226. G. F. Kirkbright and T. S. West, *Appl. Optics*, **7,** 1305 (1968). The application of separated flames in analytical flame spectroscopy.

227. G. F. Kirkbright, Paper presented to the 15th Colloquium Spectroscopicum Internationale, Madrid, 1969. The use of separated flames in atomic absorption and emission spectroscopy.

228. G. F. Kirkbright, M. Sargent and T. S. West, *Talanta*, **16,** 245 (1969). Spectroscopy in separated flames — V. The argon or nitrogen-sheathed nitrous oxide-acetylene flame in flame emission spectroscopy.

229. G. F. Kirkbright, Paper presented to the 2nd International Conference on Atomic Absorption Spectroscopy, Sheffield, 1969. The application of separated flames to atomic absorption, emission and fluorescence spectroscopy.

230. G. F. Kirkbright, A. P. Rao and T. S. West, *Anal. Letters* **2,** 465 (1969). Determination of silicon in low alloy steels by atomic fluorescence spectroscopy in a separated nitrous oxide-acetylene flame.

231. G. F. Kirkbright and M. Sargent, *Spectrochim. Acta*, **25B,** 577 (1970). The application of a piezoelectric scanning Fabry-Perot interferometer to the study of atomic line sources-I. Assembly and general application of the instrumental system.

232. G. F. Kirkbright, Paper presented to the Conference on Flame Spectroscopy, Ostrava (Czechoslovakia), 1970. The application of a piezoelectric scanning Fabry-Perot interferometer to the study of atomic line sources.

233. G. F. Kirkbright and R. M. Dagnall, *Proc. Technicon Intern. Congress*, Vol. II., New York 1970. The role of atomic fluorescence spectroscopy in analytical chemistry.

234. G. F. Kirkbright and T. S. Vetter, *Spectrochim. Acta*, **26B,** 505 (1971). The effect of inert gas shielding on temperature profiles in premixed nitrous oxide-acetylene flames.

235. G. F. Kirkbright, M. Sargent and S. Vetter, Paper presented to the 16th Colloquium Spectroscopicum Internationale, Heidelberg, 1971. Some studies of line profiles from flames and other sources using a piezoelectric Fabry-Perot interferometer.

236. G. F. Kirkbright, *Analyst*, **96**, 609 (1971). The application of non-flame atom cells in atomic absorption and atomic fluorescence spectroscopy.
237. R. Klaus, *Zeiss. Mitt. (Stuttgart)*, **4**, 26 (1966). Simple determination of zinc in the p.p.m. range by resonance fluorescence.
238. R. Klaus, *Z. Klin. Chem.*, **4**, 299 (1966). Determination of zinc and cadmium in biological material by resonance fluorescence.
239. L. Klein, *Appl. Optics*, **7**, 677 (1968). Measurements of spectral emission and absorption of a high pressure xenon arc in the stationary and the flashed modes.
240. I. Kleinmann and B. Polej, *Chem. Listy*, **65**, 1 (1971). High-frequency and microwave plasma as a radiation source for spectral analysis.
241. S. R. Koirtyohann and E. E. Pickett, *Anal. Chem.*, **38**, 1087 (1966). Light scattering by particles in atomic absorption spectrometry.
242. S. R. Koirtyohann and E. E. Pickett, *Proceedings of the 13th Colloquium Spectroscopicum Internationale*, Ottawa, p. 270, Hilger and Watts, London, 1967. The efficiency of atom formation in premixed flames.
243. S. R. Koirtyohann, *Meth. Phys. d'Analyse (G.A.M.S.)*, Special No., p. 92 (1971). Some problems in flame emission, absorption and fluorescence spectroscopy.
244. D. Kolihová and V. Sychra, Paper presented to the Conference on Flame Spectroscopy, Ostrava (Czechoslovakia), 1970. Atomic fluorescence study of elements atomized in conventional low-temperature flames by using high intensity hollow cathode lamps as excitation sources.
245. D. Kolihová, *Ph. D. Thesis*, Technical University, Prague 1971. Atomic fluorescence study of elements atomized in conventional flames by using high-intensity hollow cathode lamps as excitation sources.
246. D. Kolihová and V. Sychra, *Chem. Listy*, **66**, 93 (1972). Atomic absorption and atomic fluorescence spectrometry in biochemistry and medicine. IV. Direct determination of copper and zinc in blood serum by atomic fluorescence spectrometry.
247. D. Kolihová, Paper presented to the 1st Annual National Seminar on Flame Spectrometry, Modra (Czechoslovakia), 1971. Quenching of atomic fluorescence in various media.
248. D. Kolihová and V. Sychra, *Anal. Chim. Acta*, **59**, 477 (1972). Note on the determination of antimony by atomic fluorescence spectrometry in various premixed flames.
249. D. Kolihová and V. Sychra, *Anal. Chim. Acta*, **63**, 479 (1973). Atomic fluorescence characteristics of copper in various premixed flames.
250. Yu. I. Korovin and V. A. Kutchumov, *Zh. Prikl. Spektrosk.*, **14**, 778 (1971). Use of a hollow cathode discharge for atomization of samples.
251. E. K. Kraulinya and M. C. Arman, *Optika i Spektroskopiya*, **26**, 511 (1969). Excitation of zinc and cadmium levels in the sensitized fluorescence of Hg—Zn and Hg—Cd mixtures.
252. K. H. Krysmanski, E. Bredel and G. Boettcher, *Beitr. Plasmaphys.*, **10**, 161 (1970). Quenching of the cadmium atom resonance fluorescence with a foreign gas.
253. Kuang-Pang Li, *Ph. D. Thesis*, University of Illinois, Urbana, 1970. Fundamental consideration of the photon counting technique and its application in a comparison of flame spectroscopic techniques.
254. J. Kuhl and G. Marowsky, *Opt. Commun.*, **4**, 125 (1971). Narrow band dye laser as a light source for fluorescence analysis in subnanogram range.
255. A. F. Kuteinikov, V. A. Egorova, E. S. Konkova, I. D. Zavarueva, Y. I. Belyaev, A. V. Karyakin and A. M. Pchelintsev, *Zh. Analit. Khim.*, **25**, 1999 (1970).

Spectrographic methods for the determination of microimpurities in glassy carbon and carbon powder.

256. F. J. LANGMYHR and P. R. GRAAF, *Anal. Chim. Acta*, **21,** 334 (1959). Studies in the spectrophotometric determination of silicon in materials decomposed by hydrofluoric acid. I. Loss of silicon with hydrofluoric acid.

257. P. L. LARKINS, R. M. LOWE, J. V. SULLIVAN and A. WALSH, *Spectrochim. acta*, **24B,** 187 (1969). The use of solar-blind photomultipliers in flame spectroscopy.

258. P. L. LARKINS, *Spectrochim. Acta*, **26B,** 477 (1971). Non-dispersive systems in atomic fluorescence spectroscopy. I. A single channel system employing a solar-blind detector and an air-acetylene flame.

259. P. L. LARKINS and J. B. WILLIS, *Spectrochim. Acta*, 26**B,** 491 (1971). Non-dispersive systems in atomic fluorescence spectroscopy – II. Comparison of nitrous oxide-supported and air-supported flames.

260. R. M. LOWE, *Spectrochim. Acta*, **26B,** 201 (1971). A high-intensity hollow-cathode lamp for atomic fluorescence.

261. P. W. Y. LUNG, J. P. MATOUŠEK and G. P. THOMAS, Paper presented to the 17th Spectroscopy Symposium of Canada, Ottawa, 1970. The carbon rod atomizer in atomic absorption and atomic fluorescence.

262. B. V. L'VOV, *Atomic Absorption Spectrochemical Analysis*, Adam Hilger Ltd., London, 1970.

263. D. C. MANNING and W. SLAVIN, *At. Absorption Newsletter*, No. 8, 1962. Lithium isotope analysis by atomic absorption spectrophotometry.

264. D. C. MANNING and P. HENEAGE, *At. Absorption Newsletter*, **6,** 124 (1967). Atomic fluorescence measurements with a modified atomic absorption instrument.

265. D. C. MANNING and P. HENEAGE, *At. Absorption Newsletter*, **7,** 80 (1968). Element detection limits by atomic fluorescence spectroscopy.

266. J. M. MANSFIELD, J. D. WINEFORDNER and C. VEILLON, *Anal. Chem.* **37,** 1049 (1965). High sensitivity determination of zinc, cadmium, mercury, thallium, gallium and indium by atomic fluorescence flame spectrometry.

267. J. M. MANSFIELD, JR., *Ph. D. Thesis*, University of Florida, Gainesville, 1967. A study of line sources for use in atomic fluorescence flame spectrometry.

268. J. M. MANSFIELD, JR., M. P. BRATZEL, JR., H. O. NORGORDON, D. O. KNAPP, K. E. ZACHA and J. D. WINEFORDNER, *Spectrochim. Acta*, **23B,** 389 (1968). Experimental investigations of electrodeless discharge lamps as excitation sources for atomic fluorescence flame spectrometry.

269. J. M. MANSFIELD and J. D. WINEFORDNER, *Anal. Chim. Acta*, **40,** 357 (1968). Measurement of flow rate and pressure of gases in flame spectrometry.

270. G. B. MARSHALL and T. S. WEST, *Anal. Chim. Acta*, **51,** 179 (1970). Multi-element atomic fluorescence spectroscopy. Part I. Stimulation of atomic fluorescence of mixtures of bismuth, mercury, selenium, and tellurium; cadmium and zinc; gallium and indium by means of multi-element microwave-excited electrodeles discharge tubes.

271. G. B. MARSHALL and A. C. SMITH, Paper presented to the 3rd International Conference on Atomic Absorption and Atomic Fluorescence Spectroscopy, Paris, 1971. The determination of trace metals in high purity water using non-dispersive atomic fluorescence spectroscopy.

272. T. L. MARTIN, F. M. HAMM and P. B. ZEEMAN, *Anal. Chim. Acta*, **53,** 437 (1971). A comparative study of turbulent, premixed and separated flames in atomic fluorescence analyses, with reference to the determination of zinc.

273. J. T. MASSEY and S. M. CANNON, *J. Appl. Phys.*, **36,** 361 (1965). Constricted discharges in the rare gases. I. Spectroscopic and electrical measurements.

274. H. MASSMANN, *Spectrochim. Acta*, **23B,** 215 (1968). Atomic absorption and atomic fluorescence in a graphite cell.

275. H. MASSMANN, *Proceedings of the* **13***th Colloquium Spectroscopicum Internationale*, Ottawa, 1967, p. 284, Hilger and Watts, London. Atomic absorption and atomic fluorescence in a graphite cell.

276. J. MATOUŠEK and V. SYCHRA, *Anal. Chem.*, **41,** 518 (1969). Atomic fluorescence study of iron, cobalt and nickel.

277. J. MATOUŠEK, *Ph. D. Thesis*, Technical University, Prague, 1969. Atomic fluorescence spectrometry. II. Fluorescence media.

278. J. MATOUŠEK and V. SYCHRA, *Anal. Chim. Acta*, **49,** 175 (1970). Determination of gold by atomic fluorescence spectroscopy in premixed flames.

279. J. MATOUŠEK and V. SYCHRA, *Chem. Listy*, **63,** 1248 (1969). Atomic fluorescence spectrophotometry.

280. B. MCCARROL, *Rev. Sci. Instr.*, **41,** 279 (1970). An improved microwave discharge cavity for 2450 MHz.

281. W. J. MCCARTHY, M. L. PARSONS and J. D. WINEFORDNER, *Spectrochim. Acta*, **23B,** 25 (1967). Derivation of general expression for quantum and power efficiencies for atomic fluorescence of atoms in a hot gas with emphasis on atomic fluorescence flame spectrometry.

282. M. MCCLINTOCK and L. C. BALLING, *J. Quant. Spectr. Radiat. Transfer*, **9,** 1209 (1969). Atomic and molecular fluorescence from laser excited diatomic caesium and rubidium.

283. A. J. MCCORMACK, S. C. TONG and W. D. COOKE, *Anal. Chem.*, **37,** 1470 (1965). Sensitive selective gas chromatography detector based on emission spectrometry of organic compounds.

284. D. A. MCGILLIS and L. KRAUSE, *Can. J. Phys.*, **46,** 1051 (1968). Inelastic collisions between excited alkali atoms and molecules. IV. Sensitized fluorescence and quenching in mixtures of cesium with N_2, H_2, HD and D_2.

285. W. F. MEGGERS and F. O. WESTFALL, *J. Res. Nat. Bur. Std.*, **44,** 447 (1950). Lamps and wavelengths of mercury - 198.

286. K. D. MIELENZ, *J. Opt. Soc. Am.*, **57,** 66 (1967). Spectroscope slit images in partially coherent light.

287. B. M. MILES and W. L. WIESE, *Bibliography on Atomic Transition Probabilities.* January 1916 Through June 1969. US Dept. of Commerce, NBS Spec. Publ. 320 (1970).

288. R. L. MILLER, L. M. FRASER and J. D. WINEFORDNER, *Appl. Spectry.*, **25,** 477 (1971). A combination flame atomic fluorescence–atomic emission d.c. spectrometer for analysis of trace wear metals in jet engine oils.

289. A. C. G. MITCHELL and M. W. ZEMANSKY, *Resonance Radiation and Excited Atoms.* University Press, Cambridge, 1961.

290. D. G. MITCHELL, *Ger. Offen.* 1, 926, 456 (C. G. 01) 22, 1970. Device for atomic fluorescence spectral analysis.

291. D. G. MITCHELL and A. JOHANSSON, *Spectrochim. Acta*, **25B,** 175 (1970). Simultaneous multielement analysis using sequentially excited atomic fluorescence radiation.

292. D. G. MITCHELL, Paper presented to the 9th National Meeting of Society for Applied Spectroscopy, New Orleans, 1970. Atomic fluorescence spectroscopy: concept and design of a simultaneous multielement spectrometer.

293. D. G. MITCHELL, *Ger. Offen.* 2, 041, 744 (Cl G. 01n), 25 March 1971. Simultaneous multi-element analysis by atomic fluorescence and emission spectroscopy.

294. D. G. MITCHELL and A. JOHANSSON, *Spectrochim. Acta*, **26B,** 677 (1971). Simultaneous multielement analysis using sequentially excited atomic fluorescence radiation — II. An improved instrument for simultaneous analysis.

295. J. J. MITTELDORF and D. O. LANDON, *Appl. Optics*, **7,** 1431 (1968). Multiply diffracted light in the Czerny-Turner spectrometer.

296. E. N. MOROZOV and M. L. SOSINKY, *Optika i Spektroskopiya*, **26,** 506 (1969). Determination of the effective cross sections of second-order collisions for mercury and cadmium atoms by the methods of sensitized fluorescence.

297. V. C. MOSSOTTI and M. DUGGAN, *Appl. Optics*, **7,** 1325 (1968). Atomic emission characteristics of a premixed acetylene-nitrous oxide, total consumption flame.

298. V. G. MOSSOTTI and F. N. ABERCROMBIE, Paper presented to the 16th Colloquium Spectroscopicum Internationale, Heidelberg, 1971. Noise structure in flame spectrometry.

299. C. E. MULFORD, D. C. MANNING and W. SLAVIN, *Proceedings of the 13th Colloquium Spectroscopicum Internationale*, *Ottawa*, 1967, p. 282, Hilger and Watts, London. Atomic fluorescence and atomic absorption spectroscopy.

300. V. I. MUSCAT, T. J. VICKERS and A. ANDREN, *Anal. Chem.*, **44,** 218 (1972). A simple and versatile atomic fluorescence system for the determination of nanogram quantities of mercury.

301. V. I. MUSCAT and T. J. VICKERS, *Anal. Chim. Acta*, **57,** 23 (1971). Determination of nanogram quantities of mercury by the reduction-aeration method and atomic fluorescence spectrophotometry.

302. M. NEHASILOVÁ, *Ph. D. Thesis*, Technical University, Prague. Determination of platinum by flame spectrometry.

303. E. L. NICHOLS and H. L. HOWES, *Phys. Rev.*, **22,** 425 (1923). The photoluminiscence of flames. I.

304. E. L. NICHOLS and H. L. HOWES, *Phys. Rev.*, **23,** 472 (1924). The photoluminiscence of flames. II.

305. M. NORTON and A. GALLAGHER, *Phys. Rev. Ser.* **3A,** 3, 915 (1971). Measurements of lowest-S-state lifetimes of gallium, indium and thallium.

306. J. D. NORRIS and T. S. WEST, *Anal. Chim. Acta*, **55,** 359 (1971). Multi-element atomic fluorescence spectroscopy. Part IV. The determination of cobalt and nickel in steels by atomic fluorescence spectroscopy in a separated air-acetylene flame with a dual-element electrodeless discharge lamp.

307. M. A. NOVICE and J. VINE, *Appl. Optics*, **6,** 1171 (1967). Optical means for enhancing the sensitivity of a tri-alkali photocathode.

308. T. C. O'HAVER and J. D. WINEFORDNER, *Appl. Optics*, **7,** 1647 (1968). Equation for thermionic emission shot noise of multiplier phototubes, including dynode emission.

309. T. C. O'HAVER and J. D. WINEFORDNER, *J. Chem. Educ.*, **46,** 435 (1969). An instrument for source intensity compensation in atomic fluorescence.

310. T. C. O'HAVER and J. D. WINEFORDNER, *J. Chem. Educ.*, **46,** 241 (1969). A versatile, solid state, constant band-width, recording nanoammeter.

311. T. C. O'HAVER, *Ph. D. Thesis*, University of Florida, Gainesville, 1969. The measurement of low light intensities in atomic flame fluorescence spectrometry.

312. N. OMENETTO and G. ROSSI, *Anal. Chim. Acta*, **40,** 195 (1968). Atomic fluorescence flame spectrometry using a mercury line source.

313. N. OMENETTO and G. ROSSI, *Spectrochim. Acta*, **24B**, 95 (1969). Some observations on direct line fluorescence of thallium, indium and gallium.

314. N. OMENETTO and G. ROSSI, Paper presented to the 2nd International Conference on Atomic Absorption Spectroscopy, Sheffield, 1969. Some observations on excitation processes in atomic fluorescence spectrometry.

315. N. OMENETTO, *Met. Ital.*, **61**, 349 (1969). Considerations on the application of atomic fluorescence to analysis.

316. N. OMENETTO and G. ROSSI, *Spectrochim. Acta*, **25B**, 297 (1970). Some observations on the excitation processes in atomic fluorescence flame spectrometry.

317. N. OMENETTO, Paper presented to the Conference on Flame Spectroscopy, Ostrava (Czechoslovakia), 1970. New trends in flame analysis.

318. N. OMENETTO, P. BENETTI and G. ROSSI, *Spectrochim. Acta*, **27B**, 453 (1972). Flame temperature measurements by means of atomic fluorescence spectrometry.

319. M. L. PARSONS, W. J. MCCARTHY and J. D. WINEFORDNER, *Appl. Spectry.*, **20**, 223 (1966). Approximate half-intensity widths of a number of atomic spectral lines used in atomic-emission and atomic-absorption flame spectrometry.

320. M. L. PARSONS and J. D. WINEFORDNER, *Appl. Spectry.*, **21**, 368 (1967). Optimalization of the critical instrumental parameters for achieving maximum sensitivity and precision in flame-spectrometric methods of analysis.

321. M. L. PARSONS, W. J. MCCARTHY and J. D. WINEFORDNER, *J. Chem. Educ.*, **44**, 214 (1967). Selection of optimum conditions for spectrochemical methods. III. Sensitivity of atomic fluorescence, absorption and emission flame spectrometry.

322. M. L. PARSONS, *Ph. D. Thesis*, University of Florida, Gainesville, 1967. Derived expressions to compare atomic fluorescence flame spectrometry with other flame spectrometric methods.

323. S. J. PEARCE, L. DE GALAN and J. D. WINEFORDNER, *Spectrochim. Acta*, **23B**, 793 (1968). Experimental estimation of fluorescence power efficiencies of several atoms in three turbulent flames used in atomic fluorescence flame spectrometry.

324. A. M. PCHELINTSEV, YU. I. BELYAEV, A. V. KARYAKIN and T. A. KOVESHNIKOVA, *Zh. Analit. Khim.*, **26**, 1355 (1971). Increasing the sensitivity of the atomic fluorescence determination of cadmium, silver, thallium, mercury and bismuth in graphite (powder) by impulse selective evaporation into an argon atmosphere.

325. A. M. PCHELINTSEV, YU. I. BELYAEV, A. V. KARYAKIN and T. A. KOVESHNIKOVA, Paper presented to the 3rd International Conference on Atomic Absorption and Atomic Fluorescence Spectrometry, Paris, 1971. The increase in sensitivity of atomic fluorescence determination of cadmium, silver, thallium, mercury, and bismuth in graphite powder by impulse selective vaporization of solid samples in argon atmosphere.

326. E. E. PICKETT and S. R. KOIRTYOHANN, *Spectrochim. Acta*, **23B**, 235 (1968). The nitrous oxide-acetylene flame in emission analysis. I. General characteristics.

327. A. E. PITTS, J. C. VAN LOON and F. E. BEAMISH, *Anal. Chim. Acta*, **50**, 181, 195 (1970). The determination of platinum by atomic absorption spectrometry. I. Air-acetylene flames. II. Nitrous oxide-acetylene flames.

328. M. PLEVA, *Ph. D. Thesis*, University of New Hampshire, Durham, 1969. Practical comparison of atomic fluorescence flame spectrometry with atomic absorption flame spectrometry.

329. W. J. PRICE and J. T. H. ROOS, *Analyst*, **93**, 709 (1968). The determination of silicon by atomic absorption spectrometry, with particular reference to steel, cast iron aluminium alloys and cement.

330. W. J. Price, in *Spectroscopy*, ed. by D. R. Browning, p. 167: McGraw-Hill. London, 1969. Atomic fluorescence spectroscopy.
331. P. Pringsheim, *Fluorescence and Phosphorescence*, Interscience, New York 1949.
332. H. Prugger, *Spectrochim. Acta*, **24B,** 197 (1969). Radiance of light sources for atomic absorption and atomic fluorescence analysis.
333. H. Prugger, R. Grosskopf and R. Torge, *Spectrochim. Acta*, **26B,** 191 (1971). Spectral and time distribution of resonance radiation from hollow cathode lamps operated on the short pulse mode.
334. S. D. Rains, *U.S. Pat.* 3, 565, 567 (Cl. 356−87, G. 01j), Feb. 1971. Atomic absorption and-or atomic fluorescence spectrophotometers.
335. J. Ramírez-Muňoz, in *Flame Emission and Atomic Absorption Spectrometry*, Vol. 1, ed. by J. A. Dean and T. C. Rains, Dekker, New York, 1969, Basic principles of flame emission, atomic absorption, and fluorescence methods.
336. J. O. Rasmuson, R. N. Kniseley and V. A. Fassel, Paper presented to the Pittsburgh Conference on Analytical Chemistry and Applied Spectroscopy, 1971. Experimental and theoretical evaluation of the nitrous-oxide-acetylene flame as an atomization cell for flame spectroscopy.
337. R.C.A. Photomultiplier Tubes, Pamphlet PIT-701A (1967).
338. T. B. Reed, *J. Appl. Phys.*, **34,** 3146 (1963). High-power low-density induction plasmas.
339. R. J. Reynolds and K. Aldous, *Atomic Absorption Spectroscopy* (A Practical Guide), Griffin, London 1970.
340. S. L. Ridgway and J. B. Flato, Paper presented to the Pittsburgh Conference on Analytical Chemistry and Applied Spectroscopy, 1968. Extension of atomic absorption and flame fluorescence detection limits by noise analysis, signal conditioning and phase-sensitive detection.
341. E. W. Richards, *Spectrochim. Acta*, **22,** 158 (1966). A simple micro-wave discharge source.
342. J. W. Robinson, *Anal. Chim. Acta*, **24,** 254 (1961). Mechanism of elemental spectral excitation in flame photometry.
343. J. W. Robinson, *Develop. Appl. Spectry.*, **5,** 255 (1965). Atomic fluorescence.
344. J. W. Robinson, Chapter V, in *Atomic Absorption Spectroscopy*, Dekker, New York, 1966.
345. J. W. Robinson and J. C. Hsu, *Anal. Chim. Acta*, **43,** 109 (1968). Atomic fluorescence spectroscopy of beryllium.
346. J. P. Rodman and H. J. Smith, *Appl. Optics*, **2,** 181 (1963). Tests of photomultipliers for astronomical pulse-counting applications.
347. J. Rolfe and S. E. Moore, *Appl. Optics* **9,** 63 (1970). The efficient use of photomultiplier tubes for recording spectra.
348. G. Rossi and M. Mol, *Spectrochim. Acta*, **24B,** 389 (1969). Isotopic analysis of uranium by an optical spectral method-III. Determination of U^{235}/U^{238} ratios with a hollow cathode source and a direct reading attachment.
349. G. Rossi and N. Omenetto, *Talanta*, **16,** 263 (1969). Application of a demountable water-cooled hollow-cathode lamp to atomic-fluorescence spectrometry.
350. G. Rossi, *Spectrochim. Acta*, **26B,** 271 (1971). Isotopic analysis of uranium by an optical spectral method-IV. Study of the suitability of some uranium lines for the absolute determination of isotopic ratios.
351. P. R. Rouse, B. Brixner and J. V. Kline, *J. Opt. Soc. Am.*, **59,** 955 (1969). Optimization of a 4-m asymmetric Czerny-Turner spectrograph.

352. I. RUBEŠKA, *Chem. Listy*, **61**, 865 (1967). Progress in atomic absorption spectrophotometry.

353. I. RUBEŠKA and B. MOLDAN, *Atomic Absorption Spectrophotometry*, Iliffe, London, 1969.

354. G. W. SERIES, *Proc. Phys. Soc. London* (*At. Mol. Phys.*) **3**, 84 (1970). Theory of frequency and polarization modulation in resonance fluorescence.

355. J. SHARPE, *Document* R-P 021, EMI Electronics, Ltd. (1966). Dark current in photomultiplier tubes.

356. M. SHULL and J. D. WINEFORDNER, *Anal. Chem.* **43**, 799 (1971). Aluminium ellipse for decreasing limits of detection in atomic fluorescence flame spectrometry.

357. M. SHULL and J. D. WINEFORDNER, *Appl. Spectry.*, **25**, 97 (1971). Polarization of atomic fluorescence in flames.

358. M. D. SILVESTER and W. J. MCCARTHY, *Anal. Letters*, **2**, 305 (1969). Experimental parameters affecting the intensity of cadmium electrodeless discharge lamps.

359. M. D. SILVESTER and W. J. MCCARTHY, Paper presented to the Detroit Anachem Conference, Detroit, 1969. Atomic fluorescence: Optimization of electrodeless discharge lamp operating conditions.

360. M. D. SILVESTER and W. J. MCCARTHY, *Spectrochim. Acta*, **25B**, 229 (1970). The intensity of electrodeless discharge lamps containing cadmium, mercury, and manganese.

361. P. J. SLEVIN, K. W. BUSCH and T. J. VICKERS, *Anal. Chem.*, unpublished work. Determination of mercury at the nanogram level by means of a hot wire nondispersive atomic fluorescence system.

362. P. J. SLEVIN, V. I. MUSCAT and T. J. VICKERS, *Appl. Spectry.*, **26**, 296 (1972). An evaluation of premixed Méker-type flame as an atom reservoir for atomic fluorescence.

363. J. A. SMIT, J. M. W. MILATZ and C. SMIT, *Physica*, **20**, 39 (1954). An integrating spectrometer for measuring spectral intensities by counting of photoelectrons.

364. R. SMITH, C. M. STAFFORD and J. D. WINEFORDNER, *Anal. Chim. Acta*, **42**, 523 (1968). Some interference effects occurring in the hydrogen/argon/entrained air flame.

365. R. SMITH, C. M. STAFFORD and J. D. WINEFORDNER, *Anal. Chem.*, **41**, 946 (1969). Temperature profiles of turbulent hydrogen-diffusion flames used in atomic fluorescence spectrometry.

366. R. SMITH, C. M. STAFFORD and J. D. WINEFORDNER, *Can. Spectry.*, **14**, 38 (1969). An evaluation of atomic fluorescence flame spectrometry for the determination of wear metals in jet engine lubricating oils.

367. R. SMITH and J. D. WINEFORDNER, Paper presented to the 2nd International Conference on Atomic Absorption Spectroscopy, Sheffield, 1969. Fundamental studies on turbulent hydrogen diffusion flames used in atomic fluorescence spectrometry.

368. R. SMITH, R. C. ELSER and J. D. WINEFORDNER, *Anal. Chim. Acta*, **48**, 35 (1969). The atomic fluorescence of copper with a high-intensity hollow-cathode lamp.

369. R. SMITH, *Advan. Anal. Chem. Instrum.*, **9**, 235 (1971). Flame fluorescence spectrometry.

370. D. S. SMYLY and J. D. WINEFORDNER, Paper presented to the Pittsburgh Conference on Analytical Chemistry and Applied Spectroscopy, 1971. Measurement of free atom fraction in premixed hydrogen-oxygen-argon flames.

371. D. S. SMYLY, W. P. TOWNSEND, P. J. TH. ZEEGERS and J. D. WINEFORDNER, *Spectrochim. Acta*, **26B**, 531 (1971). Measurement of the free atom fraction in hydrogen-oxygen-argon flames using the relative absorption technique.

372. L. E. Smythe, *At. Energy Aust.*, **13**, 26 (1970). Chemical analysis by atomic fluorescence spectroscopy.

373. B. H. Soffer and B. B. McFarland, *Appl. Phys. Letters*, **10**, 266 (1967). Continuously tunable, narrow band organic dye lasers.

374. P. A. St. John, W. J. McCarthy and J. D. Winefordner, *Anal. Chem.*, **39**, 1495 (1967). A statistical method for evaluation of limiting detectable sample concentration.

375. R. A. Staab, *Ph. D. Thesis*, University of Florida, Gainesville, 1964. Atomic fluorescence flame spectrometry as a means of chemical analysis.

376. E. B. M. Steers, *Proc. Soc. Anal. Chem.*, **2**, 108 (1965). Spectroscopic source with microwave excitation.

377. R. J. Strutt, *Proc. Roy. Soc. (London)*, **91A**, 388, 511 (1914). Observations on the fluorescence and resonance radiation of sodium vapour.

378. R. J. Strutt, *Proc. Roy. Soc. (London)*, **96A**, 272 (1919). A study of the line spectrum of sodium as excited by fluorescence.

379. M. Stupavsky and L. Krause, *Can. J. Phys.*, **47**, 1249 (1969). Inelastic collisions between excited alkali atoms and molecules. VI. Sensitized fluorescence in mixture of sodium with CH_4, CD_4, C_2H_2, C_2H_4 and C_2H_6.

380. J. V. Sullivan and A. Walsh, *Spectrochim. Acta*, **21**, 721 (1965). High intensity hollow-cathode lamps.

381. V. Svoboda, *Chem. Listy*, **57**, 419 (1963). The sensitivity of detectors used in emission spectral analysis.

382. V. Svoboda, Paper presented to the 1st International Conference on Atomic Absorption Spectroscopy, Prague, Czechoslovakia, 1967. Investigations on atomic fluorescence using evaporation from a graphite crucible.

383. V. Svoboda and R. Gerbatsch, *Z. Anal. Chem.*, **242**, 1 (1968). Contribution to the definition of "limit of detection" and "limit of determination".

384. V. Svoboda, G. Nitis and J. D. Winefordner, Paper presented to the 9th National Meeting of Society for Applied Spectroscopy, New Orleans, 1970. The use of a Fabry-Perot oscillating interferometer for wavelength modulation in atomic absorption and atomic fluorescence spectrometry.

385. V. Svoboda, Paper presented to the 1st Annual National Seminar on Flame Spectrometry, Modra (Czechoslovakia), 1971. Theoretical growth curves in atomic fluorescence flame spectrometry.

386. V. Svoboda and J. D. Winefordner, Paper presented to the 3rd International Conference on Atomic Absorption and Atomic Fluorescence Spectrometry, Paris, 1971. Influence of flame shape and method of illumination upon theoretical calibration curves in atomic fluorescence flame spectrometry.

387. V. Sychra, J. Matoušek and S. Marek, *Chem. Listy*, **63**, 177 (1969). Direct determination of trace amounts of nickel in gas oils by atomic absorption and atomic fluorescence spectrometry.

388. V. Sychra, *Ph. D. Thesis*, Technical University, Prague, 1969. Atomic fluorescence spectrometry. I. Excitation sources.

389. V. Sychra and J. Matoušek, *Anal. Chim. Acta*, **52**, 376 (1970). Direct determination of nickel in gas oils and petroleum distillation residues by atomic fluorescence spectroscopy in a separated air-acetylene flame.

390. V. Sychra and J. Matoušek, *Talanta*, **17**, 363 (1970). Atomic fluorescence spectroscopy of lead.

391. V. Sychra, P. J. Slevin, J. Matoušek and F. Bek, *Anal. Chim. Acta*, **52**, 259 (1970).

The determination of palladium by atomic absorption, atomic fluorescence and thermal-emission spectroscopy in various flames.

392. V. SYCHRA, Paper presented to the 1st Annual National Seminar on Flame Spectrometry, Modra (Czechoslovakia), 1971. Non-flame technique in atomic fluorescence spectrometry.

393. V. SYCHRA and D. KOLIHOVÁ, Paper presented to the 3rd International Conference on Atomic Absorption and Atomic fluorescence Spectrometry, Paris, 1971. New atomization techniques in atomic fluorescence microanalysis.

394. V. SYCHRA, D. KOLIHOVÁ and J. P. MATOUŠEK, Unpublished results. A critical evaluation of use of high intensity hollow-cathode lamps as excitation sources in atomic fluorescence spectrometry.

395. A. SYTY, Chapter VIII in *Flame Emission and Atomic Absorption Spectrometry*, Vol. 2, ed. by J. A. Dean and T. C. Rains, Dekker, New York, 1971. Atomic fluorescence spectrometry.

396. A. TERENIN, *Z. Physik*, **31**, 26 (1925). Excitation of atoms and molecules by light. irradiation. I.

397. A. TERENIN, *Z. Physik*, **37**, 98 (1926). Excitation of atoms and molecules by light-irradiation. II.

398. K. C. THOMPSON and P. C. WILDY, Paper presented to the 2nd International Conference on Atomic Absorption Spectroscopy, Sheffield 1969. The performance and application of electronically modulated microwave tubes for atomic fluorescence and atomic absorption spectrophotometry.

399. K. C. THOMPSON and P. C. WILDY, *Analyst*, **95**, 776 (1970). The use of electronically modulated microwave-excited discharge tubes in atomic spectroscopy.

400. K. C. THOMPSON, *Spectrosc. Letters*, **3**, 59 (1970). The feasibility of determining iodine directly by atomic absorption and atomic fluorescence spectroscopy.

401. K. C. THOMPSON and G. D. REYNOLDS, *Analyst*, **96**, 771 (1971). The atomic-fluorescence determination of mercury by the cold vapour technique.

402. W. P. TOWNSEND, *Ph. D. Thesis*, University of Florida, Gainesville, Florida, 1970. Use of atomic fluorescence with a continuous source for measurement of the damping constant (a-parameter) in flames.

403. W. P. TOWNSEND, D. S. SMYLY, P. J. TH. ZEEGERS, V. SVOBODA and J. D. WINEFORDNER, *Spectrochim. Acta*, **26B**, 595 (1971). Measurements of damping constants of atoms in flames by atomic fluorescence flame spectrometry with a continuum source.

404. R. G. TULL, *Appl. Optics*, **7**, 2023 (1968). A comparison of photon counting and current measuring technique in spectrophotometry of faint sources.

405. U. ULFVARSON, *Acta. Chem. Scand.*, **21**, 641 (1967). Determination of mercury in small quantities in biological materials by a modified photometric-mercury vapour. procedure.

406. R. J. VAN BRUNT and N. R. ZARE, *J. Chem. Phys.*, **48**, 4304 (1968). Polarization of atomic fluorescence excited by molecular dissociation.

407. J. G. M. VAN DER LINDEN, *Instrum. Rev. (Leiden)*, 2ji-**5**, 34 (1967). Atomic absorption spectroscopy.

408. Z. VAN GELDER, *Appl. Spectry.*, **22**, 581 (1968).

409. C. VAN TRIGT, T. J. HOLLANDER and T. J. ALKEMADE, *J. Quant. Spectr. Radiat. Transfer*, **5**, 813 (1963). Determination of the α-parameter of resonance lines in flames.

410. C. VEILLON and J. D. WINEFORDNER, Paper presented to the Phoenix Symposium

on Flame Spectrometric Methods of Analysis, Phoenix, 1966. Flame fluorescence spectrometry.

411. C. VEILLON, J. M. MANSFIELD, M. L. PARSONS and J. D. WINEFORDNER, *Anal. Chem.*, **38,** 204 (1966). Use of a continuous source in flame fluorescence spectrometry.

412. C. VEILLON, *Ph. D. Thesis,* University of Florida, Gainesville, 1966. Sources of excitation for atomic fluorescence flame spectrometry.

413. C. VEILLON and J. Y. PARK, *Appl. Chem.*, **44,** 1473 (1972). Direct isotopic determination by atomic fluorescence spectrometry.

414. T. J. VICKERS and S. P. MERRICK, *Talanta,* **15,** 873 (1968). Determination of parts-per milliard concentrations of mercury by atomic fluorescence flame spectrometry.

415. T. J. VICKERS and R. M. VAUGHT, *Anal. Chem.*, **41,** 1476 (1969). Nondispersive atomic fluorescence analysis.

416. T. J. VICKERS, P. J. SLEVIN, V. I. MUSCAT and L. T. FARIAS, *Anal. Chem.*, **44,** 930 (1972). Factors affecting the use of nondispersive a system for atomic fluorescence flame spectrometry.

417. R. A. VITKUN, N. S. POLUEKTOV and YU. V. ZELYUKOVA, *Zh. Analit. Khim.*, **25,** 474 (1970). Atomic fluorescence determination of mercury.

418. A. WALSH, in *Atomic Absorption Spectroscopy,* ASTM STP 443, A.S.T.M., 1969. Physical aspects of atomic absorption.

419. A. WALSH, in *Atomic Absorption Spectroscopy,* ed. by R. M. Dagnall and G. K. Kirkbright, Butterworths, London, 1970. The application of new techniques to simultaneous multi-element analysis.

420. P. D. WARR, *Talanta,* **17,** 543 (1970). Use of a filter in atomic fluorescence spectroscopy.

421. P. D. WARR, *Talanta,* **18,** 234 (1971). Determination of zinc in copper by atomic-fluorescence flame spectroscopy.

422. J. O. WEIDE and M. L. PARSONS, Paper presented to the Pittsburgh Conference on Analytical Chemistry and Applied Spectroscopy, 1971. Investigation of laser excited flame fluorescence spectroscopy.

423. T. S. WEST, *Chem. Ind. (London),* **18,** 1005 (1966). Inorganic trace analysis.

424. T. S. WEST, *Analyst,* **91,** 69 (1966). Some sensitive and selective reactions in inorganic spectroscopic analysis.

425. T. S. WEST, Paper presented to the 1st International Conference on Atomic Absorption Spectroscopy, Prague (Czechoslovakia), 1967. Some aspects of atomic fluorescence spectroscopy.

426. T. S. WEST, *Endeavour,* **26,** 44 (1967). Atomic analysis in flames.

427. T. S. WEST, *US. Department of Commerce, NBS Monograph* 100, pp. 215–301 (1967). Trace characterization.

428. T. S. WEST and X. K. WILLIAMS, *Anal. Chem.*, **40,** 335 (1968). Atomic fluorescence spectroscopy of silver using a high-intensity hollow-cathode lamp as source.

429. T. S. WEST and X. K. WILLIAMS, *Anal. Chim. Acta,* **42,** 29 (1968). Atomic-fluorescence spectroscopy of magnesium with a high-intensity hollow-cathode lamp as line source.

430. T. S. WEST and X. K. WILLIAMS, *Anal. Chim. Acta,* **45,** 27 (1969). Atomic absorption and fluorescence spectroscopy with a carbon-filament atom reservoir. Part I: Construction and operation of atom reservoir.

431. T. S. WEST, *Spectrosc. Letters,* **2,** 179 (1969). Comparison of sensitivity of atomic absorption and atomic fluorescence spectroscopy.

432. T. S. WEST, *Minerals Sci. Engng.*, **2,** 31 (1970). Atomic spectroscopy in trace analysis. II. Atomic fluorescence spectroscopy.

433. T. S. WEST, in *Atomic Absorption Spectroscopy*, ed. by R. M. Dagnall and G. K. Kirkbright, Butterworths, London, 1970. Development in atom reservoirs and line sources for atomic absorption and atomic fluorescence spectroscopy.

434. T. S. WEST, *Pure Appl. Chem.*, **26,** 47 (1971). The determination of very small amounts of materials by the technique of atomic absorption and atomic fluorescence spectroscopy.

435. J. A. WHEAT, *Appl. Spectry.*, **25,** 328 (1971). Isotopic analysis of lithium by atomic absorption spectrophotometry.

436. W. L. WIESE, M. W. SMITH and B. M. MILES, *Stand. Ref. Data Ser., Nat. Bur. Stand.* (U.S.), Vol. 1 (1966), Vol. 2 (1969). Atomic Transition Probabilities.

437. P. C. WILDY and K. BECKER, *Fachz. Lab.*, **13,** 678 (1969). Atomic fluorescence spectrophotometry with a commercially available atomic absorption spectrophotometer.

438. P. C. WILDY and K. C. THOMPSON, *Analyst*, **95,** 562 (1970). Electronic modulation of electrodeless discharge tubes for use in atomic spectroscopy.

439. J. B. WILLIS, J. O. RASMUSON, R. N. KNISELEY and V. A. FASSEL, *Spectrochim. Acta*, **23B,** 725 (1968). The temperature of the nitrous oxide-acetylene flame.

440. J. B. WILLIS, V. A. FASSEL and J. A. FIORINO, *Spectrochim. Acta*, **24B,** 157 (1969). The nitrous oxide-hydrogen flame in atomic absorption and emission spectroscopy.

441. J. B. WILLIS, *Spectrosc. Letters*, **2,** 191 (1969). Comparison of detection limits in atomic fluorescence and absorption spectroscopy.

442. J. B. WILLIS, *Spectrochim, Acta*, **25B,** 487 (1970). Atomization problems in atomic absorption spectrometry-II. Determination of degree of atomization in premixed flames.

443. J. B. WILLIS, Lecture presented to the Meeting of Czechoslovak Spectroscopic Society, Prague, October 1971. Atomic spectrometry in Australia.

444. J. D. WINEFORDNER and T. J. VICKERS, *Anal. Chem.*, **36,** 161 (1964). Atomic fluorescence spectrometry as a means of chemical analysis.

445. J. D. WINEFORDNER and R. A. STAAB, *Anal. Chem.*, **36,** 165 (1964). Determination of zinc, cadmium, and mercury by atomic fluorescence flame spectrometry.

446. J. D. WINEFORDNER and R. A. STAAB, *Anal. Chem.*, **36,** 1367 (1964). Study of experimental parameters in atomic fluorescence flame spectrometry.

447. J. D. Winefordner, Paper presented to the Eastern Analytical Symposium, New York, 1964. Atomic fluorescence flame spectrometry.

448. J. D. WINEFORDNER, Paper presented to the American Chemical Society Meeting, Atlantic City, 1965. Atomic fluorescence flame spectrometry—a critical review.

449. J. D. WINEFORDNER, Paper presented to the 9th Annual Symposium on Modern Methods of Analytical Chemistry, Baton Rouge, 1966. Atomic fluorescence flame spectrometry.

450. J. D. WINEFORDNER, W. J. MCCARTHY and P. A. ST. JOHN, *J. Chem. Educ.*, **44,** 80 (1967). The selection of optimum conditions for spectrochemical methods. I. use of signal-to-noise ratio theory.

451. J. D. WINEFORDNER, M. L. PARSONS, J. M. MANSFIELD and W. J. MCCARTHY, *Anal. Chem.*, **39,** 436 (1967). Derivation of expressions for calculation of limiting detectable atomic concentration in atomic fluorescence flame spectrometry.

452. J. D. WINEFORDNER, M. L. PARSONS, J. M. MANSFIELD and W. J. MCCARTHY, *Spectrochim. acta*, **23B,** 37 (1967). Intensity of atomic fluorescence as a function of atomic concentration in atomic fluorescence flame spectrometry.

453. J. D. WINEFORDNER and J. M. MANSFIELD, *Appl. Spectry. Reviews*, **1,** 1 (1967). Atomic fluorescence flame spectrometry.

454. J. D. Winefordner and J. M. Mansfield, Chapter 13 in *Fluorescence*, ed. by G. G. Guilbaut, pp. 565, 625, Dekker, New York, 1967. Atomic fluorescence flame spectrometry.

455. J. D. Winefordner and J. M. Mansfield, Jr., Paper presented to the Pittsburgh Conference on Analytical Chemistry and Applied Spectroscopy. 1968. New developments on electrodeless discharge tubes and turbulent flame types for atomic fluorescence flame spectrometry.

456. Winefordner, *Rec. Chem. Progr.*, **29,** 25 (1968). Flame spectrometry: Atomic fluorescence vs. atomic absorption.

457. J. D. Winefordner, Paper presented to the 15th Spectroscopy Symposium of Canada, 1968. Critical evaluation of atomic emission, atomic absorption and atomic fluorescence flame spectrometry.

458. J. D. Winefordner, Paper presented to the 8th National Meeting, Anaheim California, 1969. A comparison of the fundamental similarities and differences between flame emission, atomic absorption and atomic fluorescence spectroscopy.

459. J. D. Winefordner, in *Atomic Absorption Spectroscopy*, ed. by R. M. Dagnall and G. K. Kirkbright, Butterworths, London, 1970. Non-flame cells in atomic fluorescence spectrometry.

460. J. D. Winefordner, Paper presented to the 17th Spectroscopy Symposium of Canada, Ottawa, 1970. Critical evaluation of atomic emission, atomic absorption and atomic fluorescence spectrometry.

461. J. D. Winefordner, *U. S. Govt. Res. Develop. Rep.*, **70,** 147 (1970). Experimental and theoretical investigation of flame spectrometric methods and gas chromatographic detectors.

462. J. D. Winefordner and T. J. Vickers, *Anal. Chem.*, **42,** 206 R (1970). Flame spectrometry.

463. J. D. Winefordner and R. C. Elser, *Anal. Chem.*, **43,** (4), 24 A (1971). Atomic fluorescence spectrometry.

464. J. D. Winefordner, *Accounts of Chemical Research*, **4,** 259 (1971). Atomic fluorescence flame spectrometry.

465. J. D. Winefordner and R. Smith, Chapter 11 in *Analytical Flame Spectrometry, Selected Topics*, ed. by R. Mavrodineanu, Philips Technical Library, McMillan, London, 1970. Atomic fluorescence flame spectrometry.

466. J. D. Winefordner, V. Svoboda and L. J. Cline, *CRC Critical Reviews in Analytical Chemistry*, **1,** 233 (1970). A critical comparison of atomic emission, atomic absorption, and atomic fluorescence flame spectrometry.

467. J. D. Winefordner and T. C. O'Haver, *Fluorescence News*, **5,** 1 (1970—71). Comparison of analytical curves in atomic fluorescence spectrometry with inorganic fluorometry.

468. R. W. Wood, *Phil. Mag.*, **3,** 359 (1902). On the fluorescence and absorption spectrum of sodium vapour.

469. R. W. Wood, *Phil. Mag.*, **10,** 513 (1905). The fluorescence of sodium vapour and the resonance radiation of the atoms.

470. R. W. Wood, *Phil. Mag.*, **23,** 689 (1912). Selective reflexion, scattering, and absorption by resonating gas molecules.

471. R. W. Wood and L. Dunoyer, *Phil. Mag.*, **27,** 1018 (1914). The separate excitation of the centres of emission of the D-lines of sodium and photometric investigation of the superficial resonance of sodium vapour.

472. R. W. Wood, *Proc. Phys. Soc.* (*London*), **26,** 185 (1914).

473. R. W. Wood and F. L. Mohler, *Phys. Rev.*, **11,** 70 (1918). Resonance radiation of sodium vapour excited by one of the D-lines.
474. R. W. Wood and F. L. Mohler, *Phil. Mag.*, **37,** 456 (1919). Resonance radiation of sodium vapour excited by one of the D-lines.
475. R. Wood and T. R. Wills, Paper presented to the 3rd International Conference on Atomic Absorption and Atomic Fluorescence Spectrometry, Paris, 1971. The simultaneous solvent extraction and atomic fluorescence spectroscopic determination of six elements with the Technicon Auto Analyzer.
476. C. Woodward, *Anal. Chim. Acta*, **51,** 548 (1970). Preparation and use of electrodeless discharge tubes for atomic absorption spectrophotometry of some alkali metals.
477. C. Woodward, Paper presented to the 3rd International Conference on Atomic Absorption and Atomic Fluorescence Spectrometry, Paris, 1971. Some practical applications of electrodeless discharge tubes to atomic absorption spectroscopy.
478. E. F. Worden, R. G. Gutmacher and J. G. Conway, *Appl. Optics*, **2,** 707 (1963). Use of electrodeless discharge lamps in the analysis of atomic spectra.
479. Yu. Yokoyama, *Bunko Kenkyu*, **18,** 125 (1969). Atomic fluorescence flame spectrometry.
480. A. T. Young, *Appl. Optics*, **8,** 2431 (1969). Photometric error analysis IX. Optimum use of photomultipliers.
481. O. Youngbluth, *Appl. Optics*, **9,** 321 (1970). Fatigue effects on the area sensitivity, dynode gain, and anode output for several end-on photomultipliers.
482. K. E. Zacha and J. D. Winefordner, *Anal. Chem.*, **38,** 1537 (1966). Use of an argon/hydrogen/entrained air flame with a total consumption aspirator burner and a simple instrumental system in flame emission spectrometry for trace metal determination.
483. K. E. Zacha, M. P. Bratzel, Jr., J. D. Winefordner and J. M. Mansfield, Jr., *Anal. Chem.*, **40,** 1733 (1968). Improvements in preparation and operation of electrodeless discharge lamps as high-intensity sources in atomic fluorescence flame spectrometry.
484. K. E. Zacha, M. P. Bratzel, Jr., J. D. Winefordner and J. M. Mansfield, Jr., *Anal. Chem.*, **41,** 1315 (1969). Correction. Improvements in preparation and operation of electrodeless discharge lamps as high intensity sources in atomic fluorescence flame spectrometry.
485. A. N. Zaidel and E. P. Korrenoi, *Optics and Spectroscopy*, **10,** 299 (1961). Spectral determination of the isotopic composition and concentration of lithium in solution.
486. P. J. T. Zeegers, R. Smith and J. D. Winefordner, *Anal. Chem.*, **40,** (13), 20 A (1968). Shapes of analytical curves in flame spectrometry.
487. P. J. Th. Zeegers, W. P. Townsend and J. D. Winefordner, *Spectrochim. Acta*, **24B,** 243 (1969). Estimation of free-atom fraction and dissociation energies of compounds in flames by an atomic absorption method.
488. P. J. T. Zeegers and J. D. Winefordner, *Spectrochim. Acta*, **26B,** 161 (1971). Comparison of theoretical and experimental atomic fluorescence analytical curves for magnesium.
489. M. Zelikoff, P. H. Wyckoff, L. M. Aschenbrand and R. S. Loomis, *J. Opt. Soc. Am.*, **42,** 818 (1952). Electrodeless discharge lamps containing metallic vapors.

RECENTLY PUBLISHED PAPERS

After finishing the manuscript of this book, about one hundred interesting papers have appeared concerning theory, instrumentation and application of atomic fluorescence spectrometry. The most important of these deal with emission noise measurement in a premixed $H_2-O_2-N_2$ flame [491], classification of atomic fluorescence transitions [526], pulsed sources for atomic fluorescence spectrometry [541, 546], temperature-controlled multiple element electrodeless discharge lamps [495, 496, 530], laser excitation [507, 527, 528, 529], ultrasonic nebulization [503], construction of an automated atomic fluorescence instruments [500, 501, 508, 517, 518, 520, 533], atomic fluorescence characteristics of rare-earths [527, 528], and some transition metals [529], and applications [511, 515, 519, 522, 534, 543]. For complete list of these publications, the reader is referred to the following Appendix.

APPENDIX

490. K. M. ALDOUS and B. W. BAILEY, Paper presented to the Pittsburgh Conference on Analytical Chemistry and Applied Spectroscopy, 1972. Application of photon counting techniques in atomic spectrometry.
491. C. T. J. ALKEMADE, H. P. HOOYMAYERS, P. L. LIJNSE and T. J. M. J. VIERBERGEN, *Spectrochim. Acta*, **27B**, 149 (1972). Emission noise spectrum in a premixed $H_2-O_2-N_2$ flame.
492. B. W. BAILEY, I. S. MAINES and J. M. RANKIN, Paper presented to the Pittsburgh Conference on Analytical Chemistry and Applied Spectroscopy, 1972. Investigation of light scattering as an interference in flame spectrometry.
493. W. A. BARNETT, H. L. KAHN, E. H. SIEGLER and W. BOHLER, Paper presented to the Pittsburgh Conference on Analytical Chemistry and Applied Spectroscopy, 1972. The place of atomic fluorescence in the analytical laboratory.
494. A. T. BLADES, *Can. J. Chem.*, **49**, 2476 (1971). Hydrocarbon induced fluorescence of atoms in flames.
495. R. F. BROWNER, B. M. PATEL, T. H. GLENN, M. E. RIETTA and J. D. WINEFORDNER, *Spectrosc. Letters*, **5**, 311 (1972). A device for precise temperature control of electrodeless discharge lamps.
496. R. F. BROWNER, M. E. RIETTA and J. D. WINEFORDNER, Paper presented to the Pittsburgh Conference on Analytical Chemistry and Applied Spectroscopy, 1972. Improvements in electrodeless discharge lamps radiant flux and stability by precise temperature control.
497. J. W. CAMBELL, *Appl. Optics*, **10**, 1232 (1971). Developmental solar blind photomultipliers suitable for use in the 1450–2800 Å region.
498. M. COHEN, J. INDROCASO, CH. SALZMAN and B. SENITSKY, Paper presented to the Pittsburgh Conference on Analytical Chemistry and Applied Spectroscopy, 1972. Improved electrodeless discharge lamps for atomic fluorescence spectrometry.
499. D. O. COOKE, R. M. DAGNALL and T. S. WEST, *Talanta*, **19**, 1309 (1972). Some considerations on spectral line profiles of microwave-excited electrodeless discharge lamps.
500. E. CORDOS and V. MALMSTADT, *Anal. Chem.*, **44**, 2277 (1972). Dual channel synchronous integration measurement system for atomic fluorescence spectrometry.

501. E. Cordos and V. Malmstadt, Paper presented to the Pittsburgh Conference on Analytical Chemistry and Applied Spectroscopy, 1972. Developments in excitation source, flame nebulizer, and measurement system for a new type of automated atomic fluorescence spectrometer.

502. R. M. Dagnall, M. D. Silvester and T. S. West, *Spectrochim. Acta*, **28B**, 51 (1973). Some intense sources of radiation for the alkali and alkaline earth elements.

503. M. B. Denton and H. V. Malmstadt, *Anal. Chem.*, **44**, 1813 (1972). Ultrasonic nebulization in a low-emission flame for atomic fluorescence spectrometry.

504. L. Ebdon, G. F. Kirkbright and T. S. West, *Talanta*, **19**, 1301 (1972). Determination of manganese by atomic fluorescence spectroscopy using a carbon-filament atom-reservoir.

505. M. Failes, D. J. Koop, A. R. Barringer and M. D. Silvester, Paper presented to the Pittsburgh Conference on Analytical Chemistry and Applied Spectroscopy, 1973. Double-beam flameless atomic fluorescence technique for mercury analysis.

506. S. D. Fihn, S. W. Subber and C. D. West, Paper presented to the Pittsburgh Conference on Analytical Chemistry and Applied Spectroscopy, 1972. Determination of mercury by flameless atomic fluorescence.

507. L. M. Fraser and J. D. Winefordner, *Anal. Chem.*, **44**, 1444 (1972). Laser-excited atomic fluorescence flame spectrometry as an analytical method.

508. S. R. Goode and S. R. Crouch, Paper presented to the Eleventh National SAS Meeting, Dallas, Texas, 1972. Critical evaluation of an automated non-flame atomic fluorescence spectrometer.

509. D. S. Gough and P. Hannaford, Paper presented to the 8th Australian Spectroscopy Conference, 1971. Atomic fluorescence from vapours produced by cathodic sputtering of alloys.

510. G. M. Hieftje, R. I. Bystroff and R. Lim, *Anal. Chem.*, **45**, 253 (1973). Application of correlation analysis for signal-to-noise enhancement in flame spectrometry. Use of correlation in determination of rhodium by atomic fluorescence spectrometry.

511. M. E. Hofton and D. P. Hubbard, *Anal. Chim. Acta*, **62**, 311 (1972). The determination of trace amounts of bismuth in steels by solvent extraction and atomic absorption or atomic fluorescence spectrometry.

512. D. A. Huchital, J. W. Vollmer and W. B. Barnet, Paper presented to the Pittsburgh Conference on Analytical Chemistry and Applied Spectroscopy, 1973. Electrodeless discharge lamps for atomic spectroscopy—a progress report.

513. H. G. C. Human, *Spectrochim. Acta*, **27B**, 301 (1972). The combined hollow cathode and high frequency discharge as excitation source for atomic fluorescence spectrometry.

514. F. M. Johnson and M. V. Swagel, *Appl. Optics*, **10**, 1624 (1971). Continuously tunable pulsed UV light source.

515. M. Jones, G. F. Kirkbright, L. Ranson and T. S. West, *Anal. Chim. Acta*, **63**, 210 (1973). The simultaneous determination of traces of cobalt, chromium, copper, iron, manganese, and zinc by atomic fluorescence spectrometry with preconcentration by an automated solvent extraction procedure.

516. R. Jones, C. T. Oliver and E. R. Pike, *Appl. Optics*, **10**, 1673 (1971). Experimental and theoretical comparison of photon counting and current measurements of light intensity.

517. H. V. Malmstadt and E. Cordos, *American Laboratory*, (August), p. 35 (1972). Automated atomic fluorescence spectrometer for rapid multielement determinations.

518. H. V. MALMSTADT and E. CORDOS, Paper presented to the Pittsburgh Conference on Analytical Chemistry and Applied Spectroscopy, 1972. A new type of automated atomic fluorescence spectrometric instrument for rapid multielement quantitative determination.
519. G. B. MARSHALL and A. C. SMITH, *Analyst*, **97,** 447 (1972). The determination of zinc in the feed-water to high pressure boilers by atomic fluorescence spectroscopy.
520. A. MONTASER and S. R. CROUCH, Paper presented to the Pittsburgh Conference on Analytical Chemistry and Applied Spectroscopy, 1973. Computer-controlled, non-flame atomic absorption-atomic fluorescence spectrometer.
521. V. I. MUSCAT and T. J. VICKERS, Paper presented to the Eleventh National SAS Meeting, Dallas, Texas, 1972. Advantages and limitations of non-dispersive atomic fluorescence.
522. J. D. NORRIS and T. S. WEST, *Anal. Chim. Acta*, **59,** 474 (1972). Multi-element atomic fluorescence spectroscopy. Part V. The determination of chromium and manganese in steels in a separated air-acetylene flame with a dual-element electrodeless discharge lamp.
523. J. D. NORRIS and T. S. WEST, *Anal. Chim. Acta*, **59,** 355 (1972). The atomic fluorescence characteristics and determination of chromium in an argon-separated air-acetylene flame.
524. J. D. NORRIS and T. S. WEST, *Anal. Chem.*, **45,** 226 (1973). Application of a wavelength scanning technique to multi-element determination by atomic fluorescence spectrometry.
525. N. OMENETTO, R. BROWNER, J. D. WINEFORDNER, G. ROSSI and P. BENETTI, *Anal. Chem.*, **44,** 1683 (1972). Color temperature atomic fluorescence method for flame temperature measurements.
526. N. OMENETTO and J. D. WINEFORDNER, *Appl. Spectry.*, **26,** 555 (1972). Types of fluorescence transitions in atomic fluorescence spectrometry.
527. N. OMENETTO, N. N. HATCH, L. M. FRASER and J. D. WINEFORDNER, Paper presented to the Eleventh National SAS Meeting, Dallas, Texas, 1972. Laser-excited atomic fluorescence of rare earths and other refractory transition metals in the nitrous oxide-acetylene flame.
528. N. OMENETTO, N. N. HATCH, L. M. FRASER and J. D. WINEFORDNER, *Anal. Chem.*, **45,** 195 (1973). Laser-excited atomic and ionic fluorescence of the rare earths in the nitrous oxide-acetylene flame.
529. N. OMENETTO, N. N. HATCH, L. M. FRASER and J. D. WINEFORDNER, *Spectrochim. Acta*, **28B,** 65 (1973). Laser-excited atomic fluorescence of some transition elements in the nitrous oxide-acetylene flame.
530. B. M. PATEL, R. F. BROWNER and J. D. WINEFORDNER, *Anal. Chem.*, **44,** 2272 (1972). Design and operation of temperature-controlled multiple element electrodeless discharge lamps for atomic fluorescence spectrometry.
531. E. H. PIEPMEIER, Paper presented to the 10th National SAS Meeting, 1971. Consideration of the characteristics of lasers as light sources for atomic fluorescence.
532. E. H. PIEPMEIER, *Spectrochim. Acta*, **27B,** 431 (1972). Theory of laser saturated atomic resonance fluorescence.
533. T. C. RAINS, M. S. EPSTEIN and O. MENIS, Paper presented to the Pittsburgh Conference on Analytical Chemistry and Applied Spectroscopy, 1973. Automatic correction system for light scatter in atomic fluorescence.
534. J. W. ROBINSON and Y. E. ARAKTINGI, *Anal. Chim. Acta*, **63,** 29 (1973). Study of the

application of atomic fluorescence spectrometry to the direct determination of mercury and cadmium in the atmosphere.

535. W. G. SCHRENK, S. V. VALENTE and K. E. SMITH, *Appl. Spectry.*, **26**, 108 (1972). Tuning stubs as an aid to coupling RF-energy to electrodeless discharge lamps.

536. C. J. SCHULER, C. T. PIKE and M. A. MIRANDA, *Appl. Optics*, **10**, 1689 (1971). Dye laser probing of the atmosphere using resonant scattering.

537. A. H. SOMMER, *Appl. Optics*, **12**, 90 (1973). Stability of photocathodes.

538. L. STOOKEY and R. HERRON, Paper presented to the Pittsburgh Conference on Analytical Chemistry and Applied Spectroscopy, 1972. An automated multielement solvent extraction system for trace metal determination in natural waters by atomic fluorescence spectrometry.

539. V. SVOBODA, R. F. BROWNER and J. D. WINEFORDNER, *Appl. Spectry.*, **26**, 505 (1972). Analytical curves in atomic fluorescence spectrometry.

540. C. VEILLON and M. K. MURPHY, Paper presented to the Eleventh National SAS Meeting, Dallas, Texas, 1972. The use of photon counting in atomic fluorescence spectrometry.

541. J. O. WEIDE and M. L. PARSONS, *Anal. Letters*, **5**, 363 (1972). The use of pulsed hollow cathode lamps in atomic fluorescence spectrometry.

542. T. S. WEST, Paper presented to the Pittsburgh Conference on Analytical Chemistry and Applied Spectroscopy, 1972. Multichannel atomic fluorescence and molecular chemiluminescence emission spectrometry for the analysis of technical materials.

543. T. S. WEST, Paper presented to the Pittsburgh Conference on Analytical Chemistry and Applied Spectroscopy, 1972. Practical applications of the carbon filament atom reservoir technique of atomic absorption and fluorescence analysis.

544. J. D. WINEFORDNER, R. F. BROWNER, C. J. MOLNAR and B. M. PATEL, Paper presented to the Eleventh National SAS Meeting, Dallas, Texas, 1972. Improved electrodeless discharge lamps for atomic fluorescence spectrometry with flames and graphite filaments atomizers.

545. J. D. WINEFORDNER and R. F. BROWNER, Paper presented to the Pittsburgh Conference on Analytical Chemistry and Applied Spectroscopy, 1973. Thermostated electrodeless discharge lamps for atomic fluorescence spectrometry.

546. J. D. WINEFORDNER, Paper presented to the Pittsburgh Conference on Analytical Chemistry and Applied Spectroscopy, 1973. Pulsed source for atomic fluorescence spectrometry.

547. A. T. YOUNG, *Appl. Optics*, **10**, 1681 (1971). Use of photomultiplier tubes for photon counting.

548. J. ZYNGER and S. R. CROUCH, *Appl. Spectry.*, **26**, 631 (1972). High-intensity, low-duty cycle, nanosecond light source for fluorescence lifetime studies and analytical applications.

549. T. K. AIDAROV and V. A. I'LIN, *Zh. Prikl. Spektrosk.* **17**, 924 (1972). Simple flame atomic fluorescence photometer.

550. T. K. AIDAROV, *Zh. Prikl. Spektrosk.*, **18**, 344 (1973) State of prospects for the development of simple devices using flame spectrophotometry.

551. T. K. AIDAROV, *Zh. Prikl. Spektrosk.*, **18**, 309 (1973). Use of solar-blind photoresistors in dispersion-free systems of atomic fluorescense spectrophotometry.

552. C. T. J. ALKEMADE, Paper presented to the 4th International Conference on Atomic Spectroscopy, Toronto, 1973. The foundations of flame atomic emission, absorption and fluorescence svectroscopy.

553. A. N. ATANOV, B. I. VERKHOVSKII, T. V. GASSANOVA, V. I. KANUNNIKOVA, E. E.

MAIZIL and G. I. SATARINA, *Zh. Prikl. Spektrosk.*, **17**, 577 (1972) Highly sensitive atomic fluorescence flame photometry with a continuous spectrum source.

554. Y. I. BELYAEV, T. A. KOVESHNIKOVA and V. I. VERNADSKII, Paper presented to the 17th Colloquium Spectroscopicum Internationale, Firenze, 1973. On the limits of atomic absorption and atomic fluorescence detection with selective flameless atom reservoirs for powdered samples.
555. R. F. BROWNER and D. C. MANNING, *Anal. Chem.*, **44**, 843 (1971). Utility of the separated air-acetylene flame in atomic fluorescence spectrometry.
556. R. F. BROWNER, B. M. PATEL and J. D. WINEFORDNER, Paper presented to the 17th Colloquium Spectroscopicum Internationale, Firenze, 1973. Thermostated electrodeless discharge lamps as sources for atomic absorption and atomic fluorescence spectrometry.
557. R. F. BROWNER and J. D. WINEFORDNER, *Spectrochim. Acta*, **28B**, 263 (1973). Optimization of thermostated electrodeless discharge lamps for analytical atomic spectrometry.
558. R. F. P. BECKWITH and J. D. WINEFORDNER, Paper presented to the 4th International Conference on Atomic Spectroscopy, Toronto, 1973. Electrodeless discharge lamps for atomic fluorescence spectrometry.
559. K. W. BUSCH and T. J. VICKERS, *Spectrochim. Acta* **28B**, 85 (1973) Fundamental properties characterizing low-pressure microwave induced plasmas as excitation sources for spectroanalytical chemistry.
560. M. CHENEVIER and M. LOMBARDI, *Chem. Phys. Lett.*, **16**, 154 (1972). Magnetic depolarization of atomic fluorescence in flames.
561. S. A. CLYBURN, B. R. BARTSCHMID and C. VEILLON, Paper presented to the Pittsburg Conference on Analytical Chemistry and Applied Spectroscopy, 1974. Atomic fluorescence spectrometry with a continuum source, graphite atomization and photon counting.
562. E. CORDOS and H. V. MALMSTADT, *Anal. Chem.* **44**, 2407 (1972). Programmable power supply for operation of hollow cathode lamps in an intermittent current-regulated high-intensity mode.
563. E. CORDOS and H. V. MALMSTADT, *Anal. Chem.*, **45**, 27 (1973). Characteristics of hollow cathode lamps operated in an intermittent high current mode.
564. R. M. DAGNALL, Paper presented to the 4th International Conference on Atomic Spectroscopy, Toronto, 1973. Non-dispersive atomic fluorescence spectrometry.
565. W. S. GLEASON and R. PERTEL, *Rev. Sci. Instrum.*, **42**, 1638 (1972);. High-stability electrodeless discharge lamps.
566. S. R. GOODE, A. MONTASER and S. R. CROUCH, *Appl. Spectry.*, **27** 355 (1973). Optimization of instrumental parameters in an automated non-flame atomic fluorescence spectrometer.
567. D. S. GOUGH, P. HANNAFORD and A. WALSH, *Spectrochim. Acta*, **28B**, 197 (1973). The application of cathodic sputtering to the production of atomic vapours in atomic fluorescence spectroscopy.
568. I. D. GUZEEV, E. S. BLINOVA, I. A. MAIOROV and V. V. NEDLER, *Zavodsk. Lab.*, **39**, 165 (1973). Determination of impurities in rare earth oxides by atomic absorption and atomic fluorescence methods with pulsed vaporization.
569. I. D GUZEEV, I. A. MAIOROV and V. V. NEDLER, *Zavodsk. Lab.*, **39**, 163 (1973) Comparison of possibilities for atomic absorption and atomic fluorescence methods of analysis in a region of small concentrations.
570. H. G. C. HUMAN, *Spectrosc. Letters*, **6**, 719 (1973). The improvement of signal-to-noise

ratio using a pulsed hollow cathode lamp and gated detector for atomic fluorescence measurement.

571. H. G. C. Human, Paper presented to the 17th Colloquium Spectroscopicum Internationale, Firenze, 1973. Atomic fluorescence measurements with a pulsed hollow cathode lamp as primary source.

572. J. D. Ingle, Jr. and S. R. Crouch, *Anal. Chem.*, **44**, 777 (1972). Pulse overlap effects on linearity and signal-to-noise ratio in photon counting system.

573. G. F. Kirkbright and T. S. West, *Chem. Brit.*, **8**, 428 (1972). Atomic fluorescence for chemical analysis.

574. G. F. Kirkbright, T. S. West and P. J. Wilson, *Anal. Chim. Acta*, **66**, 130 (1973). The direct determination of mercury by atomic fluorescence spectrometry in a nitrogen separated air-acetylene flame with excitation at 1849 A.

575. G. F. Kirkbright, O. E. Troccoli and S. Vetter, *Spectrochim. Acta*, **28B**, 1(1973). The application of a pieozelectric scanning Fabry-Perot interferometer to the study of atomic line sources —II. Line-widths for calcium in air-acetylene and nitrous oxide-acetylene flames.

576. G. F. Kirkbright and O. E. Troccoli, *Spectrochim. Acta*, **2*B***, 33 (1973). The application of a piezoelectric scanning Fabry-Perot interferometer to the study of atomic line sources —III. Use of the channeled spectra produced with a continuum source for studies of the absorption line-width for calcium atoms in flames.

577. K. Kitagawa and T. Takeuchi, *Anal. Chim. Acta*, **60**, 309 (1972). Spectroscopic studies of microwave-excited plasma.

578. D. Kolihová and V. Sychra, *Chem. Listy*, in press. Atomic absorption and atomic fluorescence spectrometry in biochemistry and medicine —IV. Direct microdetermination of zinc in blood serum by atomic fluorescence spectrometry.

579. J. Kuhl, G. Marowsky and R. Torge, *Anal. Chem.*, **44**, 375 (1972). Sodium analysis with a tunable dye laser.

580. M. Marinkovic and L. Stojanovic, Paper presented to the 17th Colloquium Spectroscopicum Internationale, Firenze, 1973. Determination of mercury in waters with cold vapour atomic fluorescence technique.

581. A. Montaser, S. R. Crouch and E. F. Palermo, Paper presented to the Pittsburgh Conference on Analytical Chemistry and Applied Spectroscopy, 1974. Time-division multiplexed atomic fluorescence spectrometry.

582. M. K. Murphy, S. A. Clyburn and C. Veillon, *Anal. Chem.* **45**, 1468 (1973). Comparison of lock-in amplification and photon counting with low background flame and graphite atomizers in atomic fluorescence spectrometry.

583. J. D. Norris and T. S. West, *Anal. Chem.*, **45**, 2148 (1973). Investigation of spectral overlap of the neon 359.352 nm and chromium 359.349 nm spectral lines in atomic absorption and atomic fluorescence spectrometry of chromium.

584. N. Omenetto, L. P. Hart and J. D. Winefordner, *Appl. Spectry.*, **26**, 612 (1972) Light scattering in turbulent and laminar flames.

585. N. Omenetto, P. Benetti, L. P. Hart, J. D. Winefordner and C. T. J. Alkemade, *Spectrochim Acta*, **28B**, 289 (1973). Non-linear optical behaviour in atomic fluorescence flame spectrometry.

586. N. Omenetto, L. P. Hart, P. Benetti and J. D. Winefordner, *Spectrochim. Acta*, **28B**, 301 (1973). On the shape of atomic fluorescence analytical curves with a laser excitation source.

587. N. Omenetto, P. Benetti, L. P. Hart, J. D. Winefordner and G. Rossi, Paper

presented to the 17th Colloquium Spectroscopicum Internationale, Firenze, 1973. Temperature profiles in flames by means of laser excited atomic fluorescence.

588. N. OMENETTO, Paper presented to the 4th International Conference on Atomic Spectroscopy, Toronto, 1973. The use of a tunable laser in atomic fluorescence.

589. B. M. PATEL, R. D. REEVES, R. F. BROWNER, C. J. MOLNAR and J. D. WINEFORDNER *Appl. Spectry.*, **27**, 171 (1973). Atomic fluorescence spectrometry with a graphite rod atomizer and thermostated electrodeless discharge lamps.

590. B. M. PATEL and J. D. WINEFORDNER, *Anal. Chim. Acta,* **64,** 135 (1973). Graphite rod atomization and atomic fluorescence for the simultaneous determination of silver and copper in jet engine oils.

591. E. H. PIEPMEIER, *Spectrochim. Acta,* **27***B*, 445 (1972). Influence of non-quenching collisions upon saturated resonance fluorescence.

592. F. G. PLANKEY and J. D. WINEFORDNER, Paper presented to the Pittsburgh Conference on Analytical Chemistry and Applied Spectroscopy, 1974. Multielement analysis by Hadamard transform fluorescence spectroscopy.

593. T. C. RAINS and O. MENIS, Paper presented to the 4th International Conference on Atomic Spectroscopy, Toronto, 1973. Application of atomic fluorescence spectrometry for the analysis of standard reference material.

594. T. C. RAINS, M. S. EPSTEIN and O. MENIS, *Anal. Chem.*, **46,** 207 (1974). Automatic correction system for light scatter in atomic fluorescence spectrometry.

595. W. E. RIPPETOE, V. I. MUSCAT and T. J. VICKERS, *Anal. Chem.*, **46,** 796 (1974). Determination of iron in serum by atomic fluorescence flame spectrometry.

596. S. SHIMOMURA and R. HIROTO, *Anal. Letters,* **6,** 613 (1973). Determination of trace amounts of mercury by non-dispersive atomic fluorescence method.

597. S. SHIMOMURA, Y. HAYASHI, H. MORITA and T. SHINOHARA, *Anal. Letters,* **6,** 769 (1973). Behaviour of mercury vapor upon irradiation by a mercury lamp in an inert gas.

598. R. SPILLMAN, R. KALEY, J. LOWRY and H. V. MALMSTADT, Paper presented to the Pittsburgh Conference on Analytical Chemistry and Applied Spectroscopy, 1974. New developments in an automated atomic fluorescence/atomic emission spectrometer for rapid multielement determinations.

599. K. C. THOMPSON, *Lab. Pract.*, **21,** 645 (1972). A cold vapour mercury atomic fluorescence detector.

600. C. VEILLON, Paper presented to the 4th International Conference on Atomic Spectroscopy, Toronto, 1973. Atomic fluorescence spectrometry with graphite atomization and photon counting techniques.

601. A. WALSH, *Pure Appl. Chem.*, **34,** 145 (1973). Non-dispersive systems in atomic absorption spectroscopy.

602. A. WALSH, Paper presented to the 4th International Conference on Atomic Spectroscopy, Toronto, 1973. Spectrochemical methods for the analysis of metals.

603. J. O. WEIDE and M. L. PARSONS, *Anal. Chem.*, **45,** 2417 (1973). Pulsed-mode atomic fluorescence utilizing a demountable hollow cathode excitation source.

604. T. S. WEST, *Chem. Brit.*, **7,** 378 (1971). Versatile atomic fluorescence spectroscopy.

605. T. S. WEST and M. S. CRESSER, *Appl. Spectrosc. Reviews,* **7,** 79 (1973). Atomic fluorescence spectroscopy.

606. T. S. WEST, Paper presented to the 4th International Conference on Atomic Spectroscopy, Toronto, 1973. Some recent investigations in analytical techniques of atomic spectroscopy.

607. J. B. WILLIS, Paper presented to the 4th International Conference on Atomic Spec-

troscopy, Toronto, 1973. Analysis of solid samples by atomic absorption and atomic fluorescence techniques.

608. J. D. WINEFORDNER and T. J. VICKERS, *Anal. Chem.*, **44**, (5), 150 R (1972). Flame spectrometry.

609. J. D. WINEFORDNER, Paper presented to the 17th Colloquium Spectroscopicum Internationale, Firenze, 1973. High-intensity pulsed source atomic fluorescence spectrometry.

610. J. D. WINEFORDNER, Paper presented to the 4th International Conference on Atomic Spectroscopy, Toronto, 1973. Application of atomic fluorescence spectrometry to trace elements.

Index